Handbook of Statistics

Volume 32

Computational Statistics with R

Handbook of Statistics

Series Editor

C.R. Rao
C.R. Rao AIMSCS, University of Hyderabad Campus,
Hyderabad, India

Handbook of Statistics

Volume 32

Computational Statistics with R

Edited by

Marepalli B. Rao
Division of Biostatistics and Epidemiology,
Department of Environmental Health,
University of Cincinnati, Cincinnati, Ohio, USA

C.R. Rao
C.R. Rao AIMSCS,
University of Hyderabad Campus,
Hyderabad, India

ELSEVIER

Amsterdam • Boston • Heidelberg • London • New York • Oxford
Paris • San Diego • San Francisco • Singapore • Sydney • Tokyo
North-Holland is an imprint of Elsevier

N·H

North-Holland is an imprint of Elsevier
Radarweg 29, PO Box 211, 1000 AE Amsterdam, Netherlands
The Boulevard, Langford Lane, Kidlington, Oxford OX5 1 GB, UK

ISBN: 978-0-444-63431-3
ISSN: 0169-7161

For information on all North-Holland publications
visit our website at http://store.elsevier.com/

Contents

Contributors

Numbers in Parentheses indicate the pages on which the author's contributions begin.

Hansen Bannerman-Thompson (221), Department of Environmental Health, University of Cincinnati, Cincinnati, Ohio, USA

Joshua Betz (257), Department of Biostatistics, Johns Hopkins Bloomberg School of Public Health, Baltimore, Maryland, USA

Subramanyam Kasala (177), Department of Mathematics and Statistics, University of North Carolina, Wilmington, North Carolina, USA

Palash Mallick (93), Department of Environmental Health, University of Cincinnati, Cincinnati, Ohio, USA

John Muscheli (257), Department of Biostatistics, Johns Hopkins Bloomberg School of Public Health, Baltimore, Maryland, USA

Chaitra H. Nagaraja (1), Gabelli School of Business & Graduate School of Business Administration, Fordham University, New York, USA

Karl E. Peace (339), Jiann-Ping Hsu College of Public Health, Georgia Southern University, Georgia, USA

C.R. Rao (357), Research Professor, University of Buffalo, Buffalo, New York, USA and CRRAO AIMSCS, Hyderabad, India

Marepalli B. Rao (93, 177, 221, 357), Division of Biostatistics and Epidemiology, Department of Environmental Health, University of Cincinnati, Cincinnati, Ohio, USA

Deepayan Sarkar (49), Theoretical Statistics and Mathematics Unit, Indian Statistical Institute, New Delhi, India

Kao-Tai Tsai (339), Jiann-Ping Hsu College of Public Health, Georgia Southern University, Georgia, USA

Ravi Varadhan (257), Division of Biostatistics and Bioinformatics, Department of Oncology, Johns Hopkins University School of Medicine, Baltimore, Maryland, USA

Hrishikesh D. Vinod (143), Professor of Economics, Fordham University, Bronx, New York, USA

Derek S. Young (309), Department of Statistics, University of Kentucky, Lexington, Kentucky, USA

Preface

A renaissance is currently in progress in all that is statistical. Those were the days when we used desk calculators to fit one of the Pearsonian distributions to the data collected from the field. Those were the days when we used punch cards on a computer occupying an entire large room to work out the Fisher's discriminant function on a large multivariate data. Those were the days some people felt that all Statistics does were to indulge in some uninspiring drab calculations. Those days were gone. A revolution occurred with the advent of sophisticated computers. Size of the data was no hindrance in fishing information from the data. A number of statistical software programs mushroomed. Statistics made inroads into realms not ventured before. News paper articles mine vast amounts of data to hammer out the essence of the point under discussion.

Another revolution is occurring: personal computers and open source statistical software. If established statistical software were to be regarded as brand name palliatives, the free software R is generic doing as good as the brand name ones and more. The software R is empowering data analysts to venture deeper into data analysis. All it requires for empowerment is downloading R, which takes less than 5 min mostly. The software has more than 4000 packages. The Bioconductor package has more than 6000 packages. Each is tailored to focus on certain statistical methodologies. Most packages have data sets and R codes specific to that package. Each code has documentation and examples of usage of the code. Anyone can develop a package. Figuratively, we can say that there are a million brains behind R, volunteering their time and effort to bring out a package. In any commercial software setup, a few dedicated groups of code writers are behind the software.

It is easy to immerse oneself into all that is R. All one needs is a laptop, an internet connection with R downloaded, and a decent working knowledge of Statistics. Running the examples provided in the documentation of the codes is educational and knowledge-fulfilling.

The handbook is addressed to those who have some background knowledge of statistics and some experience of usage of some statistical software and would like to explore what R can offer. We have requested experts to contribute to the handbook with the hope that the book will provide some basics and whet the appetite for more. Here is an outline of the chapters.

CHAPTER 1: INTRODUCTION TO R

Dr. Chaitra H. Nagaraja is an assistant professor at Fordham University. She uses R in her classes and also conducted workshops on R. She provides a thoughtful introduction to set the tone of the book.

CHAPTER 2: R GRAPHICS

Dr. Deepayan Sarkar is a professor at the Indian Statistical Institute, Delhi. He is an expert on Multivariate Data Visualization using the *lattice* system. He wrote a successful book on this theme (2008). He shares with us his insights on graphics in this chapter.

CHAPTER 3: GRAPHICS MISCELLANEA

Dr. Palash Mallick is a post doc at the University of Cincinnati involved in data analysis pertaining to environmental issues. Dr. M.B. Rao is a professor of statistical genetics and biomedical engineering at the University of Cincinnati. In his classes, he uses R extensively. He conducted workshops on Graphics, Statistical Genetics, Spatial Statistics, Bayesian Networks, and Shape Analysis, among others, all involving R. They want to convey their immense appreciation of range and depth that various packages in R offer in the field of graphics.

CHAPTER 4: MATRIX ALGEBRA TOPICS IN STATISTICS AND ECONOMICS USING R

Dr. Hrishikesh D. Vinod is a professor at Fordham University. He has contributed extensively to Econometrics and usage of matrices is pervasive in his field. He has penned a book (2011) on multifarious uses of matrices ably harnessing R. He shares with us his insights on matrices and R.

CHAPTER 5: SAMPLE SIZE CALCULATIONS WITH R: LEVEL 1

Dr. S. Kasala is a professor at the University of North Carolina, Wilmington. Sample size calculations are extensively sought out in our consulting work. Some rudimentary ideas behind sample size calculations are set out along with tables of the numbers using a specific R package covering some scenarios.

CHAPTER 6: SAMPLE SIZE CALCULATIONS WITH R: LEVEL 2

Dr. Hansen Bannerman-Thompson was a post doc working freelance now specializing in machine learning techniques in R. He collaborated with M.B. Rao to bring out tables of sample sizes covering some basic testing environment.

CHAPTER 7: BINOMIAL REGRESSION IN R

Dr. Ravi Varadhan (Johns Hopkins University) along with his collaborators brings out a timely chapter on Binomial Regression, which is a useful counterweight to the Logistic Regression. He is also an author of the package behind Binomial Regression.

CHAPTER 8: COMPUTING TOLERANCE INTERVALS AND REGIONS USING R

Dr. Derek S. Young is an assistant professor at the University of Kentucky. He has penned a chapter on Tolerance Limits and usage of R. This chapter falls into the realm of how R can be utilized to solve problems in a specific applied field such as Tolerance Limits.

CHAPTER 9: MODELING THE PROBABILITY OF SECOND CANCER IN CONTROLLED CLINICAL TRIALS

This chapter comes under the purview of applied statistics. Drs. Kao-Tai Tsai and Karl E. Peace, Georgia Southern University, were the authors of the chapter. Professor Peace is an emeritus professor very active in Biostatistics research.

CHAPTER 10: BAYESIAN NETWORKS

Bayesian Networks is an interesting and exciting area of statistical applications requiring a fairly large amount of computational work. R has a number of packages meeting the challenges of computational work. The editors of the volume have presented a rudimentary introduction to the field.

C.R. Rao
Marepalli B. Rao

REFERENCES

Sarkar, D., 2008. Lattice, Multivariate Visualization with R. Springer, New York.
Vinod, H., 2011. Hands-On Matrix Algebra Using R. World Scientific, Singapore.

Chapter 1

Introduction to R

Chaitra H. Nagaraja[1]
Gabelli School of Business & Graduate School of Business Administration, Fordham University, New York, USA
[1]*Corresponding author: e-mail: cnagaraja@fordham.edu*

ABSTRACT
The basics of R programming are discussed from syntax, conditional statements, and control structures, to writing functions. Fundamentals such as data cleaning, exploratory data analysis, hypothesis testing, and regression are introduced. Simulation and random generation and numerical methods are considered as well. Finally, a list of add-on packages and additional resources are listed.

Keywords: CRAN, Workspace, R syntax, Functions, Statistical analysis, Simulation, Numerical methods

1 INTRODUCTION

The statistical programming language R is a free version of the language S initially developed at Bell Laboratories in 1976 by John Chambers. Due to its open source philosophy and versatility, R is now used widely in academia, government, and industry. Its flexibility allows one to link to low-level programming languages, such as C++ and FORTRAN, scripting languages like UNIX, blend with word processors such as LaTeX, and handle large data sets with parallel computing capabilities.

Learning any new language requires an initial investment of time to internalize the syntax and intuition behind its structure. This chapter is designed to get you started in this process, along with illustrating the broad range of possible applications.

We begin with how to install R in Section 2 and introduce the command line environment. In Section 3, various R object and data types are discussed along with functions for common calculations. To help write your own programs, we focus on control statements, function construction, and error correction in Section 4. The remainder of the chapter focuses on applications of R. In Sections 5, 6, and 7 we discuss uploading and writing data files, data cleaning, and exploratory data analysis. Basic statistical inference and regression are

Handbook of Statistics, Vol. 32. http://dx.doi.org/10.1016/B978-0-444-63431-3.00001-2

TABLE 1 Workspace Functions

Operation	Function(s)
Save/load workspace objects	`load(), save(), save.image()`
Read code into workspace	`source()`
Commands history	`history(), loadhistory(), savehistory()`
Get working directory	`getwd()`
Set working directory	`setwd()`
Workspace maintenance	`gc(), ls(), rm(), rm(list=ls())`
Help	`?, help()`

discussed in Section 8. In Section 9, we describe how to set up simulation studies and sampling schemes. Numerical methods, such as finding roots and integration, are introduced in Section 10. We conclude in Section 11 with some references for R and related applications. To illustrate some of the topics, we sparingly use plots with little clarification regarding the code. Refer to Chapter 2 of this book for more on R graphics. Finally, to eliminate dependence on supplementary materials, we will use the data that comes with the base R package in our examples; see Section 2.1 for a list.

To help the reader, we have compiled tables to summarize many of the commands and R packages introduced within the text: see Table 1 for workspace functions, Table 2 for packages and functions for a wide range of statistical methods, Table 3 for miscellaneous functions, Table 4 for numerical operators, Table 5 for logical operators and conditional statements, Table 6 for data processing functions, Table 7 for summary statistics, Table 8 for basic hypothesis testing functions, Table 9 for probability distributions and random number generation, and Table 10 for numerical methods.

We now introduce the notation for expressing R code within the text of this chapter. In particular, `R_objects` for objects, `function()` for functions, "`name`" in double quotes for function arguments or variable names, an italicized *`R_package`* for add-on packages, and *`USER_INPUT`* in all capital letters for user input such as file and object names. Blocks of code will be written in `this font` and indicated as such. Consequently, when we write code in the R environment, we will begin each line with > to simulate the console (note: do not copy the > symbol); if a particular command continues to a second line, > is replaced by +. Within these blocks of code, to avoid confusion, we will not display the output, only the commands.

TABLE 2 R Packages for Statistical Analysis (*base* Indicates Function in Standard Installation of R); Other Applications: Table 3; Basic Hypothesis Testing: Table 8; Numerical Algorithms: Table 10

Method	R Package: Notes
ANOVA, ANCOVA, MANOVA	*base*: anova(), aov(), lm(), manova(), oneway.test()
Bayesian methods	*arm*; *bayess*; *mcmc*; *MCMCpack*; *rbugs*; *R2WinBUGS*
Capture–recapture	*Rcapture*
Conjoint analysis	*conjoint*
Copulas	*copula*
Cluster analysis	*base*: hclust(), kmeans(); *cluster*; *mclust*; *pvclust*
Data mining, machine learning	*ada*; *adabag*; *e1071*; *gbm*; *ipred*; *mboost*; *neuralnet*; *nnet*; *party*; *randomForest*; *rpart*; *tree*
Discriminant analysis	*MASS*: lda(), qda()
Experiments	*pcalg*; *experiment*; *stepp*
Factor analysis	*base*: factanal()
Functional data analysis	*fda*; *fda.usc*
Generalized linear models	*base*: glm(); *gam*
Hierarchical/mixed models	lme4: glmer(), lmer(), nlmer(); *hglm*; *nlme*
Instrumental variables	*AER*: ivreg(), ivreg.fit(); *sem*: tsls()
Kernel methods	*base*: density(); *kernlab*; *ks*
Loess/lasso	*basic*: loess(); *genlasso*; *glmnet*; *lars*; *lasso2*
Linear regression	*base*: AIC(), anova(), BIC(), lm(), predict.lm(), residuals.lm(), summary.lm(), plot.lm(), step(); *cars*: vif(); *MASS*: stepAIC(); *leaps*: leaps()
Missing data methods	*mi*; *mice*; *MImix*; *mitools*; *pan*
Network analysis	*igraph*; *network*; *sna*; *statnet*; *statnet.common*
Observational studies	*Matching*; *MatchIt*; *multilevelPSA*; *nonrandom*; *optmatch*; *PSAgraphics*; *rbounds*; *RItools*
Order statistics	*ismev*; *heavy*
Principal components	*base*: princomp()

Continued

TABLE 2 R Packages for Statistical Analysis (*base* Indicates Function in Standard Installation of R); Other Applications: Table 3; Basic Hypothesis Testing: Table 8; Numerical Algorithms: Table 10—Cont'd

Method	R Package: Notes
Psychometrics	*psych*; *psychotools*
Quantile regression	*quantreg*
Queueing	*queueing*
Resampling methods	*boot*; *bootES*; *bootstrap*
Signal processing, wavelets	*signal*; *signalextraction*; *tuneR*; *wavelets*; *waveslim*; *wavethresh*
Spatial statistics	*maptools*; *OpenStreetMap*; *raster*; *rgdal*; *rgeos*; *sp*; *spatial*; *SpatialPack*; *spatial.tools*
Splines	*base*: `smooth.spline()`, `splinefun()`; *polspline*; *pspline*
State space models	*dse*; *dlm*; *KFAS*
Statistical process control	*qat*; *qcc*; *qcr*
Structural equation modeling	*sem*; *SEMID*; *semdiag*; *semGOF*
Survey methods	*EVER*; *sampling*; *samplingbook*; *SamplingStrata*; *samplingVarEst*; *stratification*; *survey*;
Survival analysis	*OIsurv*; *survival*; *survivalMPL*; *survivalROC*
Time series	*base*: class `ts`, `acf()`, `ar()`, `arima()`, `arima.sim()`, `pacf()`, `predict.Arima()`, `tsdiag()`; *forecast*; *gsarima*; *rugarch*; *seasonal*; *timeSeries*; *tseries*; *tseriesChaos*

2 SETTING UP R

In the following section, we describe how to install R and begin coding. We briefly discuss how memory is managed and list additional packages that expand the capabilities of R.

2.1 Installing and Starting R

Install R from the CRAN (Comprehensive R Archive Network) Website: http://cran.r-project.org/; choose the link applicable to your computer system (e.g., Linux, Mac, Windows). R is continually updated; therefore, it is important to check the CRAN Website periodically for new versions.

TABLE 3 Miscellaneous R Functions and Packages

Method	R Package: Notes
Data processing	*plyr*
Data sets	*base*: `library(help = "datasets")`
Date/time stamp	*base*: `date()`, `timestamp()`, `proc.time()`
Debugging	*base*: `browser()`, `debug()`, `trace()`, `untrace()`, `traceback()`; *debug*
Integrate R with LaTeX	`Sweave()`, *knitr*
Parallel computing	*snow*
R colors	*base*: `col2rgb()`, `colors()`, `colours()`, `rgb()`
R color spectra	*base*: `cm.colors()`, `heat.colors()`, `rainbow()`, `terrain.colors()`, `topo.colors()`

The base package (herby referred to as base R or *base*) contains a variety of functions and operations most commonly used in data processing and statistics. Furthermore, a range of data sets are included (type `library(help = "datasets")` for more details); to make this text self sufficient, we will use these data in our examples. Specifically, we use: `airmiles` (commercial airline revenue), `attitude` (Chatterjee–Price Attitude survey data on a firm's employees), `faithful` (geyser data from Yellowstone National Park's Old Faithful), `iris` (iris plant measurements), and `ToothGrowth` (tooth growth in guinea pigs receiving Vitamin C supplements).

Opening the R program loads a console where you can type commands directly into a workspace. The symbol > below simulates the R interface; if a command continues to a second line, it is indicated by +, as in the R console. Now, try the following commands (output omitted below):

```
> # R will ignore everything after the symbol #
> x              # error: x is not a defined object yet
> x <- 4         # assign the number 4 to the object x
> y <- sqrt(x)   # assign square root of x to new object y
> x <- x+5       # save the value of x plus five in x
> x              # x now defined; print value(s) of x
```

Observe that all text after # in a given line is ignored by the computer; this feature allows us to add comments to our code. Comments are a crucial component of any programming language. It is easy to forget what your code signifies after time has passed. Do not learn this lesson the hard way. Comment your code!

TABLE 4 Basic Numerical Operations

Operation	Function name	Matrix operation	Function
Arithmetic	`+-*/`	Transpose vector/matrix	`t()`
Modulo	`%%`	Solve linear system, invert matrix	`solve()`
Exponent	`^`	Eigenvalues, eigenvectors	`eigen()`
Square root	`sqrt()`	Determinant	`det()`
Logarithm	`log()`	Decompositions	`chol, svd(), qr()`
Absolute value	`abs()`	Vector/matrix multiplication	`%*%`
Exponential	`exp()`	Diagonal matrices	`diag()`
Sum/product of vector	`sum(), prod()`	Outer product	`%o%`
Cumulative sum/product of vector	`cumsum(), cumprod()`	Cross product	`crossprod()`
Sort vector	`sort()`	Trace	`sum(diag())`
Rounding	`ceiling(), floor(),`		
	`round(), signif(),`		
	`trunc()`		
Integer division	`%/%`		
Trigonometry	`sin(), cos(), tan(),`	Miscellaneous	
	`acos(), atan(), asin()`	Infinity	`Inf, -Inf`
Combinations	`choose()`	π	`pi`
Factorial	`factorial()`	Missing data	`NA`
Complex numbers	`complex(), Re(), Im(),`	Not a number	`NaN`
	`Mod(), Arg(), Conj()`	Null (empty)	`NULL`

TABLE 5 Components for Conditional Statements; for Examples, Let `x <- 5`, `y <- c(7,4,2)`, **and** `z <- c("a","b","c")`

Symbol/Function	Description	Example	Example Output
`==`	Equal to	`x == 4`	FALSE
		`z == "c"`	FALSE FALSE TRUE
`!=`	Not equal to	`x != 4`	TRUE
		`z!="b"`	TRUE FALSE TRUE
`<, >`	Less/greater than	`x<4`	FALSE
`<=, >=`	Less/greater than or equal to	`y >= 5`	TRUE FALSE FALSE
`&`	And	`(x>4) & (x <=10)`	TRUE
`&&`	& Expression TRUE for all elements in vector	`(y<4) && (y>=6)`	FALSE
`|`	(Inclusive) or	`(z=="a") | (z=="b")`	TRUE TRUE FALSE
`||`	| Expression TRUE for at least one element in vector	`(y<4) || (y>=6)`	TRUE
`!`	Not	`!(x>4)`	FALSE
`any()`	Expression TRUE for at least one element in vector	`any(z=="b")`	TRUE
`all()`	Expression TRUE for all elements in vector	`all(x==3)`	FALSE
`is.element()`,	Are the elements of x in y?	`is.element(x,y)`	FALSE
`%in%`		`x%in%y`	FALSE

Continued

TABLE 5 Components for Conditional Statements; for Examples, Let `x <- 5`, `y <- c(7,4,2)`, and `z <- c("a","b","c")`—Cont'd

Symbol/Function	Description	Example	Example Output
`which()`	Which elements of x satisfy a certain condition	`which(z=="b")`	2
`which.min()`, `which.max()`	Position of first min/max value in a vector	`which.max(y)`	1
`complete.cases()`, `is.finite()`, `is.infinite()`, `is.na()`, `is.nan()`	Checking for missing and/or infinite data	`is.na(y)`	FALSE FALSE FALSE
`is.character()`, `is.factor()`, `is.integer()`, `is.numeric()`	Checking data types	`is.factor(z)`	FALSE
`is.data.frame()`, `is.list()`, `is.matrix()`, `is.vector()`	Checking object types	`is.vector(y)`	TRUE

After entering the five lines of code above, the workspace now has two objects, `x`, which is equal to 9, and `y`, equal to 2. Note that changing the value of `x` after `y` is defined does not alter the value of `y`. Nearly any name can be used to label an object, provided that it does not start with a number; however, contextually meaningful names help when deciphering your code. The operator `<-` assigns the value on the right to the object on the left. The equal sign (=) can technically also be used, but should be avoided as it has alternate interpretations in R. Finally, `sqrt()` denotes the square root function. In R, arguments for a function are listed within the parentheses. Typing the function name without the parentheses generally prints the code for the function. To obtain details about `sqrt()`, type `?sqrt` or `help(sqrt)`.

As you code, your workspace will be filled with objects. To keep it organized, especially if you run out of space, the following commands are helpful (also see Table 1):

```
> # define objects x, y, and z
> x <- 10
> y <- log(5)
```

```
> z <- "a"
>
> ls()              # list everything in the workspace
> rm(x, z)          # remove objects x and z, but not y
> rm(list=ls())     # remove all objects from the workspace
> gc()              # list current memory usage
```

By defining `x <- 10`, we have overwritten the previous command `x <- 4` from the previous section of code; `y` has been replaced by the natural log of 5.

TABLE 6 Input/Output and Processing Functions

Operation	Function(s)
Read data into console	`scan()`, `read.table()`, `read.csv()`; *foreign*; *xlsx*
Write to a file	`write()`, `write.table()`
Formatting output	`cat()`, `format()`, `paste()`, `round()`, `truncate()`, `ceiling()`, `floor()`
Data summaries	`colnames()`, `dim()`, `head()`, `names()`, `ncol()`, `nrow()`, `rownames()`, `tail()`
Reverse/sort vector	`rev()`, `sort()`
Sort data frame	`order()`
Subset data frame	`subset()`
Separate data frame by category	`split()`
Training and test sets	`sample()`
Categorical variables	`chartr()`, `cut()`, `grep()`, `strsplit()`, `substr()`, `tolower()`, `toupper()`
Construct/append to a data frame	`cbind()`, `data.frame()`, `rbind()`
Merge data frames	`merge()`
Computations by row or column	`apply()`
Computations for each element of a list	`lapply()`
Computations on a data frame by category	`aggregate()`
additional data processing functions in package *plyr*	

TABLE 7 Summary Statistics and Plots

Statistic (R Package)	Function	Summary Plot	Function
Mean, trimmed mean	`mean()`	Histogram	`hist()`
Weighted mean	`weighted.mean()`	Box plot	`boxplot()`
Mean absolute deviation	`mad()`	Bar graph	`barplot()`
Median	`median()`	Scatter plot/time series	`plot()`
Mode	`mode()`	Stem and leaf plot	`stem()`
Variance	`var()`	Pie graph	`pie()`
Standard deviation	`sd()`	Mosaic plot	`mosaicplot()`
Max/min	`max()`, `min()`	Dot chart	`dotchart()`
Cumulative max/min	`cummax()`, `cummin()`	Matrix plotting	`matplot()`
Range	`diff(range())`	Contour plot	`contour()`
Quantile	`quantile()`	Plot a function	`curve()`
Six number summary	`summary()`	Normal quantile plot	`qqnorm()`, `qqline()`
Covariance	`cov()`		
Correlation	`cor()`	Quantile–quantile plot	`qqplot()`, `qqline()`
Skewness (*moments*)	`skewness()`		
Kurtosis (*moments*)	`kurtosis()`		
Contingency table	`table()`		

2.2 Memory

The computing speed in R depends heavily on your computer. We will not go into much detail here regarding memory in R. Use `?Memory` for information on memory usage in R and `?.Machine` for details on the largest and smallest numerical values possible on your machine. As mentioned above, `gc()`, which stands for garbage collection, lists current memory usage. If you run out of memory, or want to clear your workspace, remove unnecessary objects using `rm()`.

TABLE 8 Hypothesis Tests

Test	Function
One-, two-sample, paired *t*-tests for means	`t.test()`
One-, two-sample *z*-tests for proportions	`prop.test()`
Wilcoxon rank sum/Mann–Whitney tests	`wilcox.test()`
Kruskal–Wallis rank sum test	`kruskal.test()`
Bartlett test for equal variances	`bartlett.test()`
χ^2-test for contingency tables, goodness-of-fit tests	`chisq.test()`
Fisher's exact test for independence	`fisher.test()`
Kolmogorov–Smirnov goodness-of-fit tests	`ks.test()`
Shapiro–Wilk test for normality	`shapiro.test()`
t-Test for correlation	`cor.test`
t-Test for regression slopes	`summary()`
Overall *F*-test, partial-*F* test for linear models	`anova()`
One-way ANOVA	`oneway.test()`
Tukey's HSD for multiple comparisons	`TukeyHSD()`
Durbin–Watson test for autocorrelation	*lmtest*: `dwtest()`

As with all programming languages, code can be written inefficiently. To reduce this issue, check to see if built-in functions can be applied and decrease your reliance on loops where possible (see Section 4.6). Using the command `date()` before and after a piece of code can help monitor the time to run a program. Furthermore, if your data sets are large, parallel computing methods may lower computing times as well. Use the package *snow*. See the CRAN Website for more details.

2.3 Saving Your Code and Workspace

Each time you open R, files are uploaded from and downloaded to a directory (e.g., a folder on your computer); `getwd()` will print this directory file path. To change the working directory, specify the file path in `setwd()` or use the toolbar at the top of your screen. In this section, we will discuss how to save and load code and workspaces (also see Table 1). For information on reading and writing data files, refer to Section 5.

There are two ways to preserve your code. If you have entered your commands directly into the console, you can save them by using `savehistory (file = "FILE_NAME.Rhistory")` where *FILE_NAME* is the name of your file

TABLE 9 List of Common Probability Distributions

Distribution (R Package)	Density	Distribution	Quantile	Sampling
Beta	dbeta()	pbeta()	qbeta()	rbeta()
Binomial	dbinom()	pbinom()	qbinom()	rbinom()
Burr (*actuar*)	dburr()	pburr()	qburr()	rburr()
Cauchy	dcauchy()	pcauchy()	qcauchy()	rcauchy()
χ^2	dchisq()	pchisq()	qchisq()	rchisq()
Exponential	dexp()	pexp()	qexp()	rexp()
F	df()	pf()	qf()	rf()
Gamma	dgamma()	pgamma()	qgamma()	rgamma()
Hypergeometric	dhyper()	phyper()	qhyper()	rhyper()
Log-normal	dlnorm()	plnorm()	qlnorm()	rlnorm()
Multivariate normal (mvtnorm)	dmvnorm()	pmvnorm()	qmvnorm()	rmvnorm()
Multivariate t (*mvtnorm*)	dmvt()	pmvt()	qmvt()	rmvt()
Negative binomial	dnbinom()	pnbinom()	qnbinom()	rnbinom()
Normal/Gaussian	dnorm()	pnorm()	qnorm()	rnorm()
Pareto (*actuar*)	dpareto()	ppareto()	qpareto()	rpareto()
Poisson	dpois()	ppois()	qpois()	rpois()
t	dt()	pt()	qt()	rt()
Uniform (continuous)	dunif()	punif()	qunif()	runif()
Weibull	dweibull()	pweibull()	qweibull()	rweibull()
Discrete random sampling				sample()
Random number generation				RNGkind()
Setting random number seed				set.seed()

TABLE 10 Table of Numerical Calculation Functions (*base* Indicates Function in Standard Installation of R)

Method	R Package: Notes
Fourier transforms	*fftw, fftwtools*
Roots of a polynomial	*base*: `polyroot()`
Roots of a function	*base*: `uniroot()`; *rootSolve*
Optimization	*base*: `optimize()`, `optimise()`, `optim()`, `constrOptim()`, `nlm()`
Numerical derivatives	*numDeriv*: `grad()`, `jacobian()`, `hessian()`
Numerical integration	*base*: `integrate()`; *cubature*: `adaptIntegrate()`
Differential equations	*deSolve*; *ReacTran*; *bvpSolve*; *rootSolve*; *sde*; *pomp*

and ".Rhistory" denotes a history file. If you just include a file name and not a file path, the file will be saved in the current working directory. To load the saved history, use `loadhistory(file = "`*FILE_NAME*`.Rhistory")`. However, it is more efficient to write your code in a separate file. A text file is sufficient, although additional features such as color coding and parenthesis matching can help. For Apple computers, a text editor is included when downloading R; alternatives include GNU Emacs, WinEdt, and Tinn-R (see Section 11 for more details). Save your R code with a ".R" file: `"`*FILE_NAME*`.R"`. To enter the code from an entire file, use `source("`*FILE_NAME*`.R")`.

All objects in your workspace can be saved to an "`.RData`" workspace file by using `save.image(file = "`*FILE_NAME*`.RData")`. To save only a subset of objects, use `save()` and specify those objects in the "`list`" argument. To load the workspace objects, use `load(file = "`*FILE_PATH*`.RData")`.

Finally, you can save your R session (i.e., what you see on your console) as a text file by choosing "Save" from the "File" menu from the toolbar menu at the top. All commands that were typed in that R session and the output printed will be saved, but cannot be uploaded back into R.

2.4 R Packages

Apart from base R, you can expand R's capabilities by installing a variety of packages written by fellow users. These bundles contain functions (i.e., routines) and/or data sets, which can be downloaded from CRAN or from the toolbar at the top of your screen. If you do not have administrative privileges on your computer, download the ZIP or TGZ package file to a local folder instead and install it through your console.

Once installed, a package is loaded using the command `require(PACKAGE_NAME)` or, equivalently, `library(PACKAGE_NAME)`. Saved workspaces retain only the objects you created; consequently, packages must be loaded again each time you open a new R session. Each package has a help manual on the CRAN Website, which lists and describes its components. These help pages can also be accessed through the console using `?` and `help()` after loading a package.

Check functions in an R package before use to ensure they are operating as you expect them to. The list of packages is dynamic; consequently, updating your computer's operating system or your version of R may cause problems. One way to circumvent this issue is to manually install an older version of a package. See the CRAN Website for instructions on creating your own R packages.

A brief list of packages for a range of statistical methods available at the time of writing is provided in Table 2; functions and packages for miscellaneous applications are in Table 3. Note that these lists are in no way exhaustive.

3 BASIC R OBJECTS AND COMMANDS

Information in R is generally one of three variable types: numeric, character string, or logical (i.e., Boolean). These bits of information are organized using R objects such as vectors, matrices, and data frames. We begin this section with these data types, followed by a discussion of R object types. We then introduce some basic mathematical operations that can be applied to numeric data. Finally, we focus on character strings and factors, which are a special way of handling categorical data. Note that R has additional object types, such as linear model objects or spatial objects, associated with more advanced data structures. To check the types, use `class()` or `mode()`. You can check for or force a specific class by using, for example, `is.numeric()` or `as.numeric()`, respectively. This applies to both data and object types (i.e., numeric, integer, double, complex, character, logical, matrix, array, data frame, list, and factor).

3.1 Numbers, Character Strings, and Logicals

Numeric values are real numbers. The largest and smallest number allowed depend on your computer; use the command `?.Machine` for more details. Infinite values are denoted by `Inf` and `-Inf` and indeterminate forms by `NaN`. R can also handle complex numbers written symbolically such as `3+4i`, where `i` is treated not as an R object name but the imaginary number $i = \sqrt{-1}$. Operations on numerical data are discussed further in Section 3.2 and listed in Table 4.

Character strings treat words (or numbers) as abstract symbols. They must always be enclosed in double quotes, otherwise R treats the "word" as the name for an R object. We consider strings, and a related object type, factors, in Section 3.4.

The third data type, logical (i.e., Boolean), is generally used in conditional statements and to subset data. You can either use `TRUE/FALSE` or simply `T/F` to denote this type. See Section 4.1 for more details.

Below are some examples. Observe that missing data is denoted by `NA`.

```
> w <- 5          # numeric
> x <- 3+4i       # complex number
> y <- "hello"    # character string
> z <- TRUE       # logical/Boolean (TRUE/FALSE, T/F)
>
> NA              # missing data
```

3.2 Scalars, Vectors, Matrices, and Arrays

The most basic R objects are vectors and matrices that are related to the mathematical conceptions of the words. While these objects can contain any data type (e.g., character, numeric), they cannot mix data types within the object itself. For example, a vector initially created with both character string and numeric elements will be automatically converted into a character string vector. Operations can be applied element-wise or for, numeric objects, to the entire object (e.g., matrix algebra).

Up until this point, all of our examples have been with scalars. To construct a column vector from scratch use the concatenate function `c()`. For numerical vectors, three additional options exist: use : for integer sequences, `seq()` for sequences with regular spacing, and `rep()` for replicating a vector a set number of times. We examine a few examples next:

```
> ### scalar object saved to object u
> u <- TRUE
>
> ### vector: c() function concatenates values
> v <- c("hello","world",NA)             # note: 3rd element missing
> w <- -1:10                             # integer sequence: -1, 0, 1,
>                                        # 2,..., 10
> x <- seq(from=-1, to=10, by=2)         # sequence in increments of 2
> y <- seq(from=-1, to=10, length=10)    # sequence from -1 to 10 with ten
>                                        # values in equal intervals:
>                                        # -1, 0.222, 1.444,...,8.777,10
> z <- rep(c(4,5), times=6)              # repeat vector c(4,5) six times
>                                        # result: 12x1 column vector
```

Once a vector is constructed, new elements can be appended by reusing `c()`. To determine the number of elements, use `length()`. In R, vectors are indexed starting from 1. To extract elements from a vector, specify which elements between square brackets `[]`:

```
> v <- c("hello","world",NA)          # note: 3rd element missing
> v <- c(v, "R")                      # append value to end of v
> length(v)                           # length of vector v
>
> v[2:4]                              # extract 2nd-4th elements
> v[c(FALSE, TRUE, TRUE, FALSE)]      # extract 2nd-4th elements
```

Each vector is essentially a column vector, as in linear algebra, and `t()` is the transpose function. To construct a matrix from individual vectors, use `rbind()` to append by row or `cbind()` to append by column. The function `matrix()` allows one to construct a matrix directly, similar to `vector()`. As with vectors, all elements of a matrix must be of the same data type. You can name or extract the column and row names using `colnames()` and `rownames()`, respectively.

Matrix dimensions, like in mathematics, are expressed as (rows×columns). The functions `dim()`, `nrow()`, and `ncol()` return the dimensions, number of rows, and number of columns, respectively. Both rows and columns are indexed beginning from 1. You can extract certain rows, columns, or elements again by using square brackets, but separating the row and column indices by a comma:

```
> ### matrix
> A <- matrix(c(4,2,7,8,4,6), nrow=3, byrow=TRUE)
> dim(A)      # dimensions of matrix A: rows x columns
> A[1,1]      # A[row_number, column_number]
> A[1,]       # first row, all columns
```

A related object is an array, a multidimensional matrix, which can be constructed with the function `array()`. Be careful when constructing an array to ensure the elements are in the correct order:

```
> ### array: multidimensional matrix
> B <- array(1:8, dim=c(2,2,2))
>
> dim(B)
> B[1,1,1]
> B[1,,]
```

There are many numerical operations in base R, many of which we list in Table 4. Most of the functions are applied element-wise to a vector or matrix on the left side of the table. On the top-right, the matrix algebra functions are applied to the object as a whole. Some miscellaneous, but useful, representations are provided at the bottom-right of the table.

Use `help()` for details on any of these functions. For functions applied to character strings, see Section 3.4; for logical operators, see Section 4.1.

3.3 Data Frames and Lists

Data frames are what statisticians generally think of as a data table where each column represents a variable. Columns can contain data of different variable types, but must be of the same length. Lists are the more general form of a data frame, where elements are not required to have equal lengths or even the same object types.

The function `data.frame()` binds vectors together, possibly with column headings. The functions `colnames()`, `names()`, and `rownames()` can be used to attach and extract column and row names for these objects. You can extract a given row, column, or specific cell from a data frame using the column/row index value or by name using the operator `$`:

```
> ### data frame (can combine variable types among
> ### columns unlike vectors and matrices)
> y <- data.frame("a"=c(1,2,3), "b"=c("hi","hi","bye"),
>            "c"=c(TRUE,TRUE,FALSE))
>
> # 3 ways to extract first column (vector) "a"
> y$a
> y[,1]
> y[,"a"]
>
> dim(y)    # dimensions of data frame, nrow() by ncol()
```

Lists are constructed very similarly using the `list()` function; however, the data is not structured like a data table as the combined objects may differ. Extraction of elements differs slightly as well. In the next example, the list `z` has two elements: a vector and a data frame.

```
> ### list (general form of data.frame--components
> ###        can be of varying object type)
> z <- list("d"=c(1,2), "e"=data.frame("v1"=c("is","hi","bye"),
+                                      "v2"=c(4, 6, 3)))
> z$d            # extract vector "d"
> z[[1]]         # extract vector "d" (first list component)
> (z[[1]])[2]    # extract element 2 of vector "d"
>
> names(z)       # extract (or assign) list names
> length(z)      # number of components in z
> unlist(z)      # remove list structure; convert to vector
```

To remove the list structure, use the function `unlist()`; this collapses all components into one vector. Note that this conversion may change the variable type of the data. Lists are extremely useful when writing functions as multiple results can be returned in the output. More details are provided in Section 4.5.

3.4 Strings and Factors

Character strings can be used to represent text and categorical variables. Some functions are: `nchar()` to calculate the number of characters, `substr()` to extract a subset of characters within a string, `paste()` to concatenate strings, `print()` and `cat()` to print output, `strsplit()` to separate a string based on a specified character (or string), `chartr()` to replace specified characters, `grep()` for pattern matching, and `tolower()/toupper()/casefold()` to change the case of the letters. These functions are useful when cleaning data (see Section 6). Some examples are given below:

```
> w <- c("hello", "this", "is", "R")
> nchar(w)                          # output: 5 4 2 1
> substr(w, start=1, stop=2)        # output: "he" "th" "is" "R"
>
> x <- paste("Hello", " what is your name?", sep=",")
> strsplit(x, split=",")            # split x before and after comma
>
> print(x)                          # print output to console
> cat("Hello", " what is your name?", sep=",")
```

Factors are an R object type for categorical data which are treated differently from character strings. They have an additional attribute, levels, that describes all possible values a factor object can take (this can exceed the number of categories which actually exist in the data). To construct a factor, use `factor()`. The argument "`levels`" can be omitted if the levels should be extracted from the data itself. If not all categories are represented within the data, use "`levels`" to specify them. In the following example, the data contains categories `a`, `b`, and `c`, but there is a fourth category `d` that, while not observed in the data, is a possible outcome for the variable. The function `levels()` extracts the levels and `table()` produces a count of each category.

```
> y <- factor(c("a","a","b","c","c"), levels=c("a", "b", "c", "d"))
> levels(y)           # extract levels: "a", "b", "c", "d"
> table(y)            # print contingency table (note 0 counts
>                     # for level "d")
>
> as.character(y)     # remove "factor" property
> as.numeric(y)       # convert to numeric labels; remove
>                     # "factor" property
```

Factors are tricky in R as it is possible to convert objects, sometimes unknowingly, into factors causing problems with analysis. The function `is.factor()` can check for this object type. To convert factors back into character strings, use `as.character()`; factors can also be converted into numerical categories by using `as.numeric()`. This issue is discussed further in Section 4.6.

One useful application of factors is to convert a quantitative variable into a categorical one. For instance, say you have ages of people and you want to use age ranges in your analysis instead. The `cut()` function can handle this conversion:

```
> z <- runif(n=20, min=0, max=100)        # generate 20 random
>                                         # numbers from 0 to 100
>
> cut(z, breaks=c(0, 25, 50, 75, 100))   # ranges & labels:
>                                         # (0,25], (25,50],...
> cut(z, breaks=c(0, 25, 50, 75, 100), labels=c("a", "b", "c", "d"))
```

The argument "`breaks`" specifies what the cut points should be; make sure to include the minimum and maximum value in this vector. If labels are not specified, then R assigns a label for each range. Note that `cut()` has some optional arguments which allow you to set additional specifications.

4 WRITING PROGRAMS

Among base R and the additional packages, there are a wide range of available functions; however, you may need to write your own code. In this section, we describe the components of such programs beginning with conditional statements, continuing to loop structures, and ending with developing your own functions. These constructions allow you to write generalized statistical methods that you can later apply to multiple data sets and settings (or perhaps turn into your own R package). On a practical note, following good form when writing functions is critical to both readability and debugging: indent for nested expressions and comment liberally.

4.1 Conditional Statements

We introduced logical (or Boolean) types in Section 3.1 that take two values: `TRUE` or `FALSE` (the abbreviations `T` and `F` work too). To generate logical vectors, we need conditional statements that determine whether the specified conditions are true. In this section, we introduce these types of expressions and show their utility when writing functions. In Section 6, we will use them to clean and process data.

Most logical operators, such as checking for equivalence (==), nonequivalence (!=), "not" (!), inclusive "or" statements (|), and finally "and" statements (&) can be applied to both numerical and character objects. Those such as <, >, <=, and >= are interpreted mathematically and are appropriate for numerical objects only. More complex expressions can be constructed by combining operators. All of the expressions to construct conditional statements, along with examples, are listed in Table 5. Note that these operators will produce an `NA` if applied to a missing data value (e.g., `NA`).

The functions in Table 5, `any()`, `all()`, and `is.element()` apply specifically to vectors and are related to operations for sets: `union()`, `intersect()`, `setdiff()`, and `setequal()`. Finally, functions such as `is.na()` or `is.factor()` can be used to check for certain object or data types.

4.2 `if/else` Statements

If/else statements and loops are control structures; they can change the direction of program flow. We start with if/else statements in this section. These have the following structure: `if(`*`CONDITION`*`){`*`PROCESS_A`*`}else{`*`PROCESS_B`*`}`. If *`CONDITION`* is true, then *`PROCESS_A`* is implemented, otherwise (if applicable) *`PROCESS_B`* is executed. The `else` component of the statement is optional. If the processes are simple (i.e., one command), then the more efficient `ifelse()` can be used: `ifelse(`*`CONDITION, COMMAND_A, COMMAND_B`*`)`. An alternative is `switch()`, which is convenient if there are multiple "else" statements.

To compute the median value of a numeric vector, we can use an if/else statement. If the vector has an odd number of elements, the median is simply the middle value, otherwise, the average of the two "middle" values must be computed. We can determine the parity of a number using the modulo operator, `%%`. Note also how we use indentation within the if/else statement to make the program easier to read.

```
> # computing the median
> x <- c(1, 3, 5, 2, 7, 8)
> n <- length(x)
>
> if(n%%2==0){
>       # n is even
>       median.x <- sum(sort(x, decreasing=FALSE)[(n/2):(n/2+1)])/2
> }else{
>       # n is odd
>       median.x <- sort(x, decreasing=FALSE)[ceiling(n/2)]
> } # end if/else
> median.x
>
> # checking the code using built-in R functions:
> median(x)
> quantile(x, probs=0.5)
>
> # solution: median is 4
```

4.3 `for` Loops

Especially useful for resampling methods and simulation studies, a `for` loop repeats a process for a preset (fixed) number of iterations. The basic format is `for(`*`INDEX`*` in `*`VECTOR`*`){`*`PROCESS`*`}`. *`INDEX`* is the variable name that is the counter for the loops, *`VECTOR`* is a vector object or a vector itself which gives the counter values to be looped through, and *PROCESS* is the set of operations to be repeated. For example, to compute the sample mean, we must start by adding the elements of a vector, best done by a loop. In the following code, we use the data `faithful` that contains observations on Old Faithful geyser activity at Yellowstone National Park in the United States. The first column lists the length of each eruption, and the second, the time until the next eruption, both in minutes. The code below utilizes a `for` loop to compute and print the sample mean of each column; observe that we again use indentation for clarity:

```
> for(i in names(faithful)){ # loop through each column of faithful
>
>     sum.x <- 0
>     for(j in 1:nrow(faithful)){ # loop through each row of column i
>         # compute cumulative sum
>         sum.x <- sum.x + faithful[j,i]
>     } # end j for loop (through rows)
>
>     # compute and print column average
>     print(paste("sample mean of column", i, "is:",
+         round(sum.x/nrow(faithful), digits=3), sep=" "))
> } # end i for loop (through columns)
>
> # checking our code:
> apply(faithful, 2, mean)
>
> # eruptions mean: 3.488
> # waiting mean: 70.897
```

We have two loops that are nested in the example above. For loop `i`, *VECTOR* is composed of character strings, the column names, whereas loop `j` is indexed by the row numbers.

The function `paste()` was used above to format the output generated by `print()` (or `cat()`). A second use is to print how many iterations have been completed. For instance, say the loop is indexed by `i` through a numerical vector:

```
> i <- 100
> if(i%%100==0) print(paste("iter",i,"is complete", sep=" "))
> # sep=" " indicates a space should separate each component in paste()
```

This prints `i` after every 100 iterations using the modulo operator `%%`. For loops with time intensive processes, generating such a statement is convenient to monitor your progress.

4.4 `while` Loops

All `for` loops can be written as `while` loops, the more general type of loop. The general format of a `while` loop is: `while(`*CONDITION*`){`*PROCESS*`}`, where *PROCESS* is repeated until *CONDITION* is false. We rewrite our inner `for` loop in the previous section with a `while` loop to compute the average length of Old Faithful eruptions (from data `faithful`):

```
> x <- faithful$eruptions
>
> # average the numbers in vector x using a while() loop
>
> sum.x <- 0
> index <- 1 # set index
> while(index <= length(x)){ # loop repeated as long as
>                            # index is less than or equal to length(x)
>
>       # compute cumulative sum
>       sum.x <- sum.x + x[index]
>
>        if(index%%50==0) print(index) # print index value (for debugging)
>
>        # increase index
>       index <- index+1
>
> } # end while loop
>
> # compute average
> sum.x/length(x)
>
> # check code using built-in R function
> mean(x)
>
> # eruptions mean: 3.488
```

Prior to executing the code in *PROCESS*, the condition `index <= length(x)` is checked. The moment `index`, the counter, is greater than the number of observations in `x`, the loop ends. Note, again, that we use indentation to highlight the body of the loop. When the number of iterations may be large or each iteration is time consuming, `print(index)` which prints the number of completed iterations is helpful (in this case, we print after every 50 iterations). See the end of Section 4.3 for another example.

For a known number of iterations, a `for` loop is easier to write; however, this number is not always known in advance. Optimization algorithms are an example of such a situation. Then, a `while` loop or a `repeat` loop with a `break` clause is necessary. As a second example, we write a `while` loop which finds the first 10 prime numbers starting from 2:

```
> num.prime <- 0                    # number of primes
> int <- 2                          # first number to check if prime
>
> prime.list <- rep(NA, times=10)   # empty vector to hold prime
>                                   # numbers as we find them
>
```

```
> while(num.prime < 10){           # check whether we've found
>                                  # the 10th prime
>
>     num.divisor <- 0             # to count number of divisors
>                                  # for int
>     for(i in 2:int){
>         if(int%%i == 0) num.divisor <- num.divisor + 1
>     } # end for loop
>
>     # if number of divisors is 1, int is prime
>     if(num.divisor == 1){
>        num.prime <- num.prime + 1
>        prime.list[num.prime] <- int
>     } # end if statement
>
>     int <- int+1                 # need to test next integer
> } # end while loop
>
> prime.list # print prime numbers
> # solution: 2, 3, 5, 7, 11, 13, 17, 19, 23, 29
```

In this example, the `while` loop repeats nearly 30 times to extract the first 10 prime numbers; we could not use a `for` loop to run this program. It is possible (and easy) to accidentally construct an infinite `while` loop, therefore, testing your code is key. The escape key is useful to end such an infinite loop.

Above, we first constructed an empty vector, `prime.list`, then filled it within the `while` loop. This is more efficient than starting with a vector of length one and continuously appending to it as the loop progresses. See Section 4.6 for more details.

A final note: it is best to avoid loops where possible as they tend to utilize more memory in R. For simple processes, vectorization can be used as a substitute. Instead of looping through a vector or columns of a matrix, for instance, writing code that can directly implement a function element-wise is more efficient. Examples of vectorized functions include `nchar()` or `log()`; in contrast, `sum()` and `det()` are applied to entire objects. The functions `apply()`, `lapply()`, and `aggregate()`, which we describe in Section 6, are essential for vectorization as well.

4.5 Functions

Functions are self-contained pieces of code creating a subenvironment from which operations are run. They allow for more flexible programs. We are already using many such functions such as `length()`, `data.frame()`, and `nchar()`. In this section, we will learn how to write our own functions.

Let us start by studying a function already available to us `matrix()`. If we read the help page for this function, we find that `matrix()` has five arguments:

"data," "nrow," "ncol," "byrow," and "dimnames." Four of these arguments have default settings. For example, if you do not specify the number of columns, "ncol," it is automatically set to 1. Some arguments may not be applicable to all uses of the function; for such a case, the default value is set to NULL as in the argument "dimnames."

By calling the function matrix() without any specified argument, R prints a 1×1 empty matrix (i.e., NA) with no row or column names. If we forget to add the parentheses and type matrix, we get the function code; for most functions in the base R package, this may not be useful, however, within your own R code or when customizing add-on packages, this can be helpful.

To call (i.e., use) a function, specify the appropriate arguments. You can define them by argument name, as in matrix x below, or not, as in matrix y. If you omit the argument name, R assumes that you have listed the arguments in the same order as described in the help page.

```
> ?matrix    # load help function
> matrix()  # constructing a matrix with all preset values
> matrix     # print function code
>
> # x and y are identical matrices
> x <- matrix(data=c(5, 6, 3, 4), nrow=2, ncol=2, byrow=FALSE)
> y <- matrix(c(5, 6, 3, 4), 2)
```

To write our own functions, we follow the format:

```
FUNCTION_NAME <- function(ARGUMENT_1=DEFAULT_1, ARGUMENT_2=
DEFAULT_2, ...){
   PROCESS
   return(OUTPUT)
} # end function
```

The arguments listed between the parentheses (possibly with default or NULL settings) are then used within the curly brackets. The return() function contains your R output; if you skip this step, R will return the results of the last command executed. If you want to include more than one object as output, specify them in list form using list() within return().

For our first example, we convert the code for computing the mean of each column of a data frame from Section 4.3 into a function. This will allow us to apply this program to any data frame, not just the Old Faithful data, faithful. We will call this function, mean.data.frame() and it will accept one argument, "data.set" with no default settings. Much of our initial program remains only with the generic "data.set" replacing the specific "faithful." At the end, we return the result as a data frame, each column representing one of the variables in data.set. To check our code, we apply the function to two data sets: faithful and attitude.

```
> mean.data.frame <- function(data.set){
>
>     # data.set: data frame with all numerical values
>
>     mean.col <- rep(NA, times=ncol(data.set))
>     n <- nrow(data.set)
>
>     for(i in 1:ncol(data.set)){ # loop through each column of data.set
>         sum.x <- 0
>             for(j in 1:n){ # loop through each row of column i
>                 # compute cumulative sum
>                 sum.x <- sum.x + data.set[j,i]
>             } # end j for loop (through rows)
>
>     mean.col[i] <- sum.x/n
>     rm(sum.x)
>   } # end i for loop (through columns)
>
>   # format output
>   result <- data.frame(t(mean.col))
>   colnames(result) <- names(data.set)
>   return(result)
> } # end function
>
> # testing function
> mean.data.frame(faithful) # Old Faithful geyser data
> mean.data.frame(attitude) # Chatterjee-Price Attitude Data
```

As with loops, it is important to use indentation and comments to make your code more readable. Furthermore, you can call other functions or even write functions within a function. It is helpful to divide a large task into smaller ones, writing a function per task. A final wrapper function can then execute the tasks in order, calling internal functions as needed. This is another technique that increases the readability of your code.

When a function is called, it opens a local environment in R. The objects defined within it are unavailable to the global environment (i.e., outside of the function) unless returned as output. However, global objects, even if not specified within the argument list, are accessible within the function environment. Apart from calling other functions within your function, it is generally inadvisable to refer directly to global objects within the local environment of the function. This limits the flexibility of the function. It is also good practice not to use similar object names within a function when they already exist as global objects; this may lead to confusion as a writer (and reader) of code along with leaving open the possibility of unintentionally altering those global objects.

As mentioned earlier, we often need to return multiple objects as output from a function; we can do this using `list()` within `return()`. In our second example, we use the code for listing consecutive prime numbers from Section 4.4. We will generalize the program to list a user-specified number of prime numbers beginning from a user-specified starting point.

Our function, `prime.number()` accepts two arguments: "`num.of.primes`" indicating how many prime numbers to list and "`start.num`" the smallest number at which to start checking for primes, the default being 2. Again, the code is essentially the same as before, but we replace "10" and "2" with the function arguments. We want to return both the vector of prime numbers and the number of values checked for the prime number property as a list. It is helpful to add names to the elements in the list to remind the user what each component represents. As with all lists, the operator `$` can be used to extract particular list elements.

```
> prime.number <- function(num.of.primes, start.num=2){
>     # num.of.primes: number of primes to find
>     # start.num: start checking for primes with this number
>
>     num.prime <- 0          # number of primes
>     int <- start.num        # first number to check if prime
>     # empty vector to hold prime numbers as we find them
>     prime.list <- rep(NA, times=num.of.primes)
>
>     while(num.prime < num.of.primes){  # kth prime found?
>
>         num.divisor <- 0         # count number of divisors for int
>         for(i in 2:int){         # check for prime number property
>           if(int%%i == 0) num.divisor <- num.divisor + 1
>         } #end for loop
>
>         # if number of divisors is 1, int is prime
>         if(num.divisor == 1){
>           num.prime <- num.prime + 1
>           prime.list[num.prime] <- int
>         } #end if statement
>
>         int <- int+1          # need to test next integer
>     } # end while loop
>
>     # results in the form of a list
>     return(list("primes"=prime.list, "num.iter"=int-start.num))
> } # end function
>
> # test function
> prime.number(num.of.primes=10)
>
```

```
> x <- prime.number(3, 4)
> x$primes    # extract primes: 5, 7, 11
> x$num.iter  # extract number of iterations
```

While writing functions, there are a few additional useful techniques. First, there are settings where all of the required arguments are not known in advance. For example, say you want to simulate values from a user-specified probability distribution. Every distribution requires a different set of parameters (e.g., for a normal distribution you need the mean μ and standard deviation σ, for Poisson, the rate λ). For your function to handle such variability, after listing all of the arguments, add an ellipsis "..." before ending the parentheses. The "..." represent the additional arguments. Whenever these arguments are required, insert "..."; the contents will be copied to those locations. When using the function, simply add these arguments in your function call in place of "....". R will know which are the extra arguments. See the function `samp.dist()` which simulates the sampling distribution of the sample mean in Section 9 for an example. (An example of a base R function with varying arguments is `list()`.)

Another useful pair of functions are `assign()` and `get()`. The first function allows you to assign names to new R objects dynamically and the second to extract existing R objects dynamically. You can also use `mget()` to obtain multiple R objects:

```
> for(i in 1:2){
>    assign(paste("vector",i,sep="."), faithful[,i])
> } # end for loop
> ls()
>
> mean(get(paste("vector",2,sep=".")))
```

For an example of using `get()` in practice, see `samp.dist()` in Section 9, where the sampling function is extracted for a user-specified distribution.

Finally, to check the run time of your code, use `date()` before and after a function call as in the example below:

```
> date()
> prime.number(num.of.primes=500)
> date()
```

4.6 Debugging and Efficiency

As programs become more complex and involve more lines of code, error checking—both of the syntactical and methodological variety—becomes harder. In this section, we outline some basic practices to reduce errors in your code; we also introduce the debugging functions within R.

Syntax errors will generally produce a warning message from R. Try the following examples:

```
> x <- c(5, 6,            # forgetting a ")"; R thinks you are continuing
>                         # to the next line
> for(i 1:6) print(i)     # forgetting "in" within the for loop
>
> y <- "a"
> y = "c"                 # want to check if the value in y is equal to "c",
>                         # instead we have replaced "a" with "c";
>                         # no error message!
```

Methodology errors, however, are much harder to spot and rarely generate an error message. These errors require careful checking of code under a variety of conditions. For example, let us apply the function `mean.data.frame()`, which we wrote in Section 4.5, to two new data sets: air quality measurements in New York (`airquality`) and plant measurements (`iris`):

```
> mean.data.frame(iris)       # output: Species column NA
> mean.data.frame(airquality) # output: Ozone and Solar.R columns are NA
```

In both cases, the output is incomplete. The function stumbles on each data set for different reasons. With `iris`, R generates warnings (type `warnings()`). If we examine the data, we see the final column, "`Species`," is a categorical variable, treated as a factor; the mean is an irrelevant statistic here, hence the errors (and warnings). (See code below.)

With `airquality`, some of the data is missing. We can see that observations have missing data and also whether the missing data occurs within specific variables only. We find that 42 of the 153 observations contain missing values for at least one of the variables; we also see that only "`Ozone`" and "`Solar.R`" have missing values; all other variables are complete. The code is provided below:

```
> head(iris)                          # first 6 rows of iris
>
> head(airquality)                    # first 6 rows of air quality
> nrow(airquality)
> # row numbers containing NAs
> sum(!complete.cases(airquality))
> # determine whether any columns have NAs
> apply(apply(airquality, 2, is.na), 2, any)
```

We did not account for either of these scenarios in `mean.data.frame()`. To do so, we update our function, now calling it `mean.data.frame2()` below. Both errors have been eliminated in our second attempt.

```
> mean.data.frame2 <- function(data.set){
>     # data.set: data frame
>     mean.col <- rep(NA, times=ncol(data.set))
```

```
>     n <- nrow(data.set)
>
>     for(i in 1:ncol(data.set)){ # loop through each column of faithful
>
>           if(is.numeric(data.set[,i])){ # column numeric?
>
>              print(paste("column",names(data.set)[i],
+                    "is numeric"))
>              sum.x <- 0
>
>              for(j in 1:n){ # loop through each row of column i
>                 # compute cumulative sum handling NAs
>                 sum.x <- sum.x + ifelse(!is.na(data.set[j,i]),
+                                         data.set[j,i], 0)
>              } # end j for loop (through rows)
>
>              mean.col[i] <- sum.x/n
>              rm(sum.x)
>           } # end if statement
>
>          if(i%%2==0) print(paste(i,"th column complete", sep=""))
>       } # end i for loop (through columns)
>
>     # format output
>     # drop categorical variable results
>     result <- data.frame(t(mean.col[!is.na(mean.col)]))
>     # drop categorical variable column names
>     colnames(result) <- names(data.set)[!is.na(mean.col)]
>     return(result)
> } # end function
>
> # testing function
> mean.data.frame(faithful)      # initial function with ideal case
> mean.data.frame2(faithful)     # updated function with ideal case
>                                # (check if answers match)
> mean.data.frame2(iris)         # with categorical data
> mean.data.frame2(airquality) # with missing data
```

There are two features in `mean.data.frame2()` that are helpful with debugging and efficiency tests. The first is `print(paste("column",names(data.set)[j], "is numeric")`; this command prints a message when the specified column is numeric indicating that the `if` condition is `TRUE`. We know that in the test case of `iris`, the first four columns are numeric and the fifth categorical. Therefore, this command should execute only four times. Printing indicators of progress throughout the function is helpful for following the flow of your program, especially if you have a lot of control structures, and to see when a function stops working. These can be commented out after you finish debugging (use #).

Another useful command is `if(i%%2==0) print(paste(i,"th column complete", sep=""))`. In this example, the column number is printed if the *i*th column is divisible by 2 (i.e., `i%%2==0`). For fast computations, this is unnecessary and slows down the execution of the program. For more complex programs, printing the iteration every so often will allow you to chart progress. Alternatively, you can print messages to a file which, if saved to a cloud, can allow you to monitor your program even if you are away from your computer (more helpful if your computer is not part of a larger server system or you cannot set up a remote desktop application).

Writing output to a file is beneficial especially if your program is memory-intensive. You can use `write.table()` to append output to an existing file as you generate results. In case your program crashes, writing to a file allows you to examine at least some of your results.

For simulated data, as will be described in Section 9, using `set.seed()` allows you to replicate your results when using random number generation. Retaining the same simulated values each time you test your function is helpful for debugging.

A common error occurs when character string vectors (or, for example, a particular column in a data frame) are incorrectly treated as factors. As they add attributes to a vector, some character string operations may execute incorrectly when used with factors. For example, selecting rows based on row names can fail if the vector of row names is treated as a factor. In such a case, using `as.character()` can fix the problem, although it is better to find the command where the problem originated (refer to Section 3.4 for more on strings and factors). Furthermore, when reading in data from a file, character string columns can automatically be converted to factors unless otherwise specified (see Section 5 for more details on avoiding this). A related issue is when factor levels are unintentionally converted into numeric values in place of the original category labels. Checking for these common problems is key if you frequently work with categorical data.

Another issue occurs with row numbers on data frames. Row numbers (row names) on a data frame are essentially attached to the original observations they label. After sorting and manipulating data, these row names will no longer be in order or match the total number of rows actually in the data set. This may cause problems when you execute commands that refer to the row number. Therefore, after the data cleaning step, it is important to renumber your data frame (i.e., `rownames(`*`DATA_FRAME`*`) <- 1:nrow(`*`DATA_FRAME`*`)`).

The techniques just described are *ad hoc*; R, however, has built-in debugging functions: `trace()`, `untrace()`, `browser()`, `traceback()`, and `debug()`. These functions allow you to step through your code to locate errors.

As mentioned earlier, to check for run times, you can use `date()` to time stamp your program. This is helpful if you to identify whether portions of your program are time intensive and may need to be more efficiently written. Loops tend to be slow in R; however, functions such as `apply()`, `lapply()`, and `aggregate()` can reduce computing times (see Section 6 for details). Vectorization of processes, as described in Section 4.4, can improve efficiency as well.

Another useful trick to reduce run times is to construct empty vectors and load them in as you progress as opposed to continually appending an element to an existing vector. We do this in our function `prime.number()` in Section 4.5 when defining the R object `prime.list` as an empty vector of length `num.of.primes`. This is more efficient than defining `prime.list <- NA` and using `prime.list <- c(int, prime.list)` instead of `prime.list[num.prime] <- int`. For more on how memory is allocated and used in R, see Section 2.2.

For complex programs, it is best to split the code into smaller functions, each with one main task. Then, at the end write a wrapper function which simply calls the sub-functions. This practice also helps with debugging as you can focus on one task in your program at a time. (Note, however, if you are having issues with memory, this suggestion may add to that problem.) Moreover, it is good practice to test your code as you progress, not after you have finished writing your program. Finally, do not use the same object name for global and local environments, as this may cause both computational and testing errors.

Text editors that color code your programs (e.g., orange for comments, green for functions, etc.) and match parenthetical/bracketed expressions can help immensely with syntax errors. Using indentation to indicate nested expressions more clearly is also a worthwhile practice. Finally, it cannot be repeated enough that commenting your code as you program is vital to both the debugging process and the reusability of your functions.

5 INPUT AND OUTPUT

Three file types can be uploaded into R: a workspace, code, and data. For the first two, refer to Section 2.3. As mentioned in the introduction, to be a self-containing text, we use the data already in R. Therefore, we start with file output before considering file input.

For example, let us use the `airmiles` data, a vector containing the number of miles people traveled on commercial U.S. airlines each year between 1937 and 1960. To write this vector to a file, use `write()`. First, specify the R object; second, the file name. (Note: make sure you know which directory your file is being written to as discussed in Section 2.3). Now, we can upload the vector using `scan()`:

```
> as.vector(airmiles)                              # vector of data
> write(as.vector(airmiles), file="airmiles.txt")  # output to a file
> scan("airmiles.txt")                             # input file as vector
```

For data frames, the function `write.table()` is used instead of `write()`. This function has numerous arguments, the most important of which are: "`x`," the data; "file," the file name; "append," whether or not the data should be added to an existing file (if FALSE, any existing file will be replaced); "sep," the character which separates values for each column within a row (common characters are a comma ",", a tab "\t", or a space " "); "row.names," whether or not row names should be written (best to set this to FALSE); and "col.names," whether the variable names of the data frame should be included or not. The "append" argument is especially useful when writing output as it is generated within a program (refer to Section 4.6).

You can then upload your data to R using read.table() where the basic arguments are the file name, the separating character, and whether the first line is a column heading. We use the Old Faithful geyser data, faithful, below as an example:

```
> head(faithful)      # print first 6 rows of faithful
> write.table(faithful, file="faithful.txt", append=FALSE, sep=",",
+          row.names=FALSE, col.names=TRUE)
> y <- read.table("faithful.txt", sep=",", header=TRUE)
> head(y)             # print first 6 rows of y
```

The function read.table() has additional arguments to specify the data type (character, logical, numerical, factor) for each column ("colClasses"), specifying the number of lines to skip in a file before beginning to read the data ("skip"), and whether character string columns should be treated as factors ("stringsAsFactors," the default is TRUE; this is generally best set to FALSE). The help page can elaborate on these function parameters. For CSV files, where variables are separated by commas (i.e., sep=","), the function read.csv() is an equivalent alternative. To upload files generated by other statistical software packages, such as Minitab, SAS, or SPSS, use the package *foreign*; for Excel files, install *xlsx*.

We end this section with a few miscellaneous functions. Before printing to a file, some formatting of the output may be required; the functions paste (), format(), and the rounding functions may be useful. R can also handle command line input using the function readline() when R is used interactively (see ?interactive for more information). To check whether your data has uploaded correctly, head(), tail(), colnames() and dim() can give you some basic information. Note that as # is the comment indicator in R, make sure the symbol does not show up in your data files. This can occur with geographical data, in particular with street addresses; remove every instance of # before uploading your data. The functions in this section are compiled in Table 6.

6 DATA PROCESSING

After reading in your data as described in Section 5, you may need to clean it. R has a number of functions to help process data. Some of these have been streamlined and improved upon in the package *plyr* (see CRAN for more details and refer to Table 6, for a list of data processing functions).

To work with vectors instead of data frames, use the function attach(). This splits the data frame into its component vectors with the column names repurposed as object names. Use detach() to undo this step.

To sort a vector, use sort(). You can sort in ascending or descending order along with returning a vector of the index using index.return=TRUE; this function can also handle missing data, NA. To sort your data frame by columns, use order() instead:

```
> # order smallest to largest
> ascend.sort <- faithful[order(faithful$eruptions),]
> # order largest to smallest
> descend.sort <- faithful[order(-faithful$eruptions),]
>
> # reassign row numbers
> rownames(ascend.sort) <- 1:nrow(ascend.sort)
```

For sorting by multiple columns, add additional columns within order() separated by commas.

In the example above, we sorted the faithful data by the column "eruptions." This operation shifts the row numbers of the data frame. To avoid problems in later stages, it is prudent to renumber the rows after cleaning your data.

To clean categorical data, commands borrowed from UNIX, such as grep(), are available. As described in Section 3.4, numerical variables can be converted into categorical variables by range using cut().

R generally ignores observations containing missing data when running analyses. To check and extract missing data, the functions is.na(), complete.cases(), and na.omit() are useful. For imputation, the packages *mi*, *mice*, *mitools*, *MImix*, and *pan* are available.

If you want to split the data into training and test sets, you can use the sample() function (see Section 9). We use the Chatterjee–Price Attitude Data here (attitude):

```
> training <- rep(TRUE, nrow(attitude)) # all observations in training set
> # randomly select 5 observations to be in the test set:
> training[sample(1:nrow(attitude), size=5, replace=FALSE)] <- FALSE
>
> # append vector training to attitude data frame
> attitude2 <- data.frame(attitude, "train.set"=training)
```

Then, use the function subset() to select observations you want to use. Alternatively, you can use the split() function to separate out the data into a list based on a category:

```
> # extract test set from attitude2
> attitude2[!attitude2$train.set,]
> # extract test set using subset()
> test.set <- subset(attitude2, train.set==FALSE)
```

```
>
> # split attitude2 into list with two components:
> #  'TRUE' (training set) and 'FALSE' (test set)
> attitude3 <- split(attitude2, f=attitude2$train.set)  # split data
> attitude3$'FALSE'                                       # extract test set
```

Note that subset() and split() can be used to subset data under any set of conditions (see Table 5 for a list of logical operators and related functions).

You can append new columns to your data frame if you, for instance, need to transform your data. Below is an example with the iris data on plant measurements:

```
> # add log(Sepal.Length) to data frame
> iris2 <- data.frame(iris, "log.Sepal.Length"=log(iris$Sepal.Length))
```

Provided that two data frames have the same columns (with the same column names), we can attach the rows together to create one larger data frame:

```
> attitude.train <- attitude[attitude2$train.set,]    # extract training set
> nrow(attitude.train)
> attitude.test <- attitude[attitude2$train.set,]  # extract test set
> nrow(attitude.test)
>
> # append data frames
> attitude.full <- rbind(attitude.test, attitude.train)
> nrow(attitude.full)
```

Note that rbind() works even if the column order is different between the data frames or if columns have different data types.

If you have two data sets with some common columns, you can match observations and combine them into one data frame using merge():

```
> x <- data.frame("a"=c(1, 6, 3), "b"=c(10, 8, 7))
> y <- data.frame("a"=c(1, 6, 4), "c"=c("yes", "no", "yes"))
>
> merge(x,y)                              # method 1
> merge(x,y, all.x=TRUE)                  # method 2
> merge(x,y, all.x=TRUE, all.y=TRUE)      # method 3
```

Method 1			**Method 2**			**Method 3**		
a	b	c	a	b	c	a	b	c
1	10	yes	1	10	yes	1	10	yes
6	8	no	3	7	<NA>	3	7	<NA>
			6	8	no	4	NA	yes
						6	8	no

Observe that the three function calls to merge() above produce different results.

With method 1, only the common rows (matched by column "a") are merged. In method 2, we specify that all observations in x should be included. Note that the third observation in x is not in y; consequently, after merging, the value in column "c" for that observation is missing. Similarly, in method 3, we specify that all observations in x and y should be included; consequently, missing values are automatically inserted where appropriate.

Finally, functions can be applied to data frames (or vectors) separately by a category using `aggregate()`. This function has the following form: `aggregate(`*`R_OBJECT`*`), by=list(`*`CATEGORY_LIST`*`, FUN=`*`FUNCTION`*`)`. In the next example, we use the `iris` data to compute the maximum value of each quantitative variable for each category in `Species`:

```
> # compute max for each species
> aggregate(iris[,1:4], by=list(iris$Species), max)
```

Here, *`CATEGORY_LIST`*, is `list(iris$Species)`; that is, the computations on `iris[,1:4]` should be completed separately for each species of iris. This must be in list form; to include more than one category, separate terms with a comma within `list()`.

The functions `subset()`, `merge()`, `apply()`, `lapply()`, and `aggregate()` can help simplify your code for both data cleaning and computational operations. By reducing your dependence on explicit loops, they also increase efficiency.

7 EXPLORATORY DATA ANALYSIS

After processing your data, exploratory data analysis is the next step. For example, we can calculate the mean of a vector with `mean()` and the standard deviation with `sd()`; however, without specifying additional arguments, they will not work with data containing missing values (`NA`). By adding "`na.rm=TRUE`" to the function call, the statistic can be calculated:

```
> mean(airquality$Ozone) # vector has NA values
> mean(airquality$Ozone, na.rm=TRUE)
>
> sd(airquality$Ozone, na.rm=TRUE)
```

For categorical data, construct a contingency table using `table()`:

```
> # counts of each species of iris in the data set
> table(iris$Species)
```

When tabulating multiple categorical variables, use a comma between each vector in the arguments for `table()`.

The functions `apply()` and `aggregate()` are helpful here too (see Section 6 for use in data processing). To compute a statistic separately for each row or column, use `apply()`:

```
> # compute average for each column in faithful data set
> apply(faithful, 2, mean)
```

For lists, `lapply()` is the analogous function.

Alternatively, to compute a statistic for a numerical variable separately for each concomitant category, use `aggregate()`:

```
> # compute standard deviation of sepal length for each
> # iris species separately
> aggregate(iris$Sepal.Length, by=list(iris$Species), sd)
```

See Table 7 for a list of functions to calculate summary statistics (numerical and categorical) and basic graphs.

8 STATISTICAL INFERENCE AND MODELING

In this section, we cover basic statistical analyses: hypothesis testing and linear regression. See Table 2 for information on more advanced methods.

8.1 Hypothesis Testing

R includes a basic set of hypothesis tests; Table 8 lists the most common ones. In the next example, we focus on the one-sample *t*-test. Say we want to test whether the average length of an Old Faithful geyser eruption is greater than 2 min:

$$H_0: \ \mu \leq 2\,\text{min}.$$
$$H_a: \ \mu > 2\,\text{min}.$$

We can answer this question using the `faithful` data with a one-sample *t*-test. The test statistic is $t = (\overline{x} - \mu)/(s/\sqrt{n})$ with $n - 1$ degrees of freedom. In R, function `t.test()` will run the test:

```
> # one-sample t-test, alternative hypothesis: ">"
> x <- t.test(faithful$eruptions, alternative="greater", mu=2,
+              conf.level=0.95)
> x
> names(x)
> x$statistic
```

We save the output in `x`. It includes, among other items, the test statistic, degrees of freedom, *p*-value, and confidence interval:

```
    One Sample t-test
data: faithful$eruptions
t = 21.498, df = 271, p-value < 2.2e-16
alternative hypothesis: true mean is greater than 2
95 percent confidence interval:
 3.373559    Inf
sample estimates:
mean of x
 3.487783
```

While the output is formatted when `x` is printed, it is actually saved in list form. The list component identifiers are generated with `names(x)`. For instance, to extract the test statistic, enter `x$statistic`.

8.2 Regression

We now introduce linear regression and diagnostic functions. For the first example, we use the `attitude` data (Chatterjee–Price Attitude Data) containing seven variables on employee performance: "`rating`" (y), "`complaints`," "`privileges`," "`learning`," "`raises`," "`critical`," and "`advance`." The variables are numerical with each representing the fraction of positive reviews for an employee on each measure. We start by examining scatterplots of all pairs of variables using `pairs()` as shown in Fig 1. From this plot, rating and complaints seem to have the strongest linear relationship which is confirmed by computing the pairwise correlations with `cor()`.

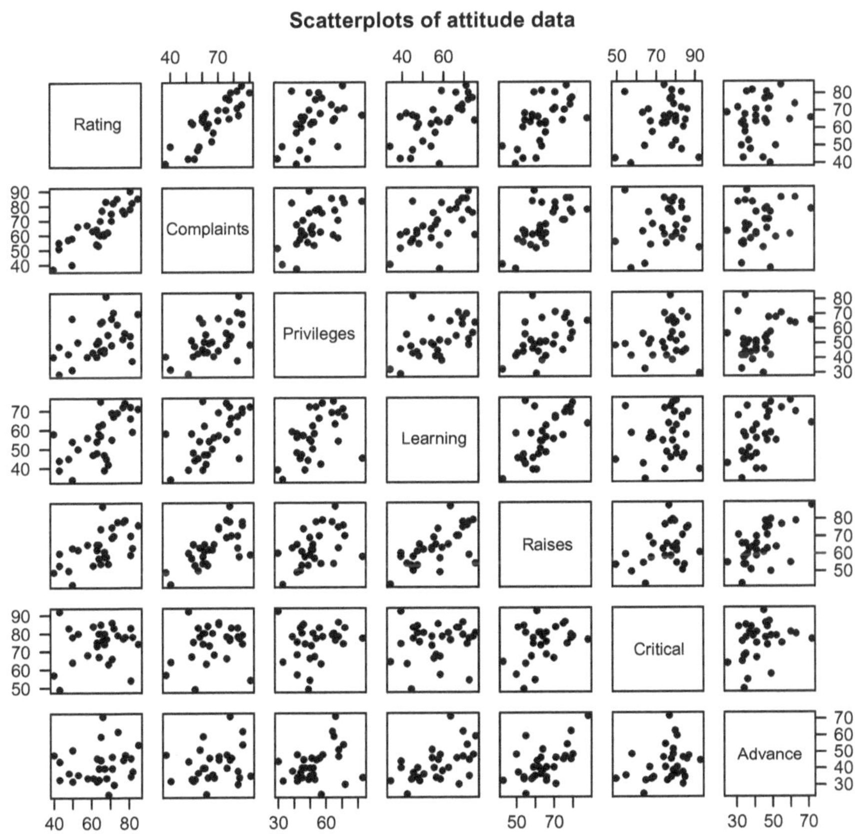

FIGURE 1 Scatterplots of all pairs of variables in `attitude` data.

Consequently, we fit:

$$y_{rating} = \beta_0 + \beta_1 x_{complaints} + \varepsilon. \quad (1)$$

Use `lm()` to run a linear regression; it returns a linear model object in list form as output. The function is called using the following format: `lm(FORMULA, data=DATA_FRAME, subset=LOGICAL_VECTOR, weights=VECTOR)`. The argument *`FORMULA`* expresses the model to be fit and takes the form: *`RESPONSE ~COVARIATE_1 +COVARIATE_2`* and so forth. The symbol "~" separates the response variable on the left with the covariates on the right. The remaining arguments are optional. If the response and covariates are not vectors in the workspace (e.g., you did not use `attach()`), specify the data frame in "`data`"; if you want to use only a subset of the observations in your regression (e.g., training and test sets) input a logical vector to the argument "`subset`"; finally, for weighted least squares, add a weight vector to "`weights`." Note that observations with any missing values among the variables selected for your model will be omitted from your analysis. The code for fitting (1) is below:

```
> # scatterplots and correlations for all pairs of variables
> pairs(attitude, las=TRUE, main="Scatterplots of Attitude Data",
+       pch=20)
> cor(attitude)
>
> # rating regressed against complaints
> model.1 <- lm(rating ~ complaints, data=attitude)
> names(model.1)          # components of linear model object model.1
>
> summary(model.1)        # coefficients, t-tests for slopes, etc.
> anova(model.1)          # overall F-test
```

The components of the linear model object, `model.1`, can be listed with `names()` and accessed as you would any list. For a summary of the coefficients and *t*-tests for slopes, use `summary()`. The output of `summary()` varies by the type of object it is applied to. Note that `summary(attitude)` and `summary(model.1)` give very different results; the first yields a six-number summary (see Table 7) while the second generates an overview of a fitted model. There are quite a few functions that behave this way in R; that is, they have multiple purposes. To get information on the use of summary for linear models, type `?summary.lm` instead of `?summary` into the command line. Finally, for the accompanying ANOVA table, use `anova()`.

For a simple regression model, we can plot the regression line on the scatterplot using `plot()` to plot the points, `coef()` to extract the regression coefficients, and `abline()` to plot those coefficients. See Fig. 2 for the graph. Like `summary()`, `plot()` yields a different result when applied to linear model objects. As we see in Fig. 3, `plot(model.1)` graphs model diagnostics (see `?plot.lm` for more details). The code is provided below:

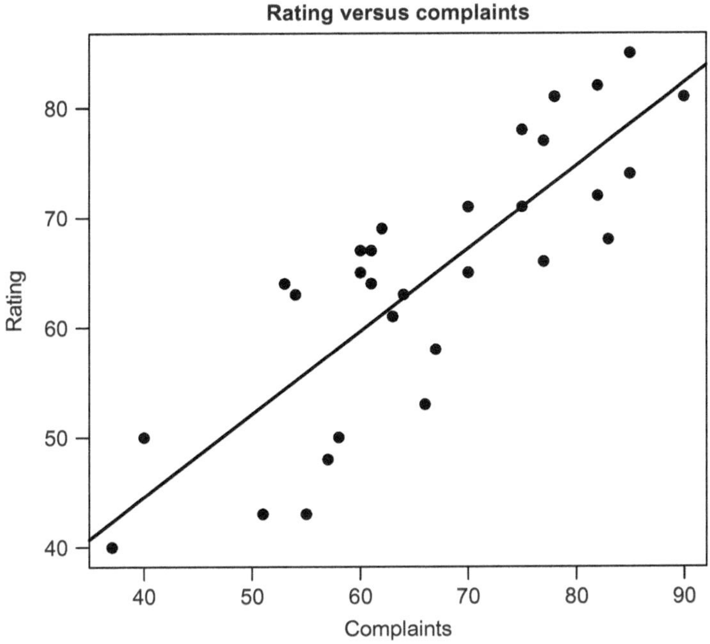

FIGURE 2 Model 1: Plotting the regression line.

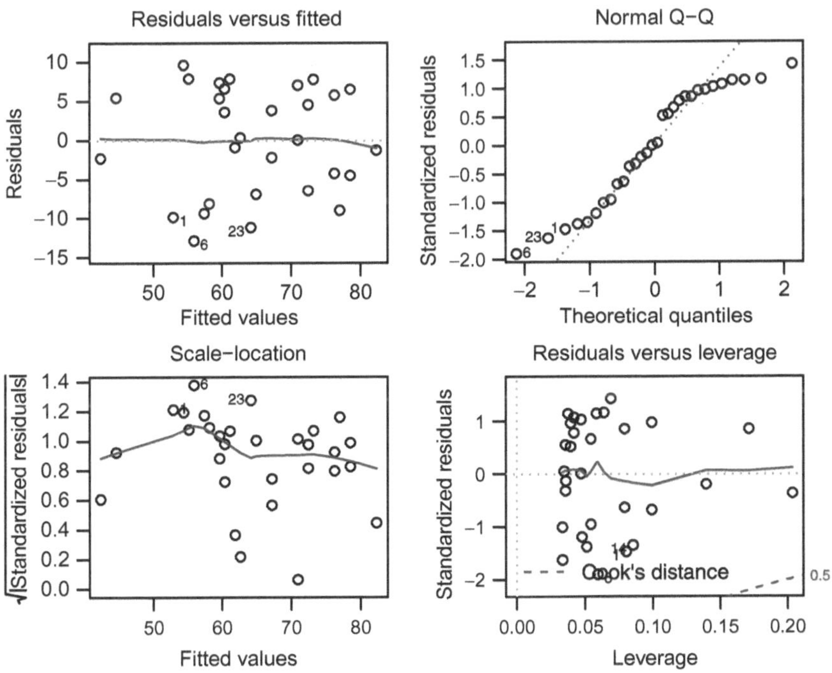

FIGURE 3 Model 1: Diagnostic plots.

```
> # scatterplot with regression line
> plot(attitude$complaints, attitude$rating, pch=19,
+        main="Rating vs. Complaints",
+        xlab="complaints", ylab="rating", las=TRUE)
> abline(coef(model.1), lwd=2)
>
> # model diagnostics
> par(mfrow=c(2,2))
> plot(model.1)
```

Residuals can be extracted one of two ways: directly from the linear model object or from `residuals()`. Other diagnostic measures such Cook's distances can be computed with `influence.measures()`, some of which appear in Fig. 3.

There are two settings to predict the response variable, $\hat{y}$: prediction for the observations used to fit the model and prediction for new observations. For the former case, as with residuals, you can extract them from the linear model object directly or use `predict()`. For new observations, construct a data frame with the relevant covariates and with column names identical to those used by the linear model. Then, use the function `predict()` with this data frame specified in the "`newdata`" argument. Note that we can also obtain standard errors along with confidence or prediction intervals at various α-levels. The code for diagnostics and fitted values are given below:

```
> # residuals, two ways
> model.1$residuals
> residuals(model.1)
>
> # other diagnostic measures
> influence.measures(model.1)
>
> # predictions for data used to fit model, two ways
> model.1$fitted.values
> predict(model.1)
>
> # predictions for new data
> attitude.data <- data.frame("complaints"=c(45, 67, 25))
> predict(model.1, newdata=attitude.data, se.fit=TRUE,
+             interval="prediction", level=0.95)
```

There are a few alternatives when expressing the regression equation. To use all of the variables as covariates, apart from the response, add a period as in `model.2` below. To fit a model without a y-intercept (not recommended), add `-1` along with the covariates as in `model.3`. Finally, you can transform your data directly in the `lm()` formula as in `model.4`. Note that nested models can be compared with partial F-tests by repurposing the function `anova()`: specify the smaller model first, then the second model separated by a comma. Code for these operations are next:

```
> # all variables used
> model.2 <- lm(rating ~ ., data=attitude)
> anova(model.1, model.2)     # partial-F test
>
> # no y-intercept
> model.3 <- lm(rating ~ -1 + complaints + privileges, data=attitude)
> # transform "privileges"
> model.4 <- lm(rating ~ complaints + log(privileges), data=attitude)
```

Until now, we have considered only numerical covariates; fortunately, categorical covariates are handled identically. However, one must be careful to check whether a column is being treated as a qualitative variable if the categories are numbers. In our second example, we use the data set `ToothGrowth` that records the length of guinea pig teeth (column "`len`") after they receive Vitamin C through orange juice or ascorbic acid ("`supp`") at one of three dosages ("`dose`"). When we examine this data, we see that the response ("`len`") is numeric, the source of the Vitamin C is categorical and listed as a character string, and the dosage is also categorical but listed in the data frame as a numerical variable.

Every statistical software has a slightly different way of expressing regression coefficients with categorical covariates. To see how R conveys the results, we fit a one-way ANOVA model with "`len`" regressed against the categorical "`supp`":

```
> # len is numerical; supp is a category (one-way ANOVA)
> model.5 <- lm(len ~ supp, data=ToothGrowth)
```

We analyze the output from this model `summary(model.5)` next:

```
Call:
lm(formula = len ~ supp, data = ToothGrowth)

Residuals:
     Min       1Q  Median      3Q     Max
-12.7633 -5.7633  0.4367  5.5867 16.9367

Coefficients:
            Estimate Std. Error t value  Pr(>|t|)
(Intercept)   20.663      1.366  15.127 <2e-16 ***
suppVC        -3.700      1.932  -1.915  0.0604 .
---
Signif. codes: 0 '***' 0.001 '**' 0.01 '*' 0.05 '.' 0.1 ' ' 1

Residual standard error: 7.482 on 58 degrees of freedom
Multiple R-squared: 0.05948,Adjusted R-squared: 0.04327
F-statistic: 3.668 on 1 and 58 DF, p-value: 0.06039
```

There are two categories of "supp": VC and OJ; however, the category OJ is dropped from the output. The *y*-intercept term, 20.663 is the sum of the intercept and the estimate for OJ. The coefficient for VC is then the *difference* between VC and OJ.

Now, let us add in "dose." If we type `lm(len ~supp + dose, data=ToothGrowth)`, R will treat "supp" as a categorical variable but "dose" as a numerical variable. To ensure that `dose` is also treated as a categorical variable, we turn it into a factor (or character string). Finally, to include interaction terms, add the lower order terms and use the multiplication operator * for the interaction effects as in `model.6` below:

```
> model.6 <- lm(len ~ supp + as.factor(dose) + supp*as.factor(dose),
+                    data=ToothGrowth)
```

To evaluate your models, `AIC()` and `BIC()` are useful. For stepwise regression, use `step()`. Test for multicollinearity with `vif()` in the package *car*. To test for autocorrelation in the residuals, use the Durbin–Watson test, `dwtest()` in the package *lmtest*. Finally, for all subsets regression, use the function `leaps()` within the package of the same name.

As listed in Table 2, R can handle more complex models. For one- or two-way ANOVA models, `lm()` can be used along with `oneway.test()` and `aov()`. Generalized linear models such as logistic regression or Poisson regression can be fit with `glm()`, generalized additive models with `gam()`. Fit mixed effects models with `lmer()` from the package *lme4*. Hierarchical generalized linear models can be fit with the *hglm* package.

9 SIMULATION

The first step in a simulation is the ability to generate random numbers. It is the basis of procedures from drawing a random sample from distributions to the bootstrap and jackknife. Internally in R, the command `.Random.seed` produces the random number generation states. The mechanics behind this function is beyond our scope; type `?.Random.seed` for the available methods (default is Mersenne–Twister).

To sample from a discrete distribution, use `sample()`. The arguments required are "x" to specify the vector to sample from (any data type); "size" for the number of draws; "replace" to indicate whether the sampling should be with or without replacement; and "prob" to indicate the probability of selection for each element in "x" (if unspecified, the distribution is assumed to be uniform). For example, let us simulate 1000 rolls of two dice, one fair and one weighted and plot the results (see Fig. 4).

```
> par(mfrow=c(1,2))
>
> # simulate and graph 1,000 rolls of a fair die
```

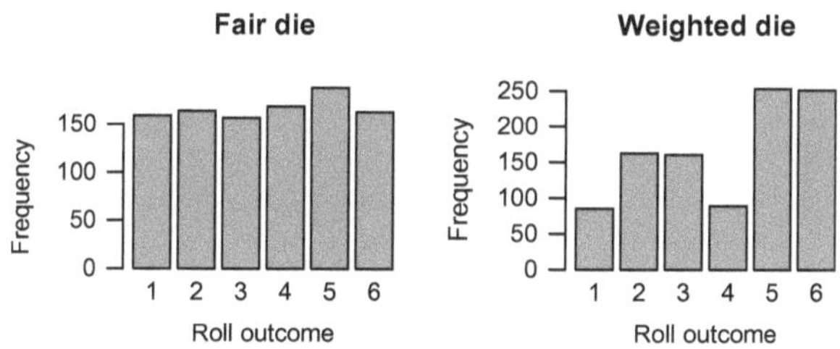

FIGURE 4 Sample simulation results for rolling dice.

```
> x <- sample(1:6, size=1000, replace=TRUE)
> barplot(table(x), main="Fair Die", las=TRUE,
+         ylab="frequency", xlab="roll outcome")
>
> # simulate and graph 1,000 rolls of a weighted die
> y <- sample(1:6, size=1000, replace=TRUE,
+            prob=c(1/12, 2/12, 2/12, 1/12, 3/12, 3/12))
> barplot(table(y), main="Weighted Die", las=TRUE,
+         ylab="frequency", xlab="roll outcome")
```

Each time we run the code above, a new set of observations are drawn. To replicate results (useful when debugging code), use `set.seed()` to set the random number generation seed.

A number of probability distributions exist within base R; add-on packages increase the available options (see Table 9 for a brief list). There are four possible functions for each distribution: density (`d`), cumulative distribution (`p`), quantile (`q`), and random draws (`r`). The normal distribution, for instance, has `dnorm()`, `pnorm()`, `qnorm()`, and `rnorm()`. For distributions not in base R or an accompanying package, you will have to use other methods such as rejection or importance sampling.

Apart from Bayesian methods, probability distribution functions are useful for simulation studies, especially when exploring asymptotic properties of statistics. In the function `samp.dist()` below, we simulate the sampling distribution of the sample mean, $\bar{x}$, from observations drawn from a given distribution. The results are plotted in Fig. 5. In the code, observe that we use the "..." notation in the function argument list. This allows any distribution to be used without having to preset all of the distribution parameters within the list of arguments for `samp.dist()`. Then "..." is passed on to any function call which requires those additional arguments (in this case, the sampling function). Furthermore, `get()` is used to obtain the sampling function for the specified distribution. (See Section 4.5 for more discussion on these two notes.)

```
> # sampling distribution for mean of a random variable
> samp.dist <- function(n, k,dist,...){
>
```

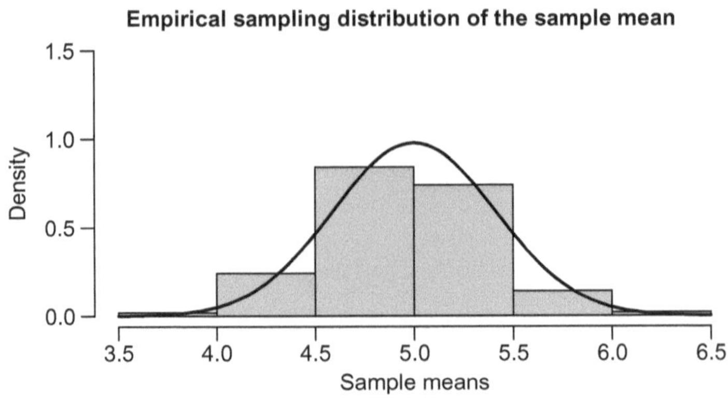

FIGURE 5 Sampling distribution simulation results.

```
>       # n: sample size; k=number of simulation runs,
>       # dist=distribution (norm, pois, etc.),
>       # ... = parameters for distribution in "dist"
>
>       # stop function if input invalid
>       if(any(c(k, n) <= 0)){return("error!")}
>
>       mean.vector <- rep(NA, times=k) # empty results vector
>
>       for(i in 1:k){
>            # extract sampling function
>            random.draws <- get(paste("r", dist, sep=""))
>              x <- random.draws(n=n,...) # simulate x_1,...,x_n
>              mean.vector[i] <- mean(x)   # compute and save x-bar
>              rm(x)                       # remove x-bar
>       } # end for loop
>
>       return(mean.vector) # function output
> } # end function
>
> # run simulation for Poisson distribution, rate parameter = 5
> z <- samp.dist(n=30, k=100, dist="pois", lambda=5)
>
> # histogram of sample means (density not frequency)
> hist(z, freq=FALSE, ylim=c(0, 1.5), las=TRUE, col="gray80",
+        main="Empirical Sampling Distribution of the Sample Mean",
+        xlab="sample means")
> # add normal density curve to histogram
> curve(dnorm(x, mean=5, sd=sqrt(5/30)), add=TRUE, lwd=2)
```

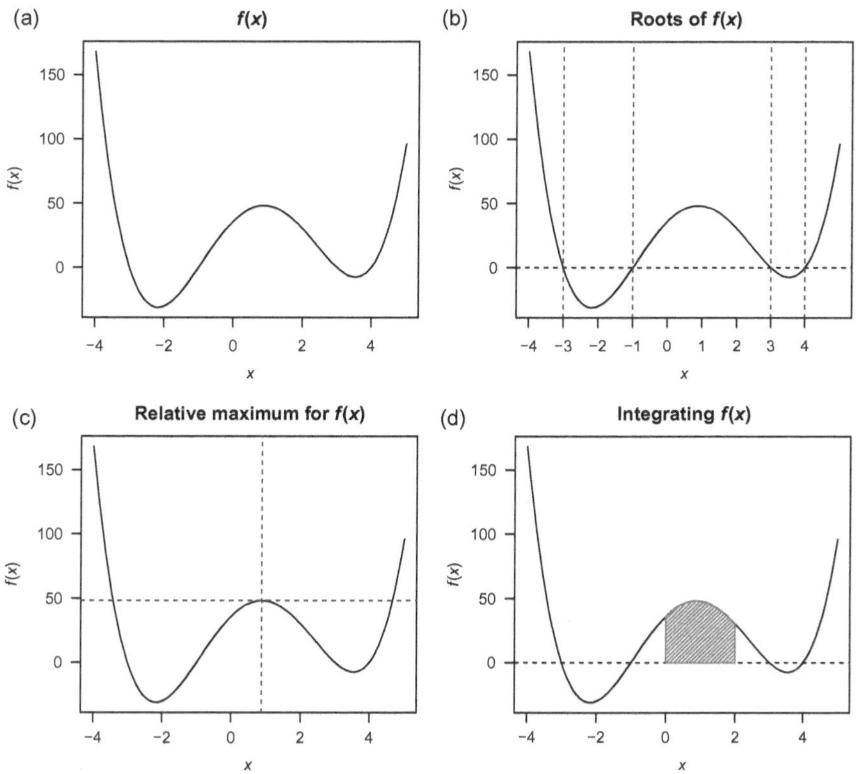

FIGURE 6 Examples of roots, relative maxima, and integration

10 NUMERICAL TECHNIQUES

We now introduce the following methods in R: finding roots of functions, optimization, derivatives, and integrals. These numerical methods are appropriate when analytical solutions are difficult or impossible to compute. Table 10 summarizes these functions.

We will use the polynomial $f(x) = x^4 - 3x^3 - 13x^2 + 27x + 36$ to demonstrate the numerical techniques in this section:

```
> f <- function(x){return(x^4-3*x^3-13*x^2+27*x+36)}
>
> # plot function
> curve(f, from=-4, to=5, main="f(x)", las=TRUE)
```

Figure 6 illustrates the output for each technique, beginning with the graph of $f(x)$.

The function `polyroot()` calculates the complex roots of a polynomial when coefficients are listed in ascending order; however, for a general one-dimensional function, `uniroot()` should be used instead.

```
> # complex roots of a polynomial:
> # enter vector of coefficients in ascending order
> polyroot(c(36, 27, -13, -3, 1))
>
> # for any function: find real root within specified range
> uniroot(f, interval=c(-2, 0))
>
> # graphing
> curve(f, from=-4, to=5, main="Roots of f(x)", las=TRUE, xaxt="n")
> abline(h=0, lty=2)
> abline(v=c(-1, -3, 3, 4), col="firebrick", lty=2)
> axis(side=1, at=-4:4)
>
> # roots at x= -1, -3, 3, 4
```

To use `uniroot()`, you must specify a search interval where the function values at the endpoints are of opposite sign; then, the function will obtain one root within the given interval. For more complex problems, the package *rootSolve* is helpful. Similar to other functions, `uniroot()` structures its output in the form of a list; use `names()` to determine the list components.

Next, we search for relative minima or maxima using the function `optimize()` (equivalently, `optimise()`). This function carries out one-dimensional optimization only. As with `uniroot()`, you must specify a search interval:

```
> # find maximum of function in specified range
> optimize(f, interval=c(0, 2), maximum=TRUE)
>
> # graphing
> curve(f, from=-4, to=5, main="Relative Maximum for f(x)",
+         las=TRUE)
> y <- optimize(f, interval=c(0, 2), maximum=TRUE)
> abline(h=y$objective, col="firebrick", lty=2)
> abline(v=y$maximum, col="firebrick", lty=2)
>
> # maximum at 48.248 when x=0.876
```

The function `optim()` can also handle maximum likelihood estimation among other optimization problems, `constrOptim()` is useful for constrained optimization, and `nlm()` for nonlinear minimization.

The final two methods we will discuss are derivatives and integrals. Let us say we want to determine the derivative of $f(x)$ at $x = 4$. Then, using the function `grad()` within the package *numDeriv*, we obtain

```
> # numerical derivative
> require(numDeriv)
> grad(f, x=4)
>
> # f'(4)= 35
```

which is equivalent to the analytical solution: $f'(x) = 4x^3 - 9x^2 - 26x + 27$, when $x = 4$. Within this package, use `jacobian()` for the Jacobian matrix and `hessian()` for the Hessian matrix to obtain partial derivatives.

To integrate over one dimension, use `integrate()`. For example, $\int_0^2 f(x)\,dx$ can be computed as follows:

```
> # integrate between lower and upper limits
> # (limits can be -Inf or Inf)
> integrate(f, lower=0, upper=2)
>
> # graphing
> curve(f, from=-4, to=5, main="Integrating f(x)", las=TRUE)
> abline(h=0, lty=2)
> x <- seq(0, 2, length=10)
> polygon(x=c(0,x,2), y=c(0,f(x),0), density=40, col="cadetblue",
+         border=TRUE) # shade in area under curve
>
> # area: 85.733
```

The limits of integration can be `-Inf` or `Inf`; insert these limits instead of a very large number for more accurate results. For multidimensional integration, use `adaptIntegrate()` in the package *cubature*.

11 ANNOTATED REFERENCES

Set Up

- The basic R package can be downloaded for free from the CRAN Website: http://cran.r-project.org/. Additional packages can be installed through your R console if you have administrative privileges on your computer or through the CRAN Website.
- For an "enhanced" version of R, with more workspace design features, install RStudio after installing R: http://www.rstudio.com; this program is also free.
- To connect R with the type-setting program LaTeX, use `Sweave()` or install and use the package *knitr*.

Text Editors

Apart from Notepad on your computer (or the built-in R text editor on Apple computers), there are other useful text editors which make reading and debugging your code easier through color coding (comments and code in different colors, etc.) and parenthesis matching (useful for nested expressions). As mentioned above, RStudio has this capability but alternatives include GNU Emacs (http://www.gnu.org/software/emacs/), RWinEdit (http://www.winedt.com, then for the R add-on go through CRAN and download package `RWinEdt`), Tinn-R (http://tinn-r.soft112.com), and Vim (http://www.vim.org). Many of these editors can be used for other programming languages as well.

Introductory Resources and Books

The following are the resources used in writing this chapter:

- Typing ? or `help()` with any function in R generates a help page describing the function arguments, some examples, and possibly a few references.
- Within the R console, if you choose "R Help" from the "Help" toolbar menu, a number of basic materials can be found, including the valuable *An Introduction to R* produced by the R Core Team.
- The CRAN Website, http://cran.r-project.org/, contains the help manual for every function (in every package) and additional documentation regarding R; the publication *The R Journal* is especially useful.
- The classic *Modern Applied Statistics with S, 4th ed.* by W.N. Venables and B.D. Ripley (Springer, 2002) is an excellent beginners guide.
- *Introductory Statistics with R, 2nd ed.* by Peter Dalgaard (Springer, 2008) is another classic. It focuses more on the statistical applications of R (summary statistics, hypothesis testing) than programming techniques.
- *Using R for Introductory Statistics, 2nd ed.* by John Verzani (Chapman & Hall/CRC The R Series, 2014) assumes the reader knows no statistics or programming.
- *Introduction to Scientific Programming and Simulation Using R, 2nd ed.* by Owen Jones, Robert Maillardet, and Andrew Robinson (Chapman & Hall/CRC The R Series, 2014), unlike Dalgaard (2008), concentrates less on statistics and more on programming and numerical algorithms.
- For a compilation of R recipes, refer to the *R Cookbook* by Paul Teetor (O'Reilly Cookbooks, 2011).
- Chapman & Hall/CRC Press has a series of books focusing on various R techniques called *The R Series* (http://www.crcpress.com/browse/series/crctherser). Similarly, Springer Press has a series titled *Use R!* (http://www.springer.com/series/6991).

Chapter 2

R Graphics

Deepayan Sarkar[1]
Theoretical Statistics and Mathematics Unit, Indian Statistical Institute, New Delhi, India
[1]Corresponding author: e-mail: deepayan.sarkar@gmail.com

ABSTRACT
Graphics is an important component of data analysis workflows, especially in interactive systems such as R. This chapter gives an overview of R's static graphics facilities. In addition to commonly used functions, we describe the underlying plotting model and how it can be exploited to customize default output. We also give a brief overview of the relatively recent Grid graphics, and the lattice and ggplot2 packages which use it to implement general purpose high-level systems.

Keywords: Graphics, Grid, Trellis, Lattice, ggplot2, Grammar of graphics

1 INTRODUCTION

R has a reputation for good graphics. Much of this reputation is based on its ability to produce publication-quality statistical displays that are static in nature. At the same time, new users who perhaps have experience with other graphics systems find the behavior of R graphics surprising and are often disappointed by the apparent lack of any sort of "dynamism." Both perceptions are true to some extent. R has an excellent (standard) graphics system that can do many things well, but is limited in other ways. As with other aspects of R, it is useful to have a sense of how the system came to be in its current state, and the standard workflow the designers expected to be followed, in order to use it effectively. It is also important to be aware of alternative graphics systems, providing various degrees of integration with R, that try to address many of the limitations of the standard system. This chapter deals entirely with the standard static graphics system, and provides pointers to some alternatives.

1.1 Origins

As with the language itself, the history of the R graphics model goes back to S. The S language was from the very beginning designed to be interactive, and graphics was naturally an essential component of the system. The model

Handbook of Statistics, Vol. 32. http://dx.doi.org/10.1016/B978-0-444-63431-3.00002-4

that the designers of S chose to adopt for graphics was the GRZ model already in use at the Bell Laboratories. This may be described as a "painter's model," where a graphic was built out of a small set of primitives such as line segments, polygons, text, etc., and later elements were drawn on top of earlier ones. There was no provision for deleting an element once it was drawn, except to start a completely new graphic.

This model allowed both input and output to be abstracted. Graphics functions that were meant for users would internally call these primitives. For output, the primitives could be implemented differently depending on the target "device," which could be Postscript or PDF files for printing, hardware devices such as pen plotters, or on-screen devices for interactive viewing. The device-specific implementations of the primitives are known as device drivers, and new drivers are even now written to support new kinds of output formats. See `?Devices` for more details.

The painter's model naturally led to a mental approach that viewed a plot as a work in progress, always with the possibility of adding something more to it. This attitude pervades much of the graphics functionality written for S in its early days and is still popular due to its simplicity, familiarity, and the availability of a large variety of graphic designs implemented using this model. We will loosely refer to this model of graphics as the *traditional graphics* model.

In the 1990s, S introduced a new approach to graphics, which was called Trellis graphics, meant to substitute the traditional graphics tools rather than to complement it. The most prominent feature of Trellis graphics was the notion of *conditioning* (also known as "small multiples" or "faceting") which allowed subsets of data to be intelligently visualized so as to enable effective comparison between those subsets. The way this was done necessitated a departure from the "work-in-progress" model and required the user to completely specify the details of the plot in one go. This was not really a departure from the painter's model in terms of implementation, but rather a change in the mental approach to plotting.

R followed S in implementing the traditional graphics model first. When it came to Trellis graphics, it took a slightly different approach that had important repercussions.

Instead of implementing Trellis graphics using the tools provided by the traditional model, as S had done, the R developers first introduced a layer of abstraction called Grid graphics. Grid was designed to be a low-level tool, which provided graphical elements as objects that could be manipulated to a considerable extent, and sophisticated viewport and layout capabilities to use these elements to construct complicated plots. Grid was used to implement two important general high-level graphics packages, as well as many other specialized packages. One was **lattice** (Sarkar, 2008), which implemented the functionality of Trellis graphics in R. The other was **ggplot2** (Wickham, 2009), which adapted ideas from Wilkinson (1999) to provide another alternative graphics system. Both systems will be discussed later in this chapter.

1.2 Principles of Data Graphics

A discussion of why graphics are important in statistical analysis and what makes good graphics good is beyond the scope of this chapter, and we refer the reader to Wickham (2013) for an excellent introduction and further references. If there is one single underlying principle, it is that good graphics should enable comparison. Starting from this principle, Cleveland and McGill (1984) performed a series of perceptual experiments that indicated, for instance, that the human eye can judge the difference in positions along a common axis better than it can judge differences in lengths of line segments, and that it is worse than either in judging quantitative differences based on a color scale. Based on such experiments, Cleveland (1985) discusses principles that when used systematically should yield effective visualizations, and these principles played an important role in the evolution of S graphics. The same principle of enabling comparison is also the basis of Trellis graphics, which grew out of ideas in Cleveland (1993). Overall, the work of Cleveland, and before him, Tukey (1977), has been extremely influential in shaping the direction and visual feel of traditional S and R graphics.

2 TRADITIONAL GRAPHICS

The core of the traditional R graphics system is the suite of functions available in the **graphics** package, with various add-on packages providing further functionality. The full list of functions can be seen, as with any other package, using

```
> library(help = graphics)
```

These functions can be roughly categorized into two groups, *high-level* and *low-level* functions. High-level functions are those that are intended to produce a complete plot by themselves. Low-level functions are those that are intended to add elements to existing plots. Of course, high-level functions are themselves built up from low-level functions.

Let us look at an example. The simplest and most common type of statistical plot is the scatterplot, which depicts bivariate numeric data as points in a Cartesian coordinate system. The high-level function that produces scatterplots is `plot()` (although that is not all `plot()` does). We use R's built-in dataset, `anscombe`, for illustration. The dataset contains Anscombe's well-known quartet of bivariate datasets (Anscombe, 1973; Tufte, 2001) that are quite different from each other, yet have the same traditional statistical summaries (mean, variance, correlation, least squares regression line, etc.). The first dataset can be plotted as follows, producing Fig. 1.

```
> plot(x = anscombe$x1, y = anscombe$y1)
```

It will be helpful if we pause here for a moment to understand how this plot could have been created using low-level functions. The plot consists of the

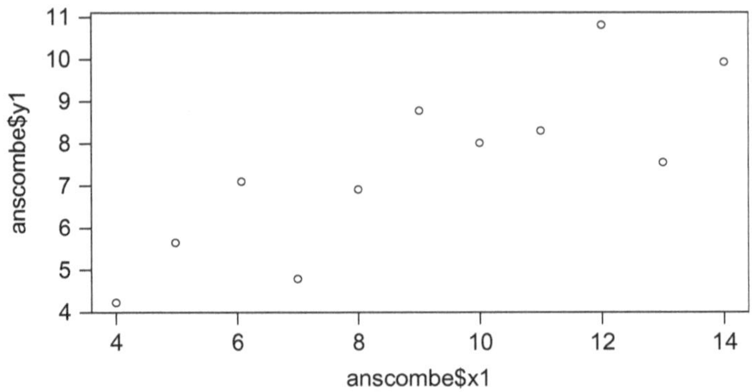

FIGURE 1 Scatterplot of Anscombe's first dataset.

points, the box surrounding the plot, the axes, and the axis labels. All these elements can be suppressed by `plot()` as follows:

```
> plot(x = anscombe$x1, y = anscombe$y1,
+         type = "n", axes = FALSE, xlab = "", ylab = "")
```

This produces a completely blank page, but performs one important task: it sets up the coordinate system for subsequent low-level calls. The extent of this coordinate system can be obtained using

```
> par("usr")
[1] 3.6000 14.4000 3.9968 11.1032
```

and is precisely the range of the data that was supplied to `plot()`, with a padding of 4% on both sides (this can be overridden by specifying the `xlim` and `ylim` arguments).

```
> range(anscombe$x1)
[1] 4 14
> range(anscombe$y1)
[1] 4.26 10.84
```

This rectangular region does not occupy the full figure area, only a part of it. This is referred to as the *plot region*. We can now draw a box around the plot region and add the data points as follows to produce the top plot in Fig. 2.

```
> box()
> points(anscombe$x1, anscombe$y1, pch = 16)
```

The area outside the plot region is known as the margin and is used for axis annotation and labels. The following low-level calls complete the plot and add a linear regression line to produce the bottom plot in Fig. 2.

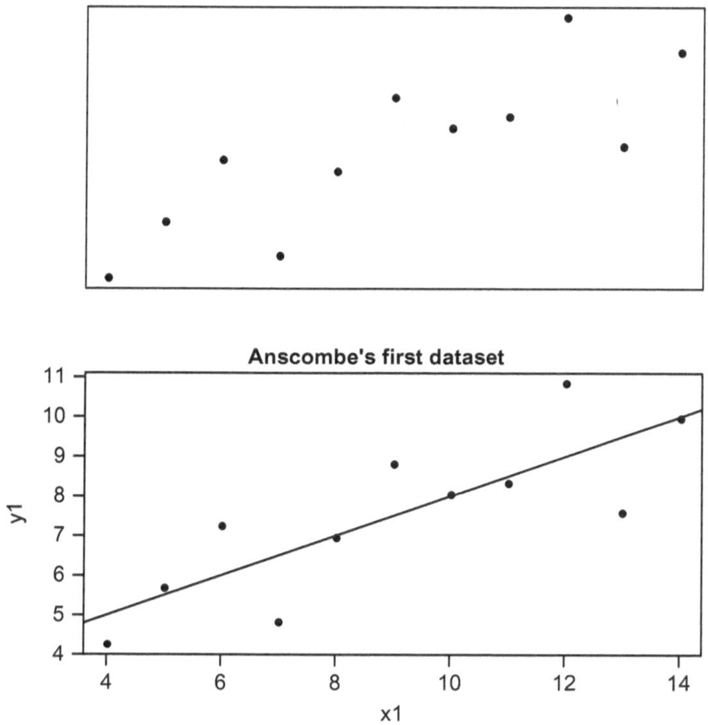

FIGURE 2 A complete scatterplot of Anscombe's first dataset built using low-level elements. The first plot represents an intermediate step when the axes and labels have not yet been added. The second plot has those elements, and adds a linear regression line.

```
> axis(side = 1)
> axis(side = 2)
> title(main = "Anscombe's first dataset", xlab = "x1",
+       ylab = "y1")
> abline(lm(y1 ~ x1, anscombe))
```

All this might seem like detail that is not relevant for routine use, but consider the following code which produces Fig. 3.

```
> plot(anscombe$x2, anscombe$y2,
+      type = "n", axes = FALSE, xlab = "", ylab = "")
> lims <- par("usr")
> rect(lims[1], lims[3], lims[2], lims[4], col = "grey80",
+      border = NA)
> abline(v = pretty(lims[1:2]), h = pretty(lims[3:4]),
+        col = "white", lwd = 2)
> axis(side = 1, col = "grey80", col.axis = "grey20")
> axis(side = 2, col = "grey80", col.axis = "grey20", las = 1)
> points(anscombe$x2, anscombe$y2, pch = 16)
```

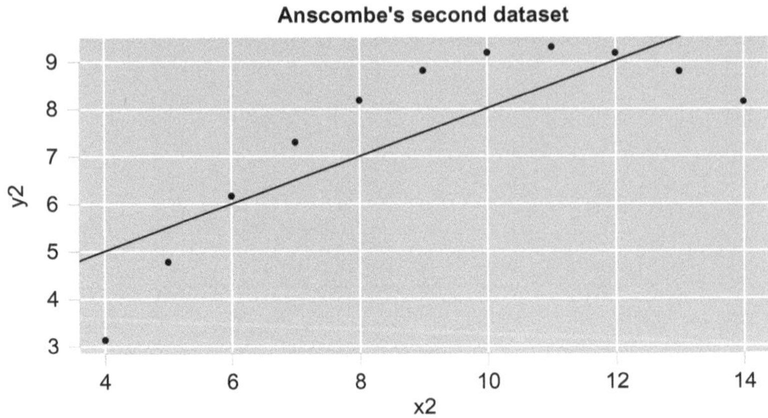

FIGURE 3 Scatterplot of Anscombe's second dataset, with various customizations.

```
> title(main = "Anscombe's second dataset",
+         xlab = "x2", ylab = "y2", col = "grey20")
> abline(lm(y2 ~ x2, anscombe), col = "grey20")
```

This kind of customization is often desirable, and with traditional graphics, making use of low-level functions as above is the canonical way to achieve them. We will not discuss the low-level structure of traditional graphics plots further, except to say that most other high-level functions in the traditional graphics suite follow this same structure. Table 1 gives a list of useful low-level functions. We refer the reader to the online documentation for more details.

2.1 The `plot()` Function

The `plot()` function is actually a generic function that can deal with the task of plotting several types of R objects. The most common method is the default method `plot.default()`, which can be used to plot paired numeric data as above, or a univariate numeric vector, which is plotted on the y-axis against its index (serial number). We refer the reader to the help pages of `plot()` and `plot.default()` for details, including the use of logarithmic scales, and a description of the very useful `type` argument, which we have already encountered and can also be used to render the data in several different ways such as joining them by lines.

Another common way to plot bivariate data is to use the formula method, which provides an interface similar to that of the various statistical modeling functions. For example, Fig. 4 is produced by

```
> plot(y3 ~ x2, data = anscombe, pch = 16)
> abline(lm(y3 ~ x3, anscombe), col = "grey20")
```

TABLE 1 Useful Low-Level Functions

`text()`	Add text to a plot
`lines()`	Add connected line segments to a plot
`points()`	Add points to a plot
`polygon()`	Polygon drawing
`polypath()`	Path drawing
`rect()`	Draw one or more rectangles
`segments()`	Add line segments to a plot
`abline()`	Add straight lines to a plot
`arrows()`	Add arrows to a plot
`axis()`	Add an axis to a plot
`box()`	Draw a box around a plot
`grid()`	Add grid to a plot
`legend()`	Add legends to plots
`rug()`	Add a rug to a plot
`title()`	Plot annotation

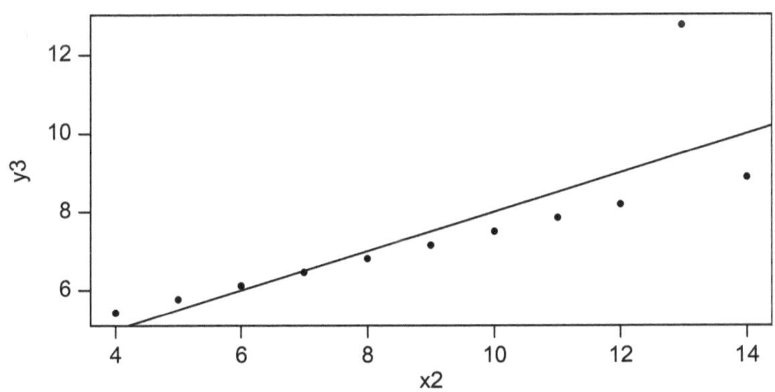

FIGURE 4 Scatterplot of Anscombe's third dataset, produced using the "formula" method of `plot()`.

The formula interface is useful because it allows a cleaner specification of the plot and leads to better default axis labels, and we will use it several times in the rest of this chapter.

The full list of currently available methods for the `plot()` generic can be listed by

```
> methods("plot")

 [1] plot.acf*          plot.data.frame*   plot.decomposed.ts*
 [4] plot.default       plot.dendrogram*   plot.density*
 [7] plot.ecdf          plot.factor*       plot.formula*
[10] plot.function      plot.hclust*       plot.histogram*
[13] plot.HoltWinters*  plot.isoreg*       plot.lm*
[16] plot.medpolish*    plot.mlm*          plot.ppr*
[19] plot.prcomp*       plot.princomp*     plot.profile.nls*
[22] plot.spec*         plot.stepfun       plot.stl*
[25] plot.table*        plot.ts            plot.tskernel*
[28] plot.TukeyHSD*

   Nonvisible functions are asterisked
```

Most of these are specialized, but a generally useful one is the method for "data.frame" objects. Figure 5, a scatterplot matrix of data on daily air quality measurements in New York over the course of a few months in 1973, is produced by the following call.

```
> plot(airquality)
```

Actually, the `plot()` method in this case simply calls `pairs()`, the function designed to draw scatterplot matrices, and the figure could also have been produced by the following call.

```
> pairs(airquality)
```

Similarly, the "factor" method for `plot()` may call `barplot()` to show a bar chart of the frequency distribution if the input is a single "factor" variable, or `boxplot()` to show box-and-whisker plots if an additional numeric variable is provided. We discuss these specialized high-level functions next.

2.2 Other Common High-Level Functions

In addition to the catch-all `plot()` function which tries to choose an appropriate plot type based on its input, there are several high-level functions for specific types of displays. Histograms are produced by the `hist()` function, for example,

```
> hist(airquality$Ozone)
```

The related kernel density plots need a two-step approach. The first step is to compute the density estimates as a "density" object, e.g.,

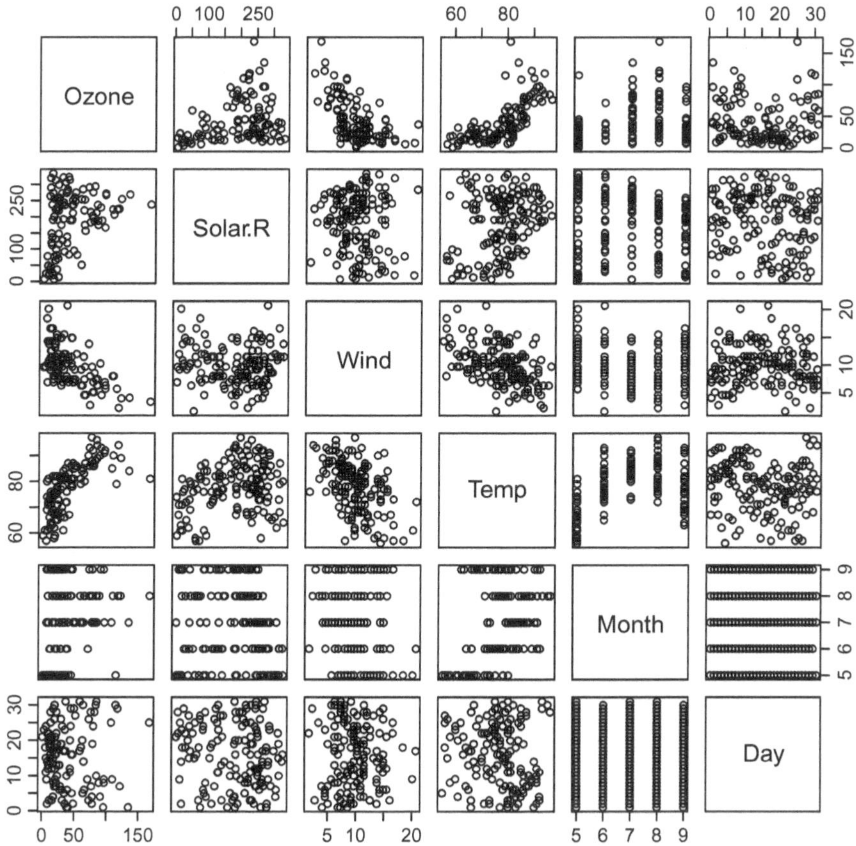

FIGURE 5 Scatterplot matrix of `airquality` data.

```
> d <- density(airquality$Ozone, na.rm = TRUE)
> str(d)
List of 7
 $ x         : num [1:512] -33.4 -33 -32.5 -32 -31.6 ...
 $ y         : num [1:512] 1.08e-05 1.24e-05 1.41e-05
               1.62e-05 1.85e-05 ...
 $ bw        : num 11.5
 $ n         : int 116
 $ call      : language density.default(x = airquality$Ozone,
               na.rm = TRUE)
 $ data.name: chr "airquality$Ozone"
 $ has.na    : logi FALSE
 - attr(*, "class")= chr "density"
```

and the second step is to plot it with the suitable `plot()` method.

```
> plot(d)
```

The outputs of the above calls are displayed in Fig. 6. Just as `density()` returns a "density" object with information about how the estimate was calculated, so does `hist()` return a "histogram" object containing details about the computations performed.

```
> h <- hist(airquality$Ozone, plot = FALSE)
> str(h)
List of 6
 $ breaks   : num [1:10] 0 20 40 60 80 100 120 140 160 180
 $ counts   : int [1:9] 37 34 14 15 9 4 2 0 1
 $ density  : num [1:9] 0.01595 0.01466 0.00603 0.00647 0.00388 ...
 $ mids     : num [1:9] 10 30 50 70 90 110 130 150 170
```

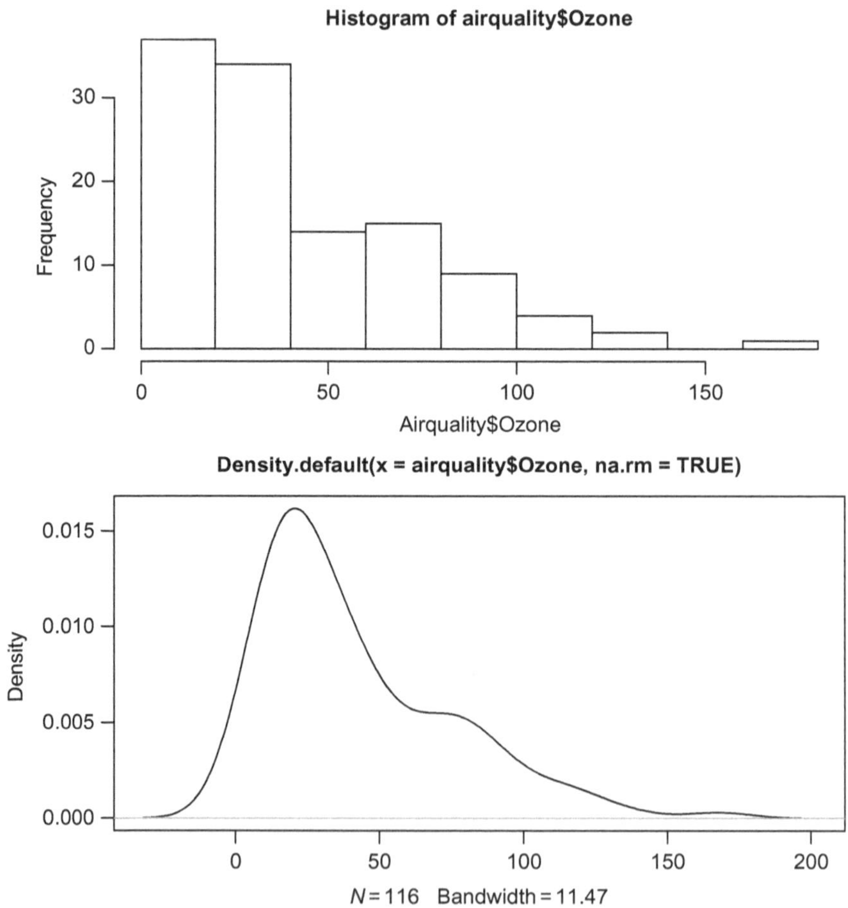

FIGURE 6 Histogram (top) and kernel density plot (bottom) of Ozone concentration data.

```
$ xname    : chr "airquality$Ozone"
$ equidist: logi TRUE
- attr(*, "class")= chr "histogram"
```

Such objects can also be plotted using the corresponding `plot()` method, but this feature is more useful because it allows for possible custom annotation on the plot. We will see an example later.

Figure 7 shows two quantile-based plots of the same dataset. The `qqnorm()` function is used to produce a normal Q–Q plot as follows:

```
> qqnorm(airquality$Ozone)
```

The `boxplot()` function is used to produce comparative box-and-whisker plots comparing the distribution of `Ozone` across months.

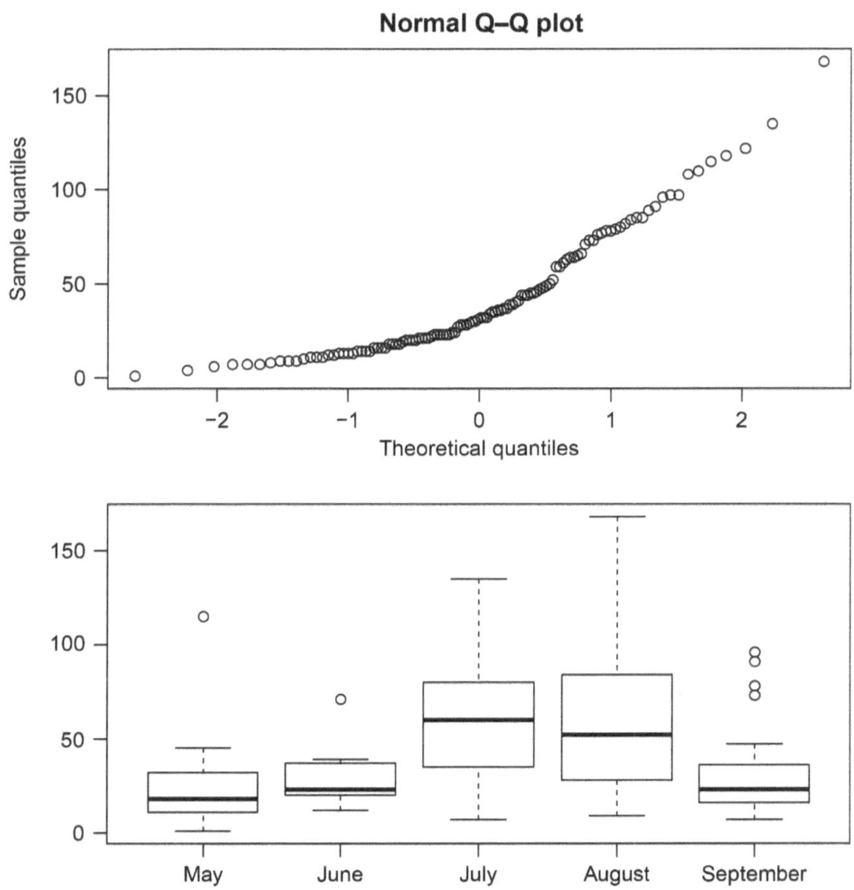

FIGURE 7 Normal Q–Q plot (top) and monthwise comparative box-and-whisker plot (bottom) of Ozone concentration.

```
> airquality$fmonth <-
+   with(airquality,
+           factor(Month, levels = 1:12, labels = month.name))
> boxplot(Ozone ~ fmonth, data = airquality)
```

As with `hist()` and `density()`, these functions also have useful return values.

The most common visualizations for tabular data are bar plots and Cleveland dot plots (Cleveland, 1985). Bar plots are produced by the `barplot()` function, and dot plots by the `dotchart()` function. Figure 8 illustrates the usage of both functions using a built-in dataset on death rates in different population subgroups in the US state of Virginia in 1940. The plots are produced by the following calls.

```
> barplot(VADeaths, beside=TRUE, horiz=TRUE,
+            legend.text=TRUE, xlim = c(0, 100))
> dotchart(VADeaths)
```

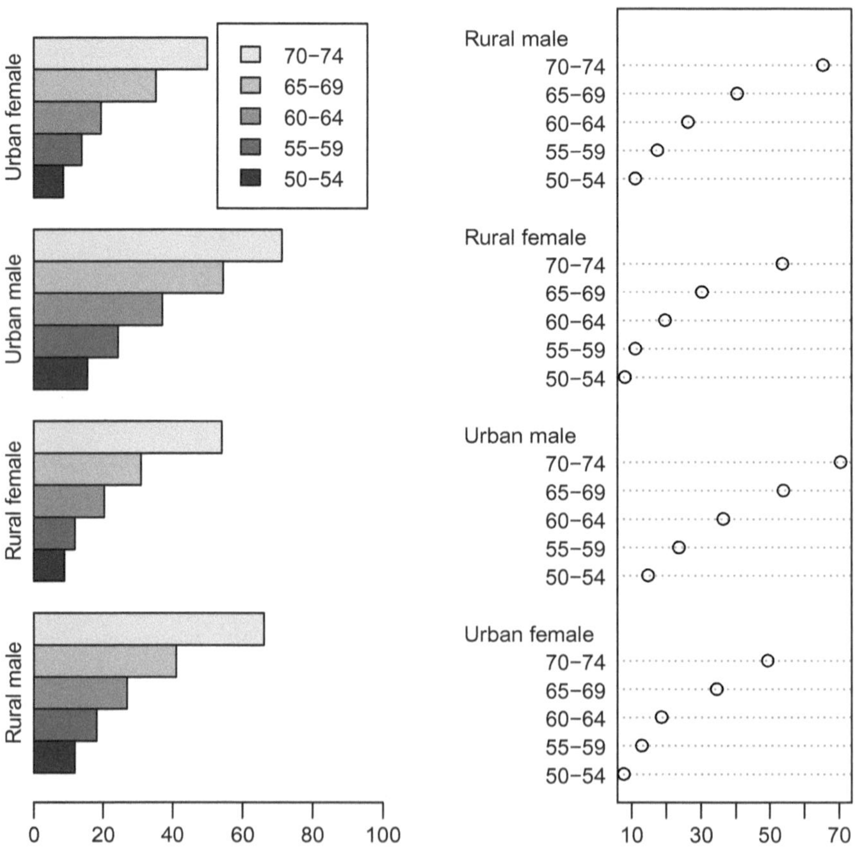

FIGURE 8 Death rates by age for different population subgroups in Virginia, 1940. The data are displayed using a bar chart (left) and a dot plot (right).

Note the use of `xlim` to add enough blank space to the plot to ensure that the legend does not overlap the bars. The legend is added by `barplot()` using the `legend()` function internally.

Another useful but less frequently used visualization for tabular data, especially with higher dimensions, is a mosaic plot (Friendly, 1994). Figure 9 illustrates mosaic plots using a well-known example of Simpson's paradox, the built-in `UCBAdmissions` data, which gives the aggregate number

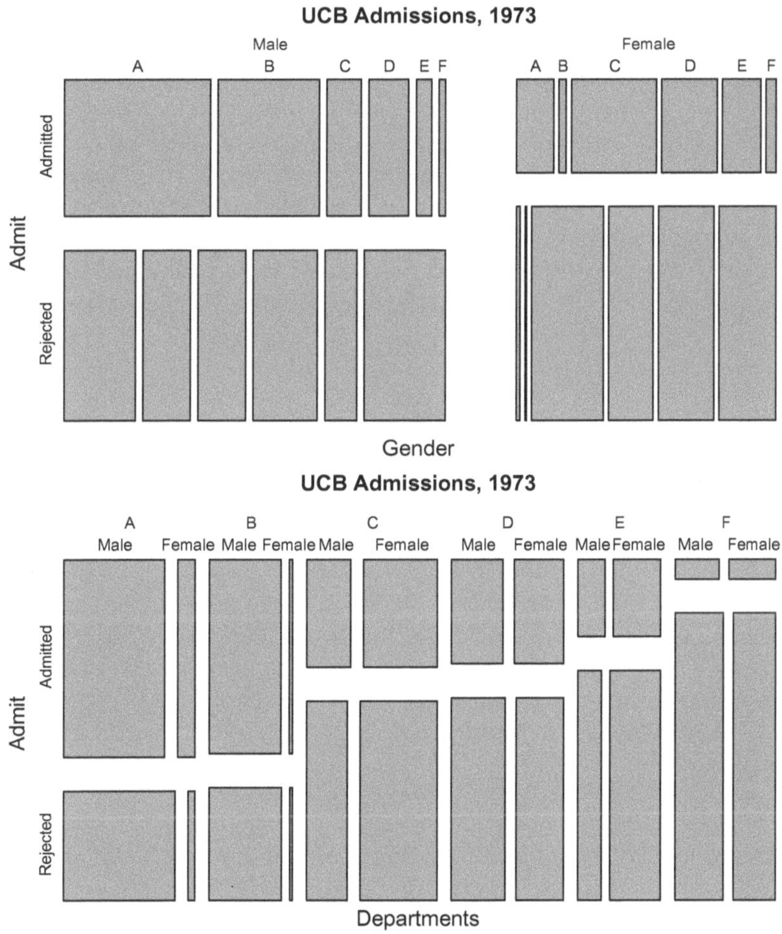

FIGURE 9 Mosaicplot of UCB admissions data. Departments are labeled as A through F. The area of each rectangle is proportional to the count it represents. The top plot suggests a strong gender bias in admissions. However, the bottom plot, which makes it easier to compare admission rates by gender across departments, shows that the per-department bias is negligible, if at all present. It just so happens that many more males than females apply to departments A and B which have comparatively lower rejection rates.

of applicants to graduate school at Berkeley for the six largest departments in 1973, classified by gender and whether they were admitted. The plots are produced by the following calls, which use `aperm()` to rearrange the dimensions of the `UCBAdmissions` array to achieve the desired hierarchy of groups.

```
> mosaicplot(aperm(UCBAdmissions, c(2, 1, 3)),
+             main = "UCB Admissions, 1973")
> mosaicplot(aperm(UCBAdmissions, c(3, 1, 2)),
+             main = "UCB Admissions, 1973")
```

2.3 Visualizations for Time Series Data

R comes with many specialized graphical designs, of which those used for time series data deserve special mention because of their ubiquity. Time series data are typically represented in R as objects of class "ts". The corresponding `plot()` method produces a standard time series plot, which is essentially a scatterplot with points joined by lines. A time series plot of the `AirPassengers` data, which records monthly totals of international airline passengers between 1949 and 1960 (Box and Jenkins, 1976), can be produced using

```
> plot(AirPassengers)
```

The estimated autocorrelation function (of the log-transformed values) is plotted by

```
> acf(log(AirPassengers), lag.max = 50)
```

and the partial autocorrelation function by

```
> pacf(log(AirPassengers), lag.max = 50)
```

The output of the three calls above are shown in Fig. 10.

In addition to these basic plots, R provides functions for various time series decompositions that can be plotted. For example, an STL decomposition (Cleveland et al., 1990) of the `AirPassengers` data, shown in Fig. 11, is produced by

```
> plot(stl(log(AirPassengers), "periodic"))
```

Other decomposition methods are implemented in the `decompose()` and `StructTS()` functions. A number of contributed R packages substantially enhance the time series analysis capabilities of R; the CRAN Task View on Time Series[1] gives a broad overview of available functionality. The **lattice** framework described later in this chapter also supports some specialized plots

1. http://cran.r-project.org/web/views/TimeSeries.html.

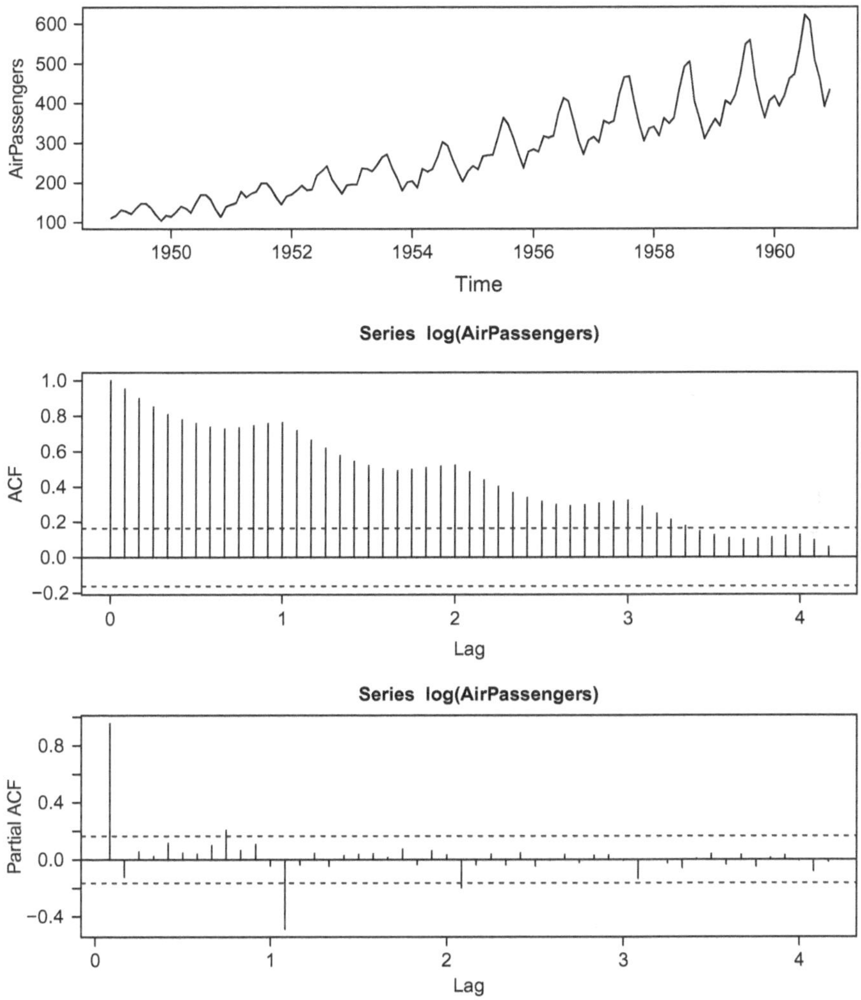

FIGURE 10 Time series plot of the built-in `AirPassengers` dataset, along with plots of estimated autocorrelation and partial autocorrelation functions.

for time series data, although these are not discussed here. For a more detailed introduction to time series analysis with R, see McLeod et al. (2012).

2.4 Customizing Plots Using Low-Level Functions

High-level graphics functions produce basic versions of common statistical graphs, but often we want to fine-tune them in different ways to get something more relevant to our purposes. We have seen a somewhat extreme example before (Fig. 3) where we built the plot almost from scratch because we wanted

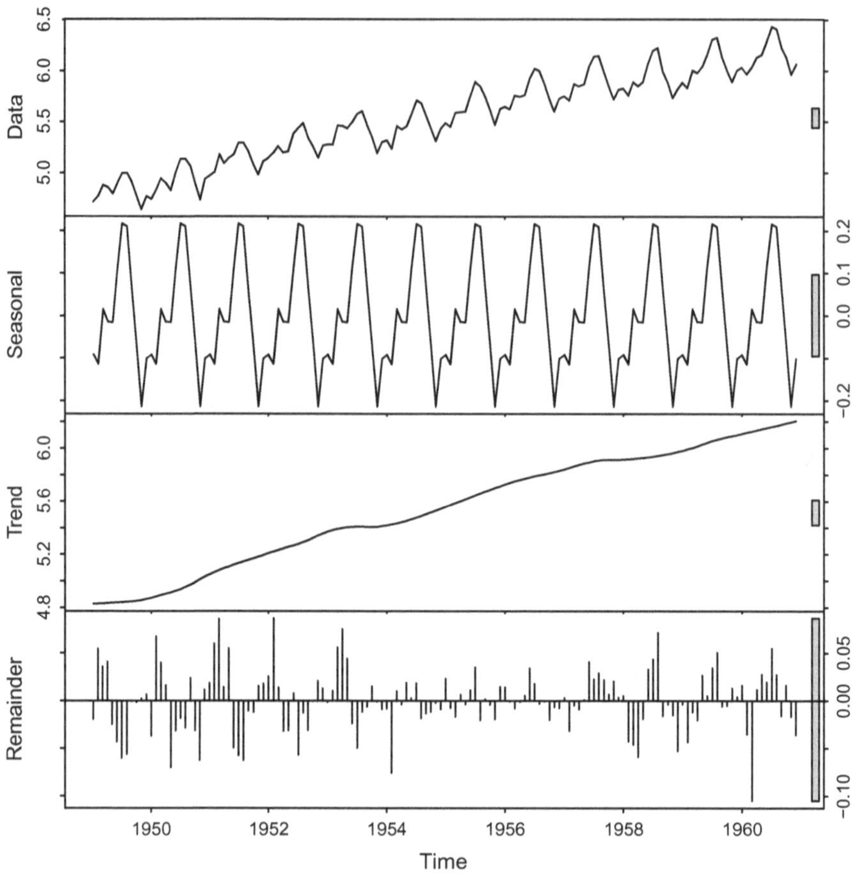

FIGURE 11 An STL decomposition of the `AirPassengers` dataset.

to customize even basic elements like the background and axes. It is more common to start with a complete high-level plot and then incrementally add components to customize it.

A common use of this feature is to add a representation of some statistical model fit using the data. For example, the following code creates the side-by-side scatterplots in Fig. 12 of Anscombe's first two datasets along with the least squares fits from a linear and quadratic model.

```
> lims <- with(anscombe, list(x = range(x1, x2), y = range(y1, y2)))
> par(mfrow = c(1, 2))
> plot(y1 ~ x1, data = anscombe, xlim = lims$x, ylim = lims$y)
> abline(lm(y1 ~ x1, data = anscombe))
> f1 <- function(x)
+     predict(lm(y1 ~ poly(x1, 2), data = anscombe),
+             newdata = list(x1 = x))
```

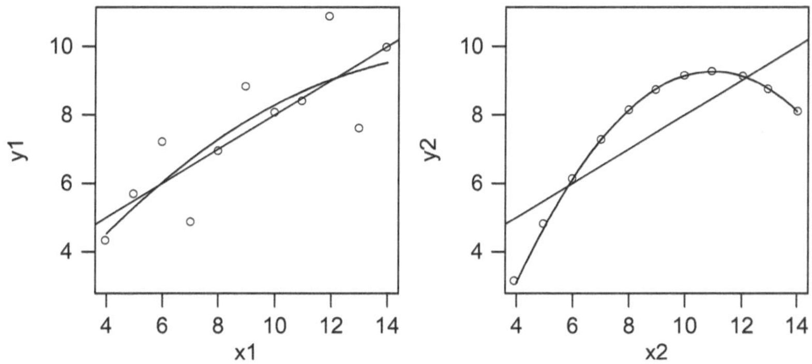

FIGURE 12 Anscombe's first two datasets with linear and quadratic fits.

```
> curve(f1, add = TRUE)
> plot(y2 ~ x2, data = anscombe, xlim = lims$x, ylim = lims$y)
> abline(lm(y2 ~ x2, data = anscombe))
> f2 <- function(x)
+     predict(lm(y2 ~ poly(x2, 2), data = anscombe),
+             newdata = list(x2 = x))
> curve(f2, add = TRUE)
```

Here, the linear fit is easily added using `abline()`, but for the quadratic fit, we need the more general `curve()` function. The second approach works for any other modeling function that has a suitable `predict()` method. The `par(-mfrow=)` construct is used to enable a "multiple figure" layout (see `?par`).

As we have remarked before, many of the high-level graphics functions such as `hist()`, `boxplot()`, `barplot()`, etc., return objects containing the results of intermediate computations. These are useful for customizations. Consider Fig. 13, which annotates a density histogram of Ozone concentration (the `airquality` data) with the bin frequencies and adds a kernel density estimate for comparison.

```
> h <- hist(airquality$Ozone, plot = FALSE)
> d <- density(airquality$Ozone, na.rm = TRUE)
> plot(h, freq = FALSE, ylim = c(0, 1.1 * max(h$density)))
> box()
> text(h$mids, h$density, labels = h$counts, pos = 3)
> lines(d)
```

Note the need to increase the y-axis range slightly to accommodate the label for the highest bin.

The return value of `barplot()`, giving the coordinates of the bar midpoints, allows similar annotation to be added to bar charts. In the following example, the age groups in the `VADeaths` data are indicated using text labels rather than a legend to produce Fig. 14.

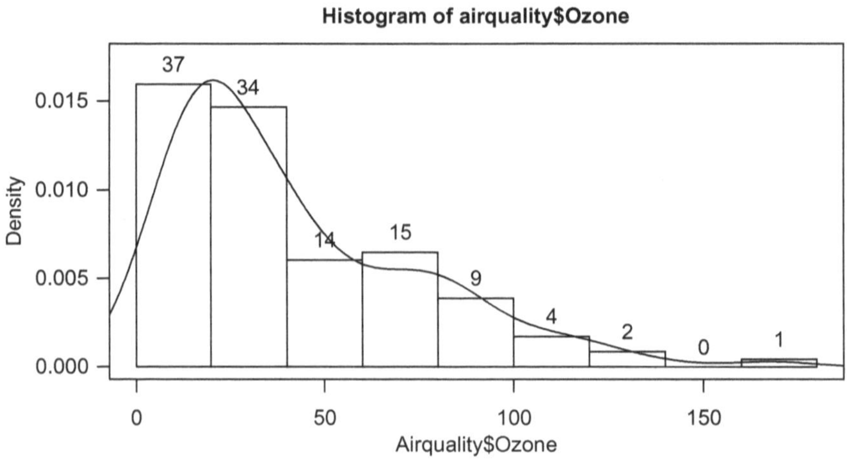

FIGURE 13 Density histogram of Ozone with a kernel density estimate and bin counts overlaid.

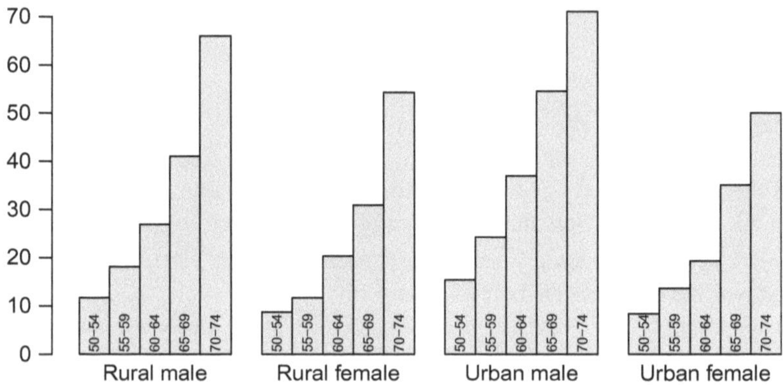

FIGURE 14 Bar chart of death rates in Virginia, 1940, with age groups indicated by text annotation inside the plot.

```
> b <- barplot(VADeaths, beside = TRUE, col = "grey90")
> text(b, 0, labels = rownames(VADeaths), srt = 90, adj = -0.1,
+       cex = .6)
```

2.5 Limitations of Traditional Graphics

The traditional graphics model is considerably limited by its design. A basic assumption in the approach is that there will be a single plot area in the center, and margins will be used for axis annotation and labels. Not all graphical designs fit into this paradigm.

One class of such examples is a graphical design known as "small multiples" (Tufte, 1990) or "conditioned plots" (Cleveland, 1993), where one page contains multiple plots grouped by some categorical variable. Here is a very poor attempt to implement such a design using traditional graphics, with output shown in Fig. 15.

```
> lims <- with(anscombe, list(x = range(x1, x2, x3, x4),
+                            y = range(y1, y2, y3, y4)))
> par(mfrow = c(2, 2), mar = c(2, 2, 1, 0))
> plot(y1 ~ x1, data = anscombe, pch = 16, xlim = lims$x, ylim = lims$y)
> abline(lm(y1 ~ x1, data = anscombe), col = "grey50")
> plot(y2 ~ x2, data = anscombe, pch = 16, xlim = lims$x, ylim = lims$y)
> abline(lm(y2 ~ x2, data = anscombe), col = "grey50")
> plot(y3 ~ x3, data = anscombe, pch = 16, xlim = lims$x, ylim = lims$y)
> abline(lm(y3 ~ x3, data = anscombe), col = "grey50")
> plot(y4 ~ x4, data = anscombe, pch = 16, xlim = lims$x, ylim = lims$y)
> abline(lm(y4 ~ x4, data = anscombe), col = "grey50")
```

This works using the multiple figure layout approach we saw earlier in Fig. 12.

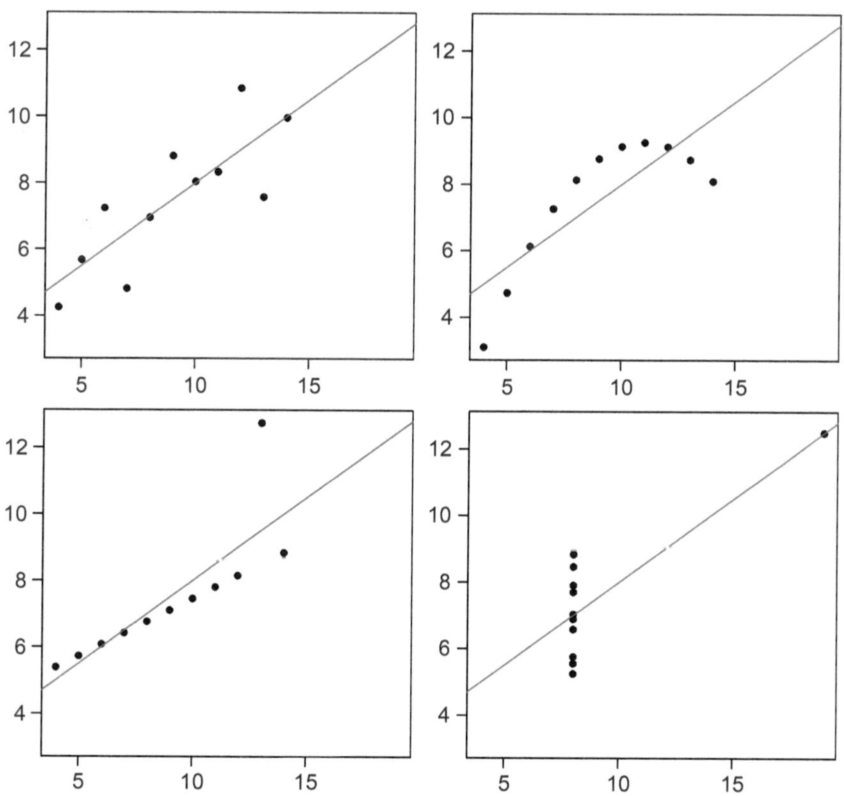

FIGURE 15 All four of Anscombe's datasets, with common axis limits.

For a more realistic example, suppose we wish to compare the distributions of `Ozone` across months, as we did using box-and-whisker plots in Fig. 7, but with density plots or histograms instead. Density estimates can be superposed in the same plot, with a little work to ensure adequate axis limits. We first use the `split()` function to divide up the `Ozone` observations by month.

```
> s <- with(airquality, split(Ozone, fmonth))
> str(s)
List of 5
 $ May      : int [1:31] 41  36 12 18 NA 28 23  19 8   NA ...
 $ June     : int [1:30] NA  NA NA NA NA NA 29  NA 71  39 ...
 $ July     : int [1:31] 135 49 32 NA 64 40 77  97 97  85 ...
 $ August   : int [1:31] 39  9  16 78 35 66 122 89 110 NA ...
 $ September: int [1:30] 96  78 73 91 47 32 20  23 21  24 ...
```

Next, we call `density()` on each component using `lapply()`.

```
> dlist <- lapply(s, density, na.rm = TRUE)
```

Each component of `dlist` now contains a "density" object with components `x` and `y` giving the estimated density function. We can compute the range needed to contain all the densities as follows:

```
> dxrng <- range(unlist(lapply(dlist, function(d) d$x)))
> dyrng <- range(unlist(lapply(dlist, function(d) d$y)))
```

We then create a blank plot (using `type="n"`) and add the densities one by one, followed by a legend.

```
> plot(dxrng, dyrng, xlab = "Ozone concentration", ylab = "Density")
> for (i in seq_along(dlist)) lines(dlist[[i]], col = i)
> legend("topright", legend = names(dlist),
+          lty = 1, col = seq_along(dlist))
```

The result is shown in the top panel of Fig. 16.

Histograms are not easily amenable to superposition, but they can be juxtaposed. Using similar tricks along with the multiple figure approach again, we obtain the bottom plot in Fig. 16.

```
> par(mfrow = c(2, 3))
> hlist <- lapply(s, hist, plot = FALSE)
> hxrng <- range(unlist(lapply(hlist, function(h) h$breaks)))
> hyrng <- range(unlist(lapply(hlist, function(h) h$density)))
> invisible(lapply(names(s),
+                  function(nm) {
+                      plot(hlist[[nm]], main = nm, freq = FALSE,
+                        xlim = hxrng, ylim = hyrng, xlab = "Ozone")
+                  }))
```

Although not very difficult, obtaining these plots is not simple either, and the results leave a lot to be desired. It is precisely the need to deal with these

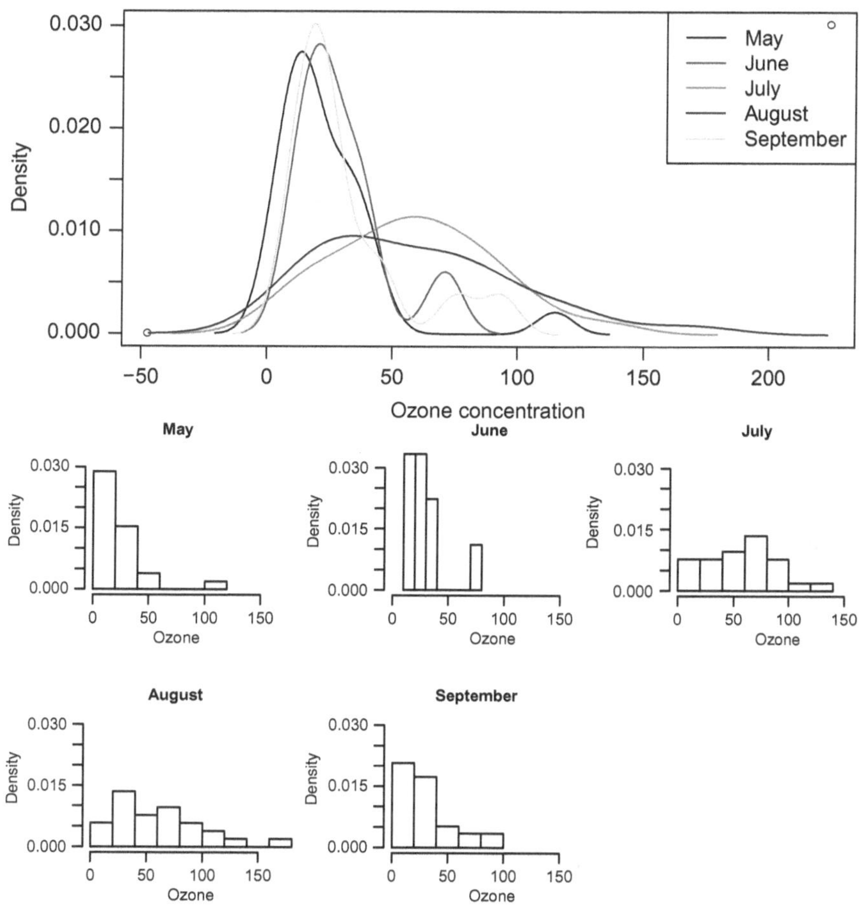

FIGURE 16 Comparison of ozone concentration by month, using superposed kernel density estimates (top) and juxtaposed histograms with common axis limits (bottom).

kinds of graphic designs that Trellis graphics as implemented in the **lattice** package, and more recently the **ggplot2** package, were developed. Before we move on to these, we will briefly discuss the low-level system these packages are based on, namely, Grid graphics.

3 GRID GRAPHICS

Grid graphics is a flexible low-level graphics toolbox. It does not provide high-level functions itself, but is used by other packages that do. We will not delve into the details of **grid**, as it is a considerable topic, and instead refer the reader to Murrell (2005). Here, we will only try to give a glimpse of its flexibility using a practical example. Our goal is to implement an *enhanced scatter plot* with the following features:

- A central plot region with a standard scatterplot.
- "Rugs" on the left and bottom boundaries showing marginal scatter.
- Density estimates on the right and top boundaries.

This is a simple design, but difficult to obtain with standard graphics.

3.1 Viewports

Viewports are a central concept in Grid graphics. They are essentially rectangular subregions of the plotting area. They can be nested within other viewports, forming a tree of viewports. The initial blank graphics area is the ROOT viewport. New viewports can be defined relative to parent either by position or in terms of a layout. Viewports are created by the `viewport()` function and made active by `pushViewport()`. The `upViewport()` function is used to navigate to the parent of the current viewport. Viewports support several coordinate systems that can be used to specify locations within the viewport. Of these, the most useful is the *native coordinate system*, which is determined by specifying *x*-axis and *y*-axis extents when creating the viewport using the arguments `xscale` and `yscale`.

3.2 Units and Primitives

Another fundamental concept is that of units of length. Grid can specify lengths in various ways. Some of these are given in Table 2.

TABLE 2 Coordinate Systems in grid

`"native"`	Native Coordinates
`"npc"`	Normalized parent coordinates
`"snpc"`	"Square" NPC
`"inches"`, `"cm"`, `"points"`, etc.	Absolute lengths
`"char"`	Character size
`"lines"`	Height of a lines of text
`"strwidth"`, `"strheight"`	Width or height of a string
`"grobwidth"`, `"grobheight"`	Width or height of A "grid object" (details later)

TABLE 3 Common Primitive Objects in `grid`

Type	Function Producing Output	Function Producing Object
Points	`grid.points`	`pointsGrob`
Lines	`grid.lines`	`linesGrob`
Text	`grid.text`	`textGrob`
Rectangles	`grid.rect`	`rectGrob`
Circles	`grid.circle`	`circleGrob`
Polygons	`grid.polygon`	`polygonGrob`
Segments	`grid.segments`	`segmentsGrob`
x-axis	`grid.xaxis`	`xaxisGrob`
y-axis	`grid.yaxis`	`yaxisGrob`
Arrows	`grid.arrows`	`arrowsGrob`

The basic low-level primitive functions in traditional graphics have analogs in **grid**. There are two versions of each function, one that produces output, and one that produces an object without actually plotting it. These objects can eventually be plotted, but they can also be queried to determine their height and width, so that appropriate space can be allocated for them. The standard primitives are listed in Table 3.

3.3 First Attempt

The data we will use for our example is the `faithful` dataset, which contains eruption times and inter-eruption intervals lengths for the famous Old Faithful geyser in Yellowstone national park. We first compute some summaries that will be required to set up the plot.

```
> str(faithful)

'data.frame':       272 obs. of  2 variables:
  $ eruptions: num 3.6 1.8 3.33 2.28 4.53 ...
  $ waiting  : num 79 54 74 62 85 55 88 85 51 85 ...

> x <- faithful$eruptions
> y <- jitter(faithful$waiting)
> xrng <- range(x)
> yrng <- range(y)
> xlim <- extendrange(xrng, f = 0.1)
> ylim <- extendrange(yrng, f = 0.1)
```

```
> xdens <- density(x)
> ydens <- density(y)
> xdens.max <- max(xdens$y)
> ydens.max <- max(ydens$y)
```

We can now use these to take a shot at our desired plot. First we create a viewport for the central plot area with a suitable native coordinate system that covers the range of the data. The location of the viewport itself is specified in terms of the parent (root) viewport. By default, the coordinate system used to specify the location of a viewport is the *normalized parent coordinate* (NPC) system, which assigns the range [0,1] to both coordinate axes.

```
> library(grid)
> pushViewport(viewport(x= 0.45, y= 0.45, width= 0.7, height= 0.7,
+                       xscale = xlim, yscale = ylim))
```

Next we add the data points using the native coordinate system and add the x- and y-axes.

```
> grid.points(x, y, default.units = "native")
> grid.xaxis()
> grid.yaxis()
```

For the rugs, we use line segments whose locations along the axes are specified using the native coordinate system, but whose lengths are specified using the NPC system to be exactly 3% of the range of the corresponding axis.

```
> grid.segments(x0 = unit(x, "native"), y0 = unit(0, "npc"),
+               x1 = unit(x, "native"), y1 = unit(0.03, "npc"))
> grid.segments(x0 = unit(0, "npc"), y0 = unit(y, "native"),
+               x1 = unit(0.03, "npc"), y1 = unit(y, "native"))
```

Finally, we can add axis labels, and then the densities by creating two new viewports and plotting the densities inside.

```
> upViewport(1)
> grid.text("Eruption times (minutes)", x = 0.45, y = 0,
+           just = c("centre", "bottom"))
> grid.text("Time to next eruption (minutes)", x = 0, y = 0.45,
+           rot = 90, just = c("centre", "top"))
> pushViewport(viewport(x= 0.45, y= 0.85, width= 0.7, height = 0.1,
+                       xscale = xlim, yscale = c(0, xdens.max),
+                       clip= "on"))
> grid.polygon(x = xdens$x, y = xdens$y, default.units = "native",
+              gp = gpar(fill = "lightgrey", col = "transparent"))
> upViewport(1)
> pushViewport(viewport(x= 0.85, y= 0.45, width= 0.1, height= 0.7,
+                       xscale = c(0, ydens.max), yscale = ylim,
+                       clip= "on"))
```

```
> grid.polygon(x = ydens$y, y = ydens$x, default.units = "native",
+              gp = gpar(fill = "lightgrey", col = "transparent"))
> upViewport(1)
```

We finish by adding a main label on top.

```
> pushViewport(viewport(x=0.5, y=0.95, width=1, height=0.1))
> grid.text("Old faithful eruptions", x = 0.5, y = 0.5)
```

The final result of these calls can be seen in Fig. 17.

This approach can be generalized to work with other datasets by writing a function that performs all the individual steps. However, there are several improvements that could be desired. Instead of allocating space for the axes, main title, etc., as proportions of the total figure area as we have done, we would prefer to use only as much space as is required, and no more. This can be done by using an alternative method where viewports are specified using *layouts*, where the figure area is divided into rows and columns whose heights and widths can be specified more flexibly. We refrain from a detailed discussion of this approach for brevity, but such advanced features of **grid** are used by the **lattice** and **ggplot2** packages, which we briefly discuss next.

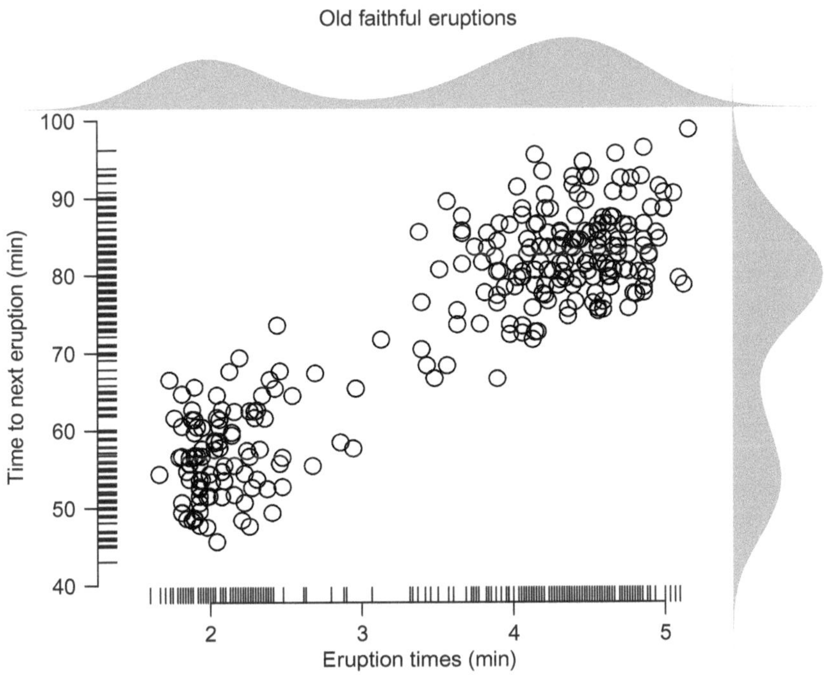

FIGURE 17 An enhanced scatter plot of eruption times and waiting times in the `faithful` data. In addition to the point scatter, density plots and rugs indicate marginal distributions.

4 LATTICE

lattice is an add-on package that implements Trellis graphics in R. It is a powerful high-level data visualization system with an emphasis on multivariate data. To illustrate the typical usage of **lattice**, let us revisit the ozone concentration data and see how we can use it to draw histograms. The function that draws histograms is `histogram()`, and a standard histogram can be created using

```
> library(lattice)
> histogram(~ Ozone, data = airquality)
```

Note that the variable `Ozone` is specified in a formula. All high-level functions in **lattice** support the use of formulas as the primary interface. The formula language allows us to elegantly specify *conditioning variables* that are used to subdivide the data into subgroups that are plotted in "panels" juxtaposed to enable comparison. To see the distribution of `Ozone` across different months, we can simply use

```
> histogram(~ Ozone | fmonth, data = airquality)
```

The output of both the calls above are shown in Fig. 18. Note that without any specific instructions in the call, the panels share the same data ranges, and in fact the axes are labeled only along the boundaries of the whole plot instead of each panel. This is the default behavior in all **lattice** plots and is intended to make optimal use of available space while making comparison easier.

A similar design with kernel density estimates instead of histograms can be created using the `densityplot()` function.

```
> densityplot(~ Ozone | fmonth, data = airquality, as.table = TRUE)
```

The `as.table` argument is used to order the panels from top-left (as in a table) rather than the default bottom-left (as in a Cartesian coordinate system). While juxtaposition with common axes makes comparison easier, even more effective comparison is enabled by direct superposition. As noted earlier, this is difficult to do with conventional histograms, but easier using kernel density estimates. This is achieved using a `groups` argument, which is supported by some but not all of the high-level functions in **lattice**.

```
> densityplot(~ Ozone, data = airquality, groups = fmonth,
+             plot.points = FALSE, auto.key = list(columns = 2))
```

The output of both calls above is given in Fig. 19. We leave it as an exercise to the reader to determine the effect of the extra arguments `plot.points` and `auto.key`.

4.1 Overview

The **lattice** package provides a high-level system for statistical graphics that is essentially independent of traditional R graphics. It is modeled on the Trellis

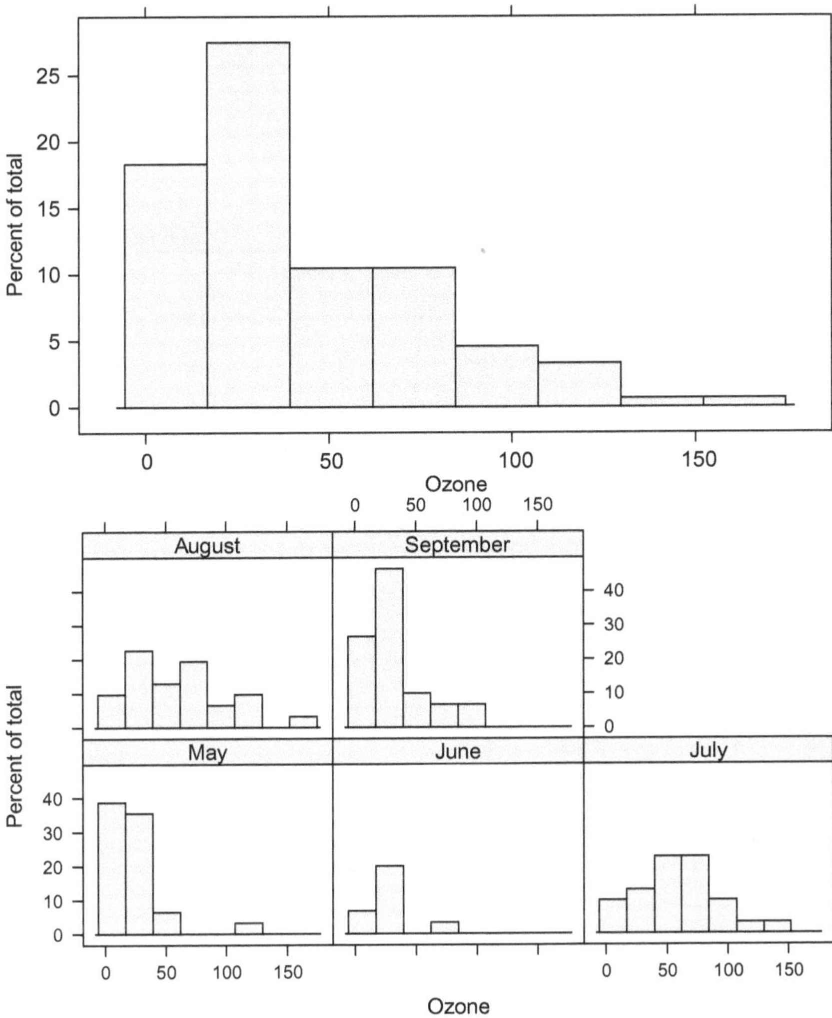

FIGURE 18 Histograms of `Ozone` concentration created using **lattice**.

suite in S (Becker et al., 1996) and uses the **grid** package (Murrell, 2005) as the underlying implementation engine. Trellis displays are defined primarily by the *type* of graphic and the *role* different variables play in it. Each supported graphic design is associated with a corresponding *high-level* function (`histogram`, `densityplot`, etc.). Possible roles depend on the type of display, but typical ones are:

- primary variables: those that define the primary display (e.g., `Ozone` in the preceding examples),

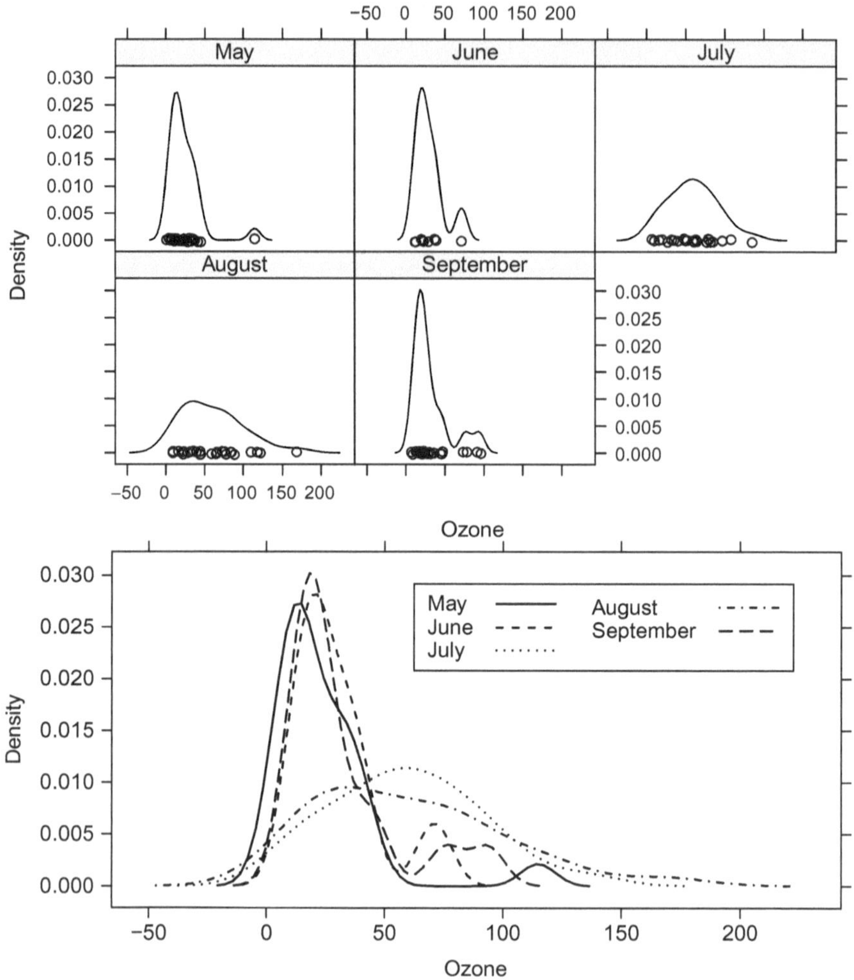

FIGURE 19 Density plots of Ozone concentration by month, juxtaposed (top) and superposed (bottom).

- conditioning variables: divides data into subgroups, each of which is presented in a different panel (e.g., `fmonth`), and
- grouping variables: subgroups are contrasted within panels by superposing the corresponding displays (e.g., `fmonth` in the last example).

The high-level display types available in **lattice** and the corresponding graphic design are given in Table 4.

New high-level functions can be written to represent further visualization types; some examples are available in the **latticeExtra** package (Sarkar and Andrews, 2013).

TABLE 4 High-Level Functions Available in the lattice Package

Function	Default Display
`histogram()`	Histogram
`densityplot()`	Kernel density plot
`qqmath()`	Theoretical quantile plot
`qq()`	Two-sample quantile plot
`stripplot()`	Stripchart (comparative 1-D scatterplots)
`bwplot()`	Comparative box-and-whisker plot
`dotplot()`	Cleveland dot plot
`barchart()`	Bar plot
`xyplot()`	Scatterplot and time series plot
`splom()`	Scatterplot matrix
`contourplot()`	Contour plot of surfaces
`levelplot()`	False-color level plot of surfaces
`wireframe()`	Three-dimensional perspective plot of surfaces
`cloud()`	Three-dimensional scatterplot
`parallelplot()`	Parallel coordinates plot

Although visualization is an art, it can benefit greatly from a systematic, scientific approach. The design of Trellis graphics was based on the work of Cleveland (1985, 1993), who demonstrated that it is possible to come up with general rules that can be applied to design more effective graphs. One of the primary goals of Trellis graphics is to provide tools that make it easy to apply these rules so that the burden of compliance is shifted from the user to the software to the extent possible. Some obvious examples of such rules are the following:

- Use as much of the available space as possible
- Force direct comparison by superposition (grouping) when possible
- Encourage comparison when juxtaposing (conditioning): use common axes and add common reference objects such as grids.

These design goals have some technical implications; for example, nonwastage of space requires the complete display to be known when plotting begins, so, the incremental approach common in traditional R graphics (e.g., adding a main title after the main plot is finished) is no longer feasible. **lattice** deals with such situations by following an object-based paradigm: plots are represented as regular R objects, and incremental updates are performed by modifying such objects and replotting them.

Although rules are useful, any serious graphics system must also be flexible. **lattice** is designed to be flexible, but obviously there is a trade-off between flexibility and ease of use for the more common tasks. **lattice** tries to achieve a balance using the following structure. A display is constructed as a combination of various elements. The defaults of all the elements are coordinated to provide meaningful results, but for finer control each element can be controlled by the user independently of the others. The main elements are:

- the primary (panel) display,
- axis annotation,
- strip annotation (describing the conditioning process), and
- legends (typically describing the grouping process).

In each case, additional arguments to the high-level calls can be used to activate common variants, and full flexibility is allowed through arbitrary user-defined functions. This is particularly useful for controlling the primary display through panel functions. Most nontrivial use of **lattice** involves manipulation of one or more of these elements.

4.2 Common High-Level Functions

As noted in Table 4, most of the traditional graphics functions we encountered earlier in this chapter have **lattice** analogues. Several of these are intended to visualize the distribution of a continuous random variable. We have already seen histograms and density plots, which are both estimates of the probability density function. Another useful display is the normal Q–Q plot, which is related to the distribution function $F(x) = P(X \leq x)$. Normal Q–Q plots can be produced by the **lattice** function `qqmath()`. Figure 20 is produced by

```
> qqmath(~ Ozone | fmonth, airquality, grid = TRUE, layout = c(5, 1))
```

The `layout` argument controls the layout of panels in columns, and rows (and possibly pages). Normal Q–Q plots plot empirical quantiles of the data against quantiles of the normal distribution (or some other theoretical distribution). They can be regarded as an estimate of the distribution function F, with the probability axis transformed by the normal quantile function. They are designed to detect departures from normality; for a good fit, the points lie approximate along a straight line. The `grid` argument adds a common reference grid to each panel that makes comparison easier.

Comparative box-and-whisker plots are produced by the **lattice** function `bwplot()`. For example,

```
> bwplot(fmonth ~ Ozone, airquality)
```

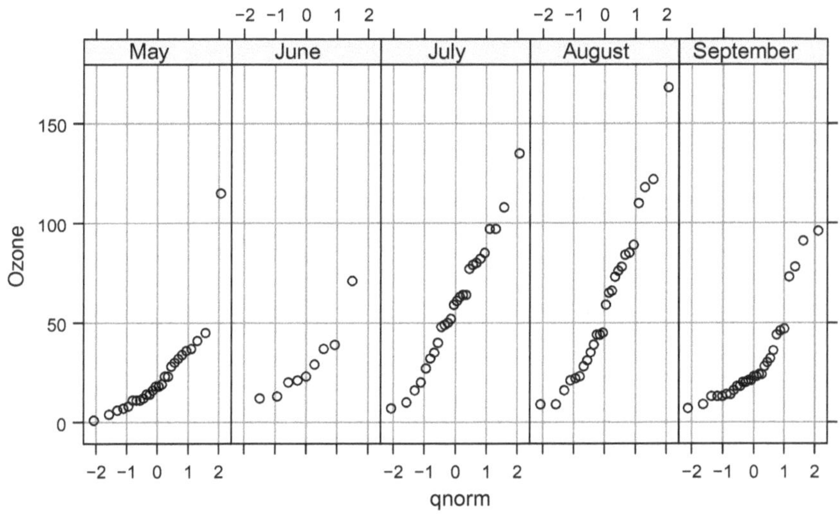

FIGURE 20 Normal Q–Q plot of Ozone concentration by month, produced by **lattice**.

It is also possible to juxtapose box-and-whisker plots of more than one variable, using a special form of the formula. In this case, it does not make sense to require a common axis range for the resulting panels, and we use the `scales` argument in the following call to allow independent ranges in the x-axis.

```
> bwplot(fmonth ~ Ozone + Solar.R, airquality,
+           outer = TRUE, scales = list(x = "free"), xlab = NULL)
```

The results of both calls are shown in Fig. 21.

4.3 Bar Charts and Dot Plots for Tabular Data

Bar charts and Cleveland dot plots are created using the `barchart()` and `dotplot()` functions, respectively. We go back to the `VADeaths` dataset for illustration. `VADeaths` is a matrix, but we need to convert it into a data frame to use the **lattice** formula interface. This may be done using

```
> VADeathsDF <- as.data.frame.table(VADeaths, responseName = "Rate")
> str(VADeathsDF)
```

```
'data.frame': 20 obs. of 3 variables:
  $ Var1: Factor w/ 5 levels "50-54","55-59",..: 1 2 3 4 5 1 2 3 4 5 ...
  $ Var2: Factor w/ 4 levels "Rural Male","Rural Female",..: 1 1 1 1 1
          2 2 2 2 2 ...
  $ Rate: num 11.7 18.1 26.9 41 66 8.7 11.7 20.3 30.9 54.3 ...
```

A bar chart can be now produced by the following call:

```
> barchart(Rate ~ Var1 | Var2, VADeathsDF, layout = c(4, 1),
+           scales = list(x = list(rot = 45)))
```

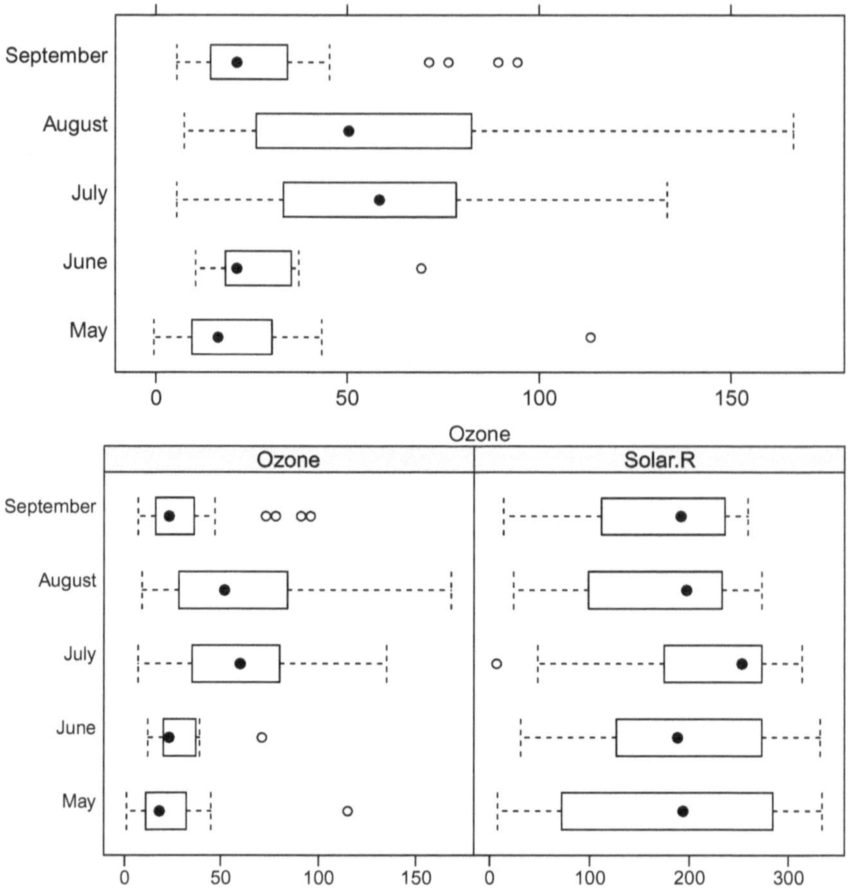

FIGURE 21 Box-and-whisker plots produced by **lattice**. The top plot shows Ozone concentration alone, and the bottom plot shows Ozone concentration as well as solar radiation.

This plot is potentially misleading because a strong visual effect in the plot is the comparison of the areas of the shaded bars, which do not mean anything. This problem can be addressed by making the areas proportional to the values they encode, by ensuring that the bars start at 0.

```
> barchart(Rate ~ Var1 | Var2, VADeathsDF, layout = c(4, 1), origin =0,
+           scales = list(x = list(rot = 45)))
```

The results of both calls are shown in Fig. 22.

A better design is to altogether forego the bars, which distract from the primary comparison of the endpoint positions, and instead use a dot plot.

```
> dotplot(Rate ~ Var1 | Var2, VADeathsDF, layout = c(4, 1),
+           scales = list(x = list(rot = 45)))
```

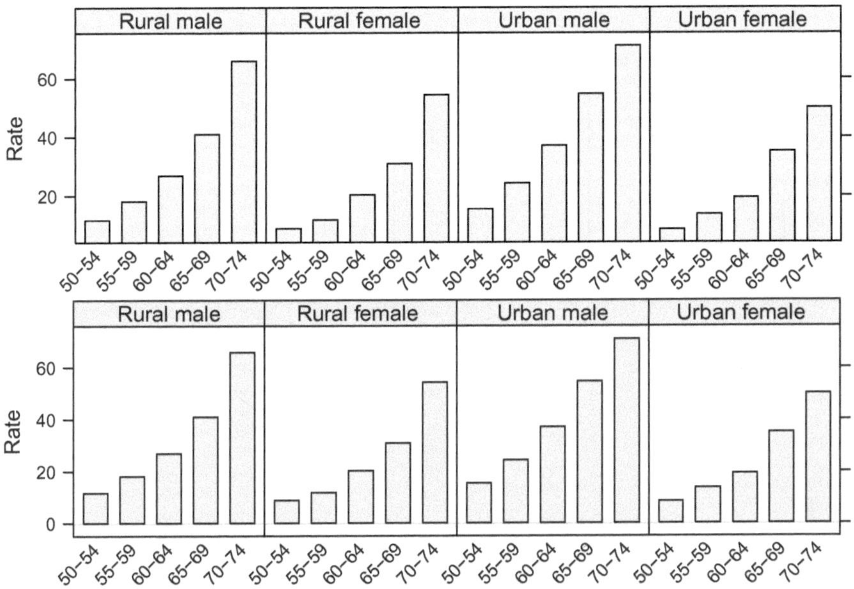

FIGURE 22 Bar charts of death rates by age and population subgroups. The bottom plot ensures that the bars start at zero so that the area of each bar is proportional to the quantity being plotted.

In this particular example, the display is more effective if we use `Var2` as a grouping variable and join the points within each group.

```
> dotplot(Rate~ Var1, data= VADeathsDF, groups= Var2, type= "o",
+           auto.key= list(columns = 2, points = TRUE,
+           lines = TRUE))
```

Results of the calls are shown in Fig. 23. It is clear from the grouped plot that the pattern of death rate by age is virtually identical for urban and rural females, with an increased rate for rural males and a further increase for urban males. This interaction is difficult to see in the preceding conditional plots.

High-level **lattice** functions are actually generic functions, with specific methods doing the actual work. All the examples we have seen so far use the "formula" methods, that is, the method called when the first argument is a formula. Because `barchart()` and `dotplot()` are frequently used for multi-way tables stored as arrays, **lattice** also includes suitable methods that bypass the conversion to a data frame that would be required otherwise. For example, the following call is an alternative way to reproduce the bottom plot in Fig. 23.

```
> dotplot(VADeaths, type= "o", ylab= "Rate", horizontal= FALSE,
+           auto.key= list(columns= 2, points= TRUE, lines= TRUE))
```

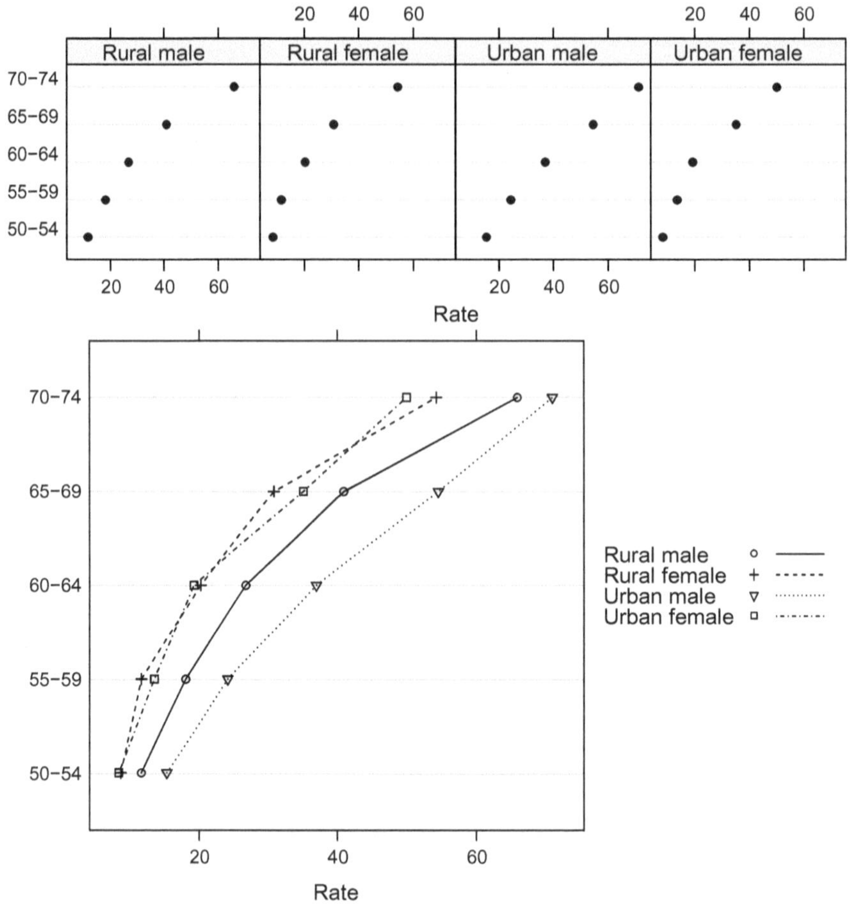

FIGURE 23 Dot plots of death rates by age and population subgroups, juxtaposed (top) and superposed (bottom). The bottom plot is a variant of Fig. 4.3 in Sarkar (2008).

Methods available for a particular generic can be listed using

```
> methods("dotplot")
[1] dotplot.array* dotplot.default* dotplot.formula* dotplot.
matrix*
[5] dotplot.numeric* dotplot.table*
  Nonvisible functions are asterisked
```

The special features of the methods, if any, are described in their respective help pages; for example, `?dotplot.matrix` for the example above.

4.4 Scatterplots and Custom Displays

The **lattice** function to draw scatterplots is `xyplot()`. Going back to Anscombe's quartet of datasets, we can create the equivalent of Fig. 1 with the following code (output not shown):

```
> xyplot(y1 ~ x1, data = anscombe)
```

To use `xyplot()` to show all four datasets together, we need to rearrange the data to make it amenable to use with the formula interface. This can be done as follows:

```
> anscombe.long <-
+     with(anscombe, data.frame(x = c(x1, x2, x3, x4),
+                               y = c(y1, y2, y3, y4),
+                               which = gl(4, nrow(anscombe))))
```

A conditional plot can now be created using

```
> xyplot(y ~ x | which, data = anscombe.long)
```

Suppose that we wanted to customize this plot by adding linear and quadratic regression lines as in Fig. 12. The idea is very similar to what we did earlier, in the sense that we need to fit the models and add lines or curves representing the fits. The difference is that with **lattice** this cannot be done after the basic plot is already drawn. Instead, we must provide a function that encodes the procedure to be followed to draw each panel. This function is known as the *panel function* and is supplied as the `panel` argument to high-level **lattice** calls. In this example, such a function might look as follows:

```
> panel.lq <- function(x, y, ...)
+ {
+     panel.grid()
+     panel.xyplot(x, y, ...)
+     fm1 <- lm(y ~ x)
+     panel.abline(fm1)
+     try({
+         fm2 <- lm(y ~ poly(x, 2))
+         f <- function(x) predict(fm2, newdata = list(x = x))
+         panel.curve(f)
+     }, silent = TRUE)
+ }
```

This function uses **lattice** analogues of traditional graphics functions such as `abline()` and `curve()`. The `panel.xyplot()` function represents the default panel function for `xyplot()`. The quadratic model fit is wrapped inside `try()` because the fit fails for the fourth dataset which has only two unique values. Figure 24 is produced using this panel function as follows:

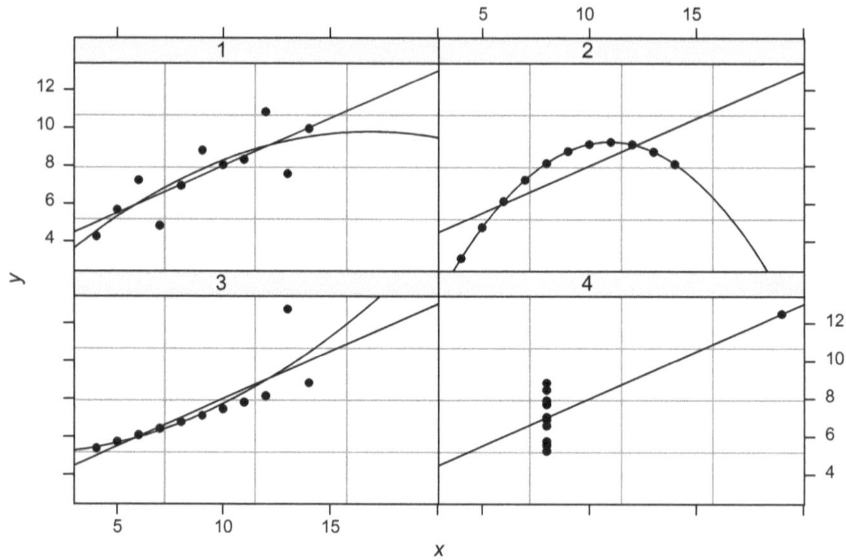

FIGURE 24 Scatterplots of Anscombe's datasets along with linear and quadratic fits.

```
> xyplot(y ~ x | which, data = anscombe.long, pch = 16,
+          as.table = TRUE, panel = panel.lq)
```

Three-dimensional scatterplots are supported by **lattice** through the `cloud()` function.[2] Figure 25 uses the `quakes` dataset to show a scatterplot of latitude, longitude, and depth of earthquake epicenters and is produced by the following code:

```
> cloud(depth ~ lat * long, data = quakes,
+          zlim = rev(range(quakes$depth)),
+          screen = list(z = 105, x = -70), panel.aspect = 0.75,
+          xlab = "Longitude", ylab = "Latitude", zlab = "Depth")
```

Static three-dimensional scatterplots are not very useful because of the strong effect of "camera" direction. Unfortunately, **lattice** does not allow interactive manipulation of the viewing direction. For an alternative interface that does allow such dynamic manipulation, see the **rgl** package (Adler and Murdoch, 2014).

4.5 The "trellis" Object

One important feature of **lattice** that makes it different from traditional R graphics is that high-level functions do not actually plot anything. Instead, they return an object of class "trellis", that needs to be `print()`-ed or `plot()`-ed. R's automatic printing rule means that in most cases, the user does not see any

2. For a traditional graphics analogue, see the **scatterplot3d** package (Ligges and Mächler, 2003).

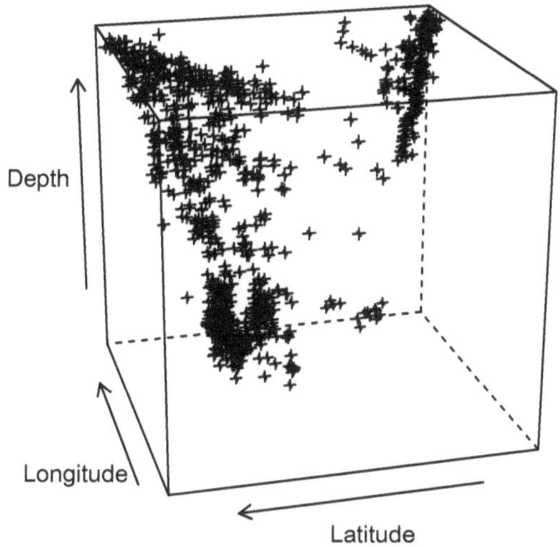

FIGURE 25 A three-dimensional scatterplot of earthquake epicenter locations.

difference in behavior. Here is one example where we use optional arguments of the `plot()` method for "trellis" objects to display two plots side by side.

```
> dp.wp <-
+     dotplot(WorldPhones/1000, groups = FALSE, layout = c(1, NA),
+             xlab = "Number of telephones (millions)")
> dp.wp.log <-
+     dotplot(WorldPhones/1000, groups = FALSE, layout = c(1, NA),
+             scales = list(x = list(log = TRUE,
+                                    equispaced.log = FALSE)),
+             xlab = "Number of telephones (millions)")
> plot(dp.wp,     split = c(1, 1, 2, 1))
> plot(dp.wp.log, split = c(2, 1, 2, 1), newpage = FALSE)
```

The result is shown in Fig. 26.

5 ggplot

Traditional graphics and **lattice** graphics both have the same basic approach. Plot functions are written to implement specific designs, usually those that have already been established as being useful. Customization is achieved through a procedural approach. The **ggplot2** package takes a different approach. It defines a "layered grammar" for defining graphical designs as a composition of various independent components, which can be used by the user to describe a plot which is not necessarily from a set of pre-specified plot types. Plots consist of one or more layers, usually a layer rendering the raw data, together with annotations and summaries. Each layer is associated with a dataset. The main components of the grammar are:

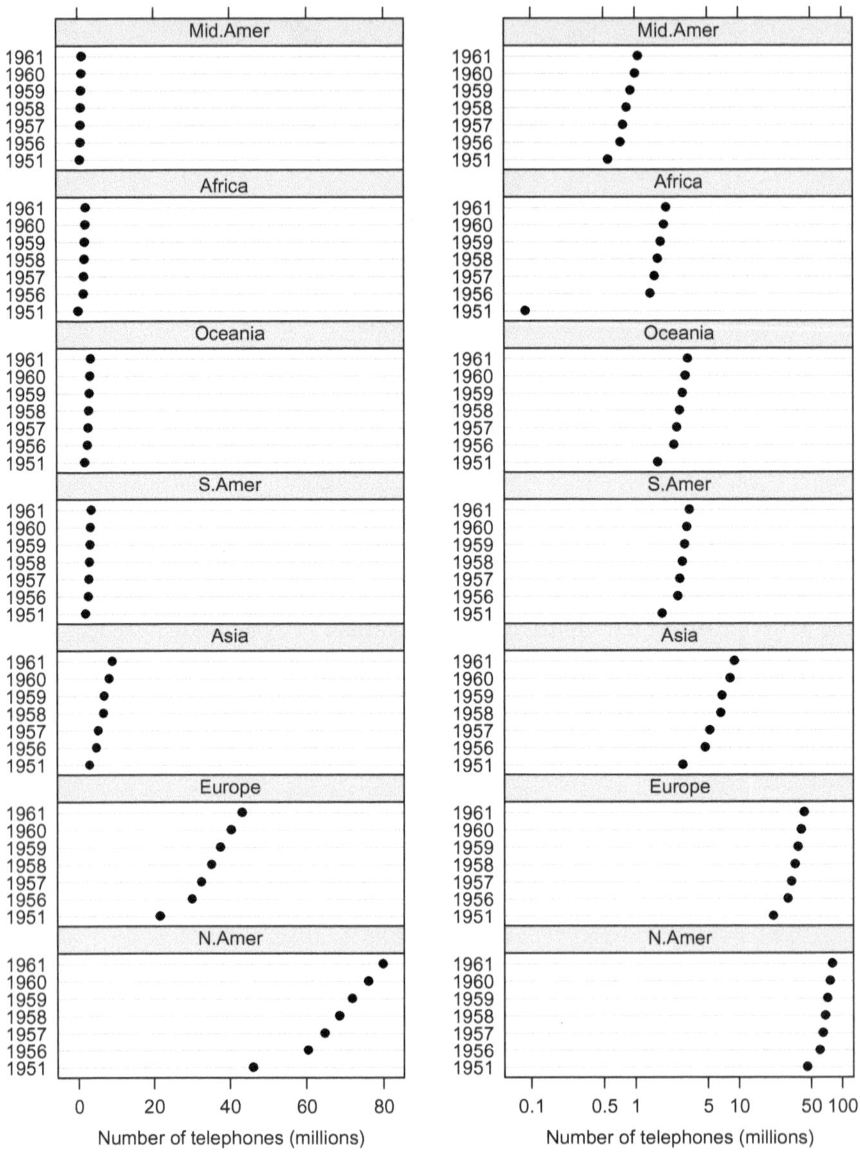

FIGURE 26 Side-by-side plots of the estimated number of telephones in various parts of the world (McNeil, 1977), raw and log-transformed, created using the `plot()` method for "trellis" objects.

- A set of *aesthetic mappings* that map data values to some aspect in the displayed graph, such as coordinate positions, color, shape, size, group, etc.
- A geometric type that is used to render the mapped data, e.g., by points, lines, polygons, or something more complex such as a box-and-whisker plot.

- Statistical transformations that are applied to the data beforehand, such as binning for histograms, or computation of kernel density estimates.
- Scales that give a visual indication of the aesthetic mappings, e.g., axis annotation for position mapping, legends for mapping to color, size, etc.
- Faceting (conditioning in Trellis terminology) to produce small multiples.

The **ggplot2** package is a comprehensive system that provides a powerful alternative to the traditional graphics approach. The system is described in detail by Wickham (2009). Here, we provide a few simple examples to give a flavor of its novel approach.

Consider again the `airquality` dataset, and suppose we wish to produce histograms and kernel density plots conditioned on `fmonth`. As the data source and faceting are common to both plots, we can capture this common information and store it in a variable using the following call:

```
> library(ggplot2)
> p <- ggplot(airquality, aes(x = Ozone)) + facet_wrap(~ fmonth)
```

The `aes()` function specifies the aesthetic mapping, saying in this case that the `Ozone` variable should be mapped to x-coordinates in the plot. `p` is not yet a valid plot, because it does not yet have any layers. To create a scatterplot, we can add a point "geom" (short for geometric type), but to do so, we also need to specify a variable that will map to the y-coordinates. Figure 27 is produced by the following code:

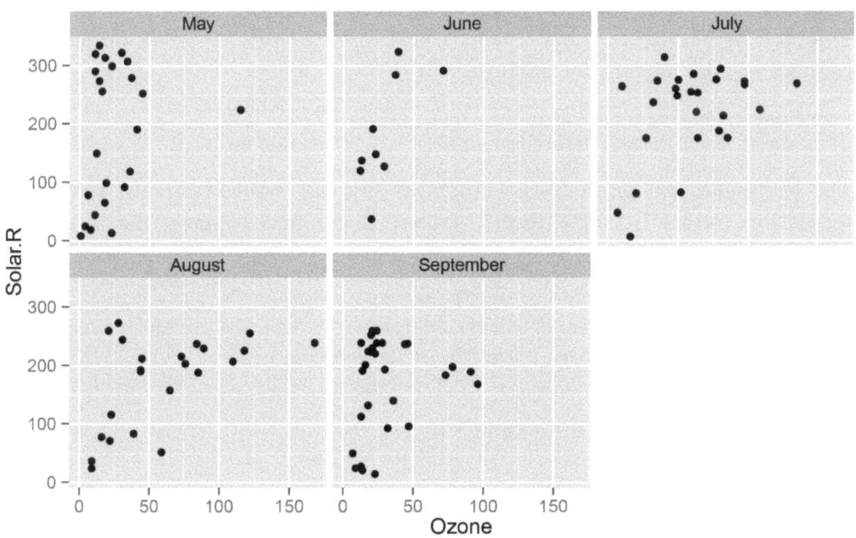

FIGURE 27 Scatterplot of solar radiation and ozone concentration by month, produced using **ggplot2**.

```
> p + geom_point(aes(y = Solar.R))
```

For a histogram, adding a geom is not enough, because we first need to transform the data by binning it. For this, we add a new "stat" layer created by `stat_bin()`, where we can also specify the geom to be used for rendering the transformed data.

```
> p + stat_bin(binwidth = 20, geom = "bar")
```

The `"bar"` geom is in fact the default, and does not need to be specified explicitly. Kernel density plots can be similarly produced by using `stat_density()`.

```
> p + stat_density(geom = "path")
```

In this case, the `"path"` geom asks the density to be rendered as a curve instead of the default of a filled area. The output of both calls is shown in Fig. 28.

Given that these are all standard graphical designs, the calls above are comparatively verbose. The `qplot()` function offers a simpler interface for common designs. The examples in Figs. 27 and 28 can be reproduced by the following code:

```
> qplot(x = Ozone, y = Solar.R, data = airquality, facets = ~ fmonth)
> qplot(x = Ozone, data = airquality, facets = ~ fmonth,
+         stat = "bin", binwidth = 20)
> qplot(x = Ozone, data = airquality, facets = ~ fmonth,
+         stat = "density", geom = "line", binwidth = 20)
```

The previous examples of course only use one layer. The full potential of the layered approach of **ggplot2** is apparent when we need to add more elements to a plot. The following call creates a scatterplot of the `faithful` data with a LOESS fit and rugs on the margin.

```
> qplot(x = jitter(eruptions), y = jitter(waiting), data = faithful) +
+         stat_smooth(method = "loess") + geom_rug()
```

The following call produces a version of Fig. 24, with linear and quadratic regression model fits (the fourth dataset is omitted because the quadratic model fit fails for it):

```
> anscombe.123 <- subset(anscombe.long, which != "4")
> qplot(x = x, y = y, data = anscombe.123, facets = ~ which) +
+         stat_smooth(method = "lm", se = FALSE) +
+         stat_smooth(method= "lm", formula = y ~ poly(x, 2), se = FALSE)
```

The results for both calls are shown in Fig. 29.

6 FURTHER READING

R graphics is an extensive topic that is impossible to cover in any depth in a single chapter. An important aspect we have glossed over is the system of

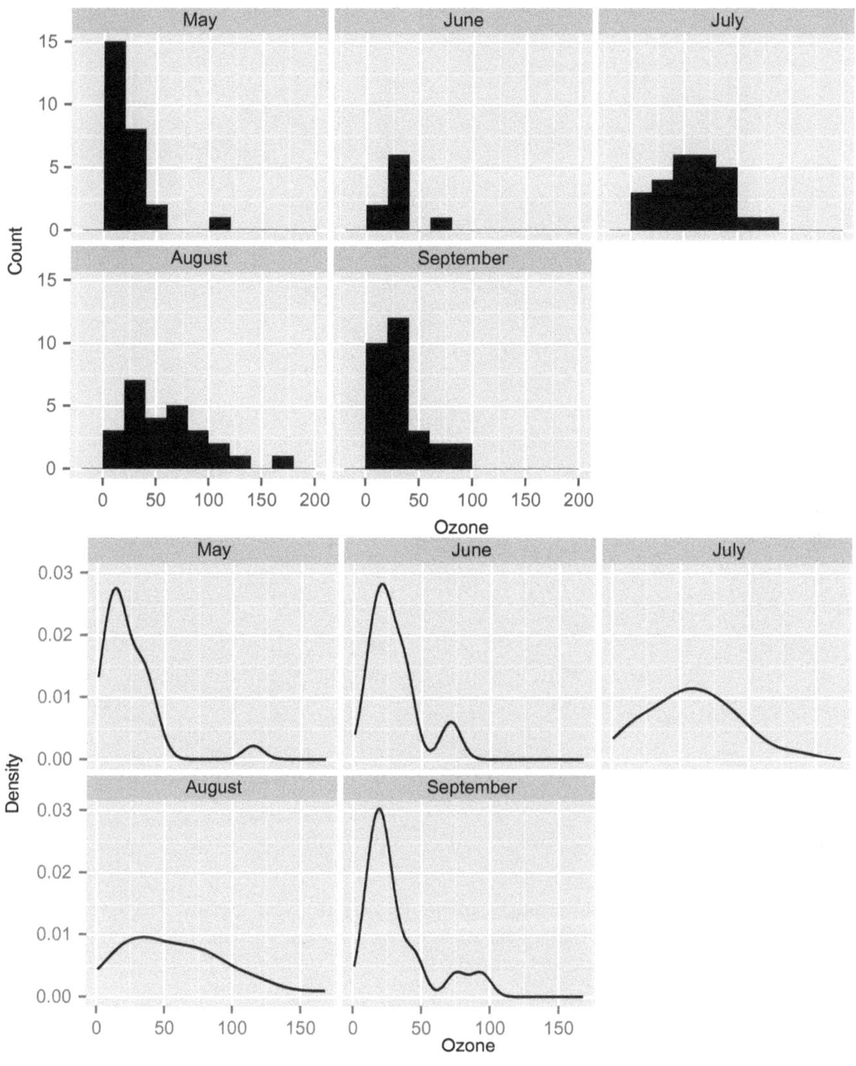

FIGURE 28 Histograms and density plots of ozone concentration by month created by **ggplot2**.

customizable parameters (see `?par` for traditional graphics, `?gpar` for Grid graphics, `?trellis.par.get` for **lattice**, and `?ggtheme` for **ggplot2**). Such parameters may include colors, line types, plotting characters, fonts, etc., and even the amount of space to be used for margins. For a fuller discussion of the traditional R graphics system and Grid graphics, see Murrell (2005). The **lattice** package is discussed by Sarkar (2008), and **ggplot2** by Wickham (2009). Various other specialized displays are implemented in add-on packages, notably **vcd** for categorical data (Hornik et al., 2006). The

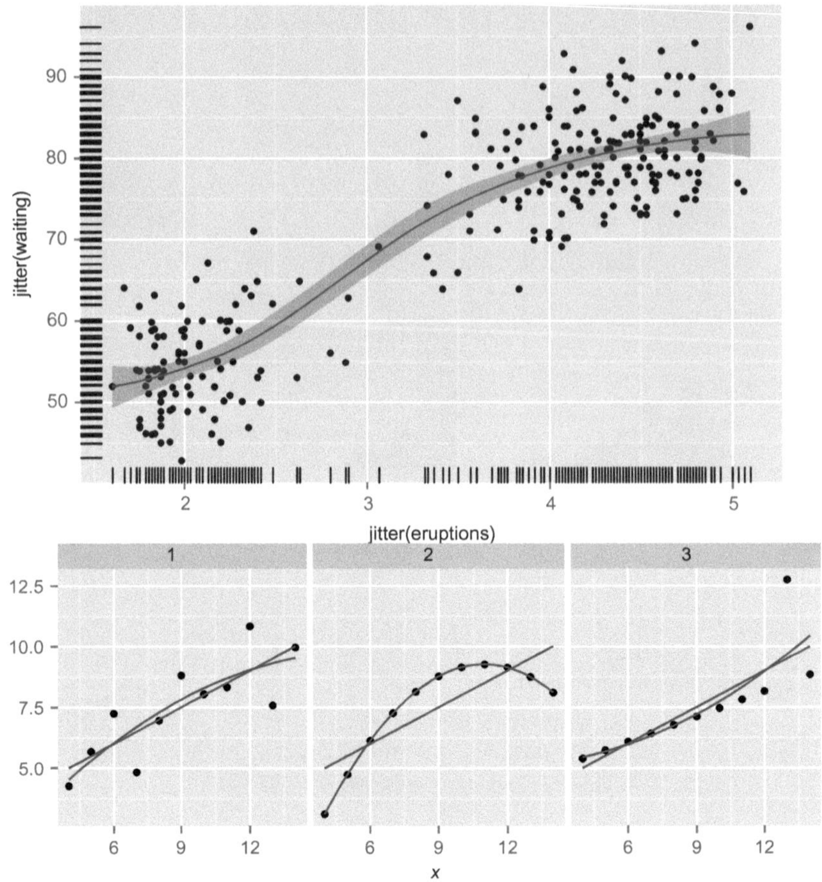

FIGURE 29 Scatterplots with various customizations created using **ggplot2**. The top plot shows a scatterplot of the `faithful` data with a LOESS fit and rugs. The bottom plot shows three of Anscombe's datasets with linear and quadratic fits.

CRAN Task View on Graphics[3] is a good starting point for pointers to many useful packages and projects.

The "mainstream" graphics utilities described above are all primarily meant for static graphics. Although both the traditional graphics and grid graphics models allow for some interaction, the need to support such interaction on a wide variety of output devices makes such interaction fairly limited. To address this shortcoming, several alternative graphics systems have been developed over the years that allow various degrees of dynamic manipulation and interaction. These are usually less portable, in the sense that they may not work on all platforms supported by R, and may be more difficult to install

3. http://cran.r-project.org/web/views/Graphics.html.

even on the platforms where they do. Of these, the **rgl** package provides mechanisms for display and interactive manipulation of three-dimensional plots using OpenGL, and the **misc3d** package builds on it to provide high-level tools (Feng and Tierney, 2008). The **rggobi** package provides an interface to GGobi, a powerful interactive graphics system (Swayne et al., 2003).

REFERENCES

Adler, D., Murdoch, D., 2014. rgl: 3D visualization device system (OpenGL), R package version 0.93.1098.

Anscombe, F.J., 1973. Graphs in statistical analysis. Am. Stat. 27, 17–21.

Becker, R.A., Cleveland, W.S., Shyu, M.J., 1996. The visual design and control of trellis display. JCGS 5, 123–155.

Box, G.E.P., Jenkins, G.M., 1976. Time Series Analysis: Forecasting and Control. Holden-Day Inc., San Francisco.

Cleveland, W.S., 1985. The Elements of Graphing Data. Wadsworth, Monterey, CA.

Cleveland, W.S., 1993. Visualizing Data. Hobart Press, Summit, NJ.

Cleveland, W.S., McGill, R., 1984. Graphical perception: theory, experimentation, and application to the development of graphical methods. J. Am. Stat. Assoc. 79, 531–554.

Cleveland, R.B., Cleveland, W.S., McRae, J.E., Terpenning, I., 1990. STL: a seasonal-trend decomposition procedure based on loess. J. Off. Stat. 6, 3–33.

Feng, D., Tierney, L., 2008. Computing and displaying isosurfaces in R. J. Stat. Softw. 28, 1–24.

Friendly, M., 1994. Mosaic displays for multi-way contingency tables. J. Am. Stat. Assoc. 89, 190–200.

Hornik, K., Zeileis, A., Meyer, D., 2006. The strucplot framework: visualizing multi-way contingency tables with vcd. J. Stat. Softw. 17, 1–48.

Ligges, U., Mächler, M., 2003. Scatterplot3d—an R package for visualizing multivariate data. J. Stat. Softw. 8, 1–20.

McLeod, A.I., Yu, H., Mahdi, E., 2012. Time series analysis with R. In: Rao, T.S., Rao, S.S., Rao, C.R. (Eds.), Handbook of Statistics, Time Series Analysis: Methods and Applications, vol. 30. Elsevier, Oxford, UK.

McNeil, D.R., 1977. Interactive data analysis. Wiley.

Murrell, P., R Graphics, 2005. Chapman & Hall/CRC. Raton, FL, Boca. ISBN: 1-584-88486-X.

Sarkar, D., 2008. Lattice: Multivariate Data Visualization with R. Springer, New York. ISBN: 978-0-387-75968-5.

Sarkar, D., Andrews, F., 2013. lattice Extra: Extra Graphical Utilities Based on Lattice. R package version 0.6–26.

Swayne, D.F., Temple Lang, D., Buja, A., Cook, D., 2003. GGobi: evolving from XGobi into an extensible framework for interactive data visualization. Comput. Stat. Data Anal. 43, 423–444.

Tufte, E.R., 1990. Envisioning Information. Graphics Press, Cheshire, CT.

Tufte, E.R., 2001. The Visual Display of Quantitative Information, second ed. Graphics Press, Cheshire, CT.

Tukey, J.W., 1977. Exploratory Data Analysis. Addison-Wesley, Menlo Park, CA.

Wickham, H., 2009. ggplot2: Elegant Graphics for Data Analysis. Springer, New York.

Wickham, H., 2013. Statistical graphics. Encyclopedia Environ, 6.

Wilkinson, L., 1999. The Grammar of Graphics. Springer, New York.

Chapter 3

Graphics Miscellanea

Palash Mallick and Marepalli B. Rao[1]
Department of Environmental Health, University of Cincinnati, Cincinnati, Ohio, USA
[1]*Corresponding author: e-mail: Marepalli.rao@uc.edu*

ABSTRACT
In this chapter, we provide a basic introduction to graphics using R complementing the chapter on R Graphics by Dr. Sarkar.

Keywords: Plot, Par, Multiple plots, Scatter plots, Regression, Multiple regression, Residuals, Time series plots, Pie chart, Box plots, xy plots, Curves, LOWESS, Sunflower plots, Violin plots, Bean plots, Bubble plots, Surface plots, Chernoff faces, Maps, Choropleth maps

1 INTRODUCTION

A picture is worth more than a thousand words. Data collection is an integral part of any research activity. Presenting the data in an appropriate graphical mode is absolutely vital in enhancing research presentation. This chapter describes a variety of ways to present data using R. We assume the reader is familiar with some basic R commands in order to enjoy this chapter. The R commands can be typed directly in the R console or in an R editor. Typing in an R editor facilitates saving and future use. An R editor can be opened from an R console by clicking "File" and "New Script." One can click on "Open Script" to recall a saved R program file. Another way to use R is by utilizing R studio. A number of graphics are presented along with their commands and output. There are a number of excellent books on R graphics. Notably, Murrell (2011), Chang (2012), Wickam (2009), Sarkar (2008), and Cook and Swayne (2007) have contributed substantially to graphics literature among many others. New packages on graphs are also coming on to the scene periodically. Some of the examples are inspired by these contributions. Some of the examples in this chapter come from these sources.

2 THE *PLOT()* COMMAND

The plot command in R is very versatile with a veritable collection of entrancements. The command is used to provide a scatter plot of bivariate data

Handbook of Statistics, Vol. 32. http://dx.doi.org/10.1016/B978-0-444-63431-3.00003-6

$(x_1, y_1), (x_2, y_2), \ldots, (x_n, y_n)$ on two quantitative variables, x and y. The core inputs for the command are `[>plot (x, y, type =...)]`. The data $(x_1, x_2, \ldots, x_n)$ on the variable x and data $(y_1, y_2, \ldots, y_n)$ on y. We will illustrate the use of the command using artificial data along with the `par` command.

2.1 Features that Can Be Included in a Scatter Plot

Here are couple of most used features on scatter plot using the `plot()` command,

`type`: determines the type of the output graph. type = "p," the symbol "p" stands for points and it is a default choice. Other choices are:
- `"l"` for **l**ines,
- `"b"` for **b**oth,
- `"c"` for the lines part alone of `"b"`,
- `"o"` for both "**o**verplotted,"
- `"h"` for "**h**istogram" like (or "high-density") vertical lines,
- `"s"` for stair **s**teps, and
- `"n"` for no plotting.

`xlab`: This is for the label on the x-axis.
`ylab`: This is for the label on the y-axis.
`pch`: `pch` argument is for the plotting character. There is a choice of 26 characters. One can improvise the character with one's own choice.
`col`: stands for color of the plot. There is a choice of 667 colors.
`las`: The `las` argument can be used to orient the numbers on the axes (`las = 0 - 3`, with `las = 0` being the default).
`xlim`: The limits of x-axis can be set by `xlim` = c(minimum, maximum).
`ylim`: The `ylim` argument has a similar connotation.
`main`: This facilitates a main title for the graph.
`sub`: This facilitates a subtitle for the graph.

The titles can be colored. For example, the input, `col.main = "red,"` paints the main title red. The size of the numbers on the axes can be controlled.

`cex`: determines the size of the plotting character.
`cex.lab`: determines the size of the text labels on the axes.
`cex.axis`: determines the size of the numbers on the tick marks.

2.1.1 *par()* Command

The `par()` command facilitates accommodation of several graphs in a single frame.

For our example, we create two folders of data named x and y.

```
> x <- c(1, 2, 3, 4, 5, 6, 7, 8, 9, 10)
> y <- c(12, 15, 16, 18, 21, 22, 28, 31, 33, 35)
```

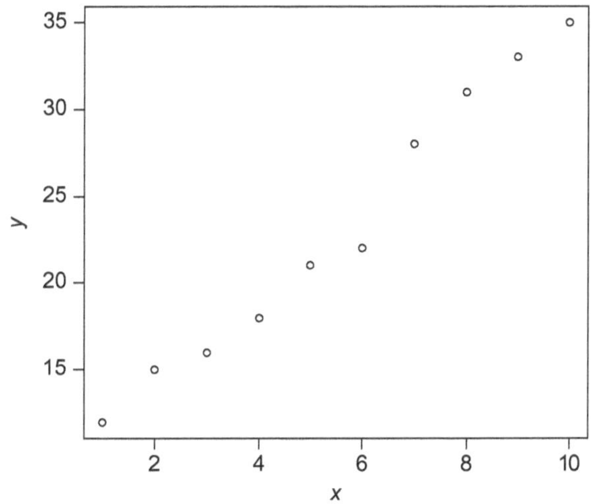

FIGURE 1 Plot command with points only. Here the labels of axes, orientation of the numbers, choice of plotting character, and sizes of the numbers on the axes are all at default settings. (las = 0; cex = 1; cex.lab = 1; cex.axis = 1.)

We want to create the default scatter diagram of the data in *x* and *y* (Fig. 1).

```
> plot(x, y)
```

We want to present six scatter diagrams with different specifications in a single frame. For this, we use the `par()` command. We also want to create spaces at the bottom worth two lines and at the top four lines. We use the input `oma = c(bottom, left, top, right)` inside the par command.

```
> par(mfrow = c(3,2), oma = c(2, 0, 4, 0))
```

A blank screen appears. There is room for six graphs and four lines of space at the top. We provide R commands graph-by-graph.

First graph,

```
> plot(x, y, type = "l", pch = 16, main = "Scatter Diagram",
+       sub = "Heart Data", xlab = "Samantha", ylab = "Ermintrude",
+       col = "black", las = 0, cex.lab = 1, cex.axis = 1,
+       cex = 1.0)
```

Second graph,

```
> plot(x, y, type = "b", pch = 16, main = "Scatter Diagram",
+       sub = "Heart Data", xlab = "Samantha", ylab = "Ermintrude",
+       col = "black", las = 2, cex.lab = 1, cex.axis = 1,
+       cex = 1.0)
```

Third graph,

```
> plot(x, y, type = "h", pch = 16, main = "Scatter Diagram",
+        sub = "Heart Data", xlab = "Samantha", ylab = "Ermintrude",
+        col = "black", las = 1, cex.lab = 1.2, cex.axis = 1.0,
+        cex = 1.0)
```

Fourth graph,

```
> plot(x, y, type = "s", pch = 16, main = "Scatter Diagram",
+        sub = "Heart Data", xlab = "Samantha", ylab = "Ermintrude",
+        col = "black", las = 1, cex.lab = 1.2, cex.axis = 1.1,
+        cex = 1.0)
```

Fifth graph,

```
> plot(x, y, type = "o", pch = 4, main = "Scatter Diagram",
+     sub = "Heart Data", xlab = "Samantha", ylab = "Ermintrude",
+     col = "black", las = 2, cex.lab = 1.2, cex.axis = 1.2,
+     cex = 1.5)
```

Sixth graph,

```
> plot(x, y, type = "c", pch = 16, main = "Scatter Diagram",
+     sub = "Heart Data", xlab = "Samantha", ylab = "Ermintrude",
+     col = "red", las = 2, cex.lab = 1.2, cex.axis = 1.2, cex = 1.5)
```

Title

```
title(main = "Examples of Scatter Plots", outer = T)
```

Here is the final output (Fig. 2)

3 SCATTER PLOTS

3.1 Regression Analysis with Scatter Plots

Scatter plots are very informative for presenting data. Also, it can be useful for displaying regression line on it. To demonstrate this, we will use the `cats` data from the `MASS` package. `cats` data contains the body weight and heart weight of male and female cats. To use this data, we activate the `MASS` package and download the `cats` data.

```
> data(cats)
```

Get documentation and other information on the data.

```
> ? cats
> dim(cats)
[1] 144 3
> head(cats)
```

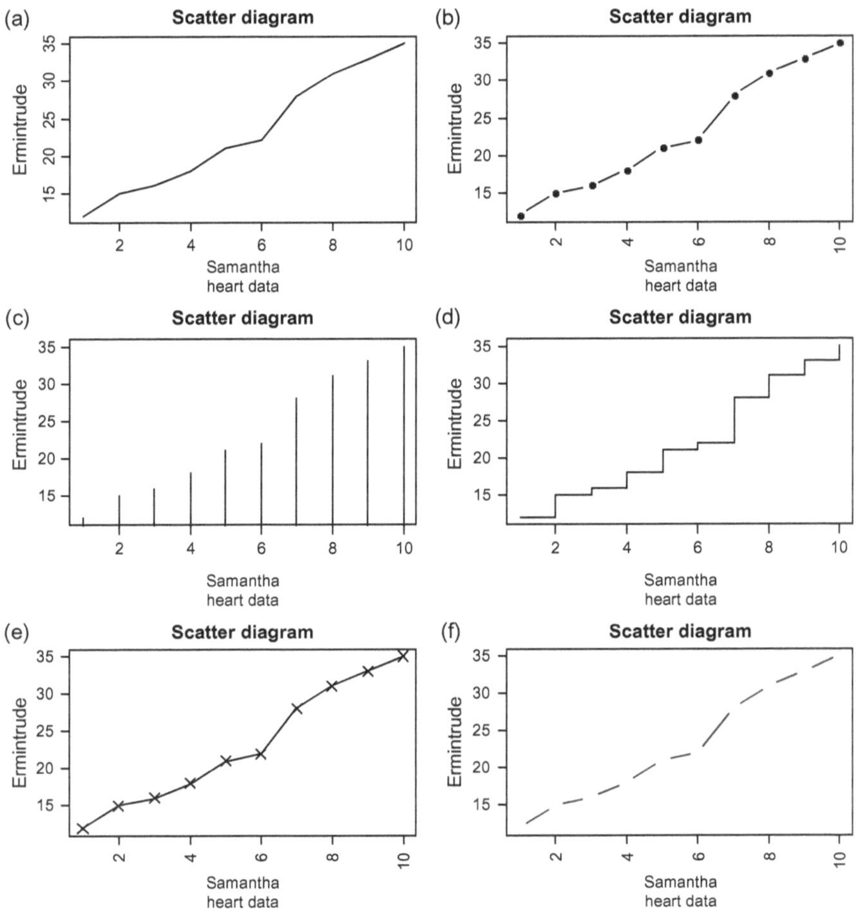

FIGURE 2 Six graphs in a single frame using the *par* command. All the graphs have same x and y value but different arguments. (a) `type` = "`l`"; (b) `type` = "`b`" with different `las` value; (c) `type` = "`h`" with different `las` and `cex.lab` values; (d) `type` = "`s`" with different `las`, `cex.lab`, and `cex.axis` values; (e) `type` = "`o`" with different `las`, `cex.lab`, `cex.axis`, `cex`, and `pch` values; and (f) `type` = "`c`" with different `las`, `cex.lab`, `cex.axis`, `cex`, and `color` values.

```
    Sex  Bwt   Hwt
1     F  2.0   7.0
2     F  2.0   7.4
3     F  2.0   9.5
4     F  2.1   7.2
5     F  2.1   7.3
6     F  2.1   7.6
```

We want to see the connection between the body weight and heart weight gender wise. Let us segregate the data gender wise and get basic information on the data.

```
> catsMale <- subset(cats, cats$Sex == "M")
> dim(catsMale)
[1] 97 3
> head(catsMale)
     Sex  Bwt  Hwt
48     M  2.0  6.5
49     M  2.0  6.5
50     M  2.1  10.1
51     M  2.2  7.2
52     M  2.2  7.6
53     M  2.2  7.9
> catsFemale <- subset(cats, cats$Sex == "F")
> dim(catsFemale)
[1] 47 3
> head(catsFemale)
    Sex  Bwt  Hwt
1     F  2.0  7.0
2     F  2.0  7.4
3     F  2.0  9.5
4     F  2.1  7.2
5     F  2.1  7.3
6     F  2.1  7.6
> summary(catsMale)
Sex          Bwt               Hwt
F: 0    Min.   : 2.0   Min.   : 6.50
M: 97   1st Qu.: 2.5   1st Qu.: 9.40
        Median : 2.9   Median : 11.40
        Mean   : 2.9   Mean   : 11.32
        3rd Qu.: 3.2   3rd Qu.: 12.80
        Max.   : 3.9   Max.   : 20.50
> summary(catsFemale)
Sex           Bwt                Hwt
F: 47   Min.   : 2.000   Min.   : 6.300
M: 0    1st Qu.: 2.150   1st Qu.: 8.350
        Median : 2.300   Median : 9.100
        Mean   : 2.360   Mean   : 9.202
        3rd Qu.: 2.500   3rd Qu.: 10.100
        Max.   : 3.000   Max.   : 13.000
```

The goal is to get a scatter plot of *body weight* versus *heart weight* gender wise in a single graph along with that regression lines. First, we will plot the male data. This is because the ranges of heart weights and body weights of males encompass those for females. (If this is not the case, we may have to set the inputs `xlim` and `ylim`.)

```
> plot(catsMale$Bwt, catsMale$Hwt, type = "p", pch = 16,
+        col = "red", xlab = "Body Weight", ylab = "Heart Weight",
+        main = "Scatter Plot of Heart Weight versus Body Weight
+        Gender-Wise", sub = "Cats")
```

Now, we want to superimpose the female scatter plot onto the male one. The `plot()` command will not accept another `plot()` command. Therefore, we need to use the `points()` command. We will also choose a different plotting character and color. All other existing specifications cannot be changed.

```
> points(catsFemale$Bwt, catsFemale$Hwt, type = "p", pch = 10,
+          col = "blue")
```

We also need to add a legend. We need to spell out which plot is for the males and which one is for females. The first input of the command is about location. We have choice: `topright; bottomright; topleft; bottomleft; top; left; right; bottom; center`. (We can also implant the legend at any location of the graph by specifying the *x*- and *y*-coordinates).

```
> legend("bottomright", legend = c("Males", "Females"),
+          pch = c(16, 10), col = c("red", "blue"))
```

Now, we want to fit a simple linear regression model with *body weight* as the response and *heart weight* as the predictor. To do this, we need to follow these steps:

```
> MaleLine <- lm(catsMale$Hwt ~ catsMale$Bwt)
> abline(MaleLine, lwd = 2, col = "red")
```

In the same way, we need to plot the regression line for female cats.

```
> FemaleLine <- lm(catsFemale$Bwt ~ catsFemale$Hwt)
> abline(FemaleLine, lwd = 2, col = "blue")
```

Add a legend about the regression lines.

```
> legend("topleft", legend = c("Hwt = -1.184 + 4.313*Bwt",
+          "Hwt = 2.981 + 2.636*Bwt"), pch = c(16, 10),
+          col = c("red", "blue"))
```

also, add a legend about the R^2 values.

```
> legend("top", legend = c("R^2 = 63%", "R^2 = 28%"),
+          pch = c(16, 10), col = c("red", "blue"))
```

Here is the final output (Fig. 3).

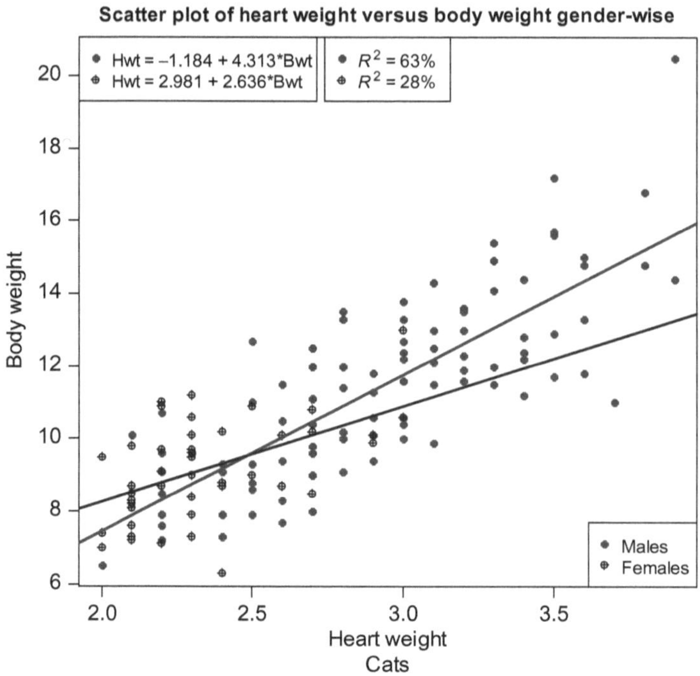

FIGURE 3 Scatter plot of heart weight versus body weight of male and female cats with regression lines for each genders.

Sometimes the data points may have a special meaning. We want to identify them on the graph. To give an example of this, we will use the data `mammals` from the same `MASS` package. `mammals` data consists of body weight (in kg) and brain weight (in gm) for 62 species of mammals.

```
> data(mammals)
> dim(mammals)
[1] 62 2
> head(mammals)
                      Body   brain
Arctic fox           3.385    44.5
Owl monkey           0.480    15.5
Mountain beaver      1.350     8.1
Cow                465.000   423.0
Grey wolf           36.330   119.5
Goat                27.660   115.0

> ?mammals
```

Let us simply plot body weight versus brain weight (Fig. 4).

```
> plot(mammals$body, mammals$brain)
```

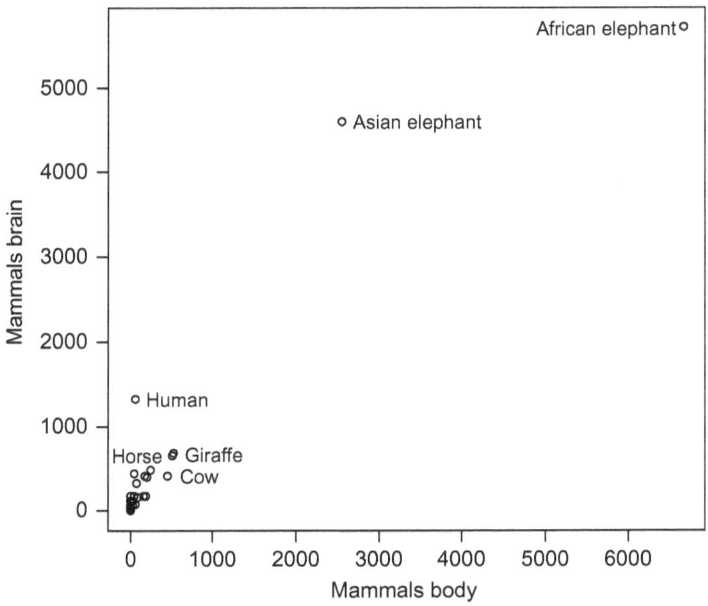

FIGURE 4 Scatter plot of brain weight versus body weight of 62 species of mammals.

The graph is chaotic. Most of the data is in the bottom left-hand corner. There is enormous variation in body and brain weights. Which animals are responsible for such a lopsided graph? To identify them use,

```
> identify(mammals$body, mammals$brain,
+            labels = row.names(mammals))
[1]   4   19   22   28   32   33
```

This command leads to a time clock that can be dragged onto the graph and clicked upon the points of interest on the graph. The elephants seem to be outliers. There may be a connection between the body weight and brain weight. When such enormous range is present, one can level the field by taking logarithms.

So, we will fit a simple linear regression model with log of *brain weight* as response and log of *body weight* as predictor.

```
> MammalsReg <- lm(log(mammals$brain) ~ log(mammals$body))
```

We get the scatter plot on log scale and the regression line on the same graph.

```
> plot(log(mammals$body), log(mammals$brain), pch = 16,
+       col = "red", xlab = "Log Body Weight",
+       ylab = "Log Brain Weight", main = "Scatter Plot Body W
+       eight versus Brain Weight", sub = "62 Mammals")
```

```
> identify(log(mammals$body), log(mammals$brain),
+           labels = row.names(mammals))
```

Warning: no point within 0.25 in.

```
[1]  2  11  19  32  33  34  35  36  55
```

```
> abline(MammalsReg, lwd = 1, col = "blue")
```

Add the legend about the regression line and R^2 value.

```
> legend("bottomright", legend = "log Brain Weight = 2.13 +
+          0.75*log Body Weight", col = "blue")
> legend("topleft", legend = "R-squared = 92.08%")
```

The final output (Fig. 5) is as follows:

3.2 Multiple Regression Analysis with Scatterplot Matrices

We have several quantitative variables measured on the sample subjects. Technically, one can develop a *scatter plot* between any two variables. We want to put all these scatter plots in a single frame. The command is `pairs()`.

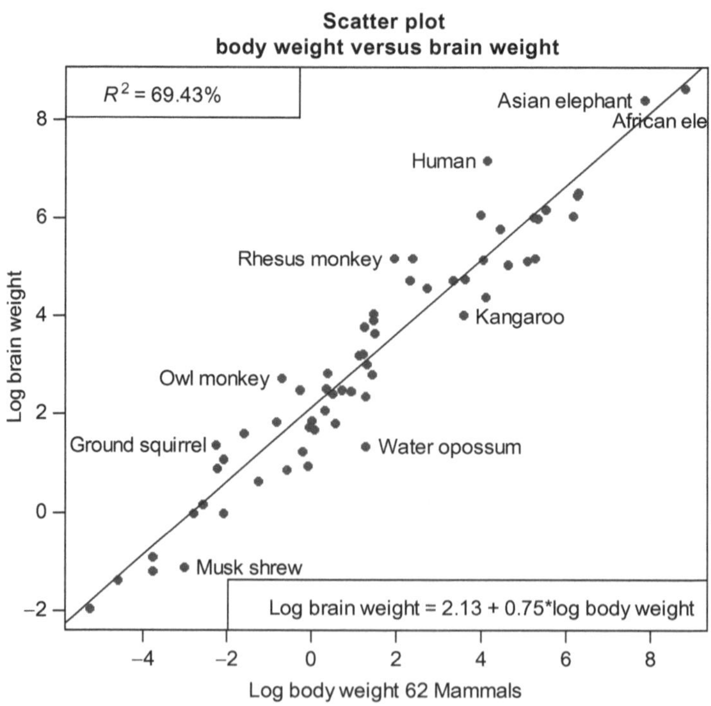

FIGURE 5 Scatter plot of log of brain weight versus log of body weight of 62 species of mammals with regression line.

Let us download the data swiss. The data swiss is available in the package datasets. The data was collected at a time when Switzerland was entering a period known as the demographic transition, i.e., its fertility was beginning to fall from the high level typical of underdeveloped countries. This data pertains to 47 French-speaking provinces of Switzerland at about 1888. The variables are common standardized fertility measure, percentage of males involved in agriculture as occupation, percentage of draftees receiving the highest mark on army examination, percentage of those with education beyond primary school, and percentage of catholics and number of live births who live less than 1 year.

Let us have a look at the data.

```
> data(swiss)
> ?swiss
> dim(swiss)
[1] 47 6
> head(swiss)

             Fertility Agriculture Examination Education Catholic
Courtelary        80.2        17.0          15        12     9.96
Delemont          83.1        45.1           6         9    84.84
Franches-Mnt      92.5        39.7           5         5    93.40
Moutier           85.8        36.5          12         7    33.77
Neuveville        76.9        43.5          17        15     5.16
Porrentruy        76.1        35.3           9         7    90.57

              Infant.Mortality
Courtelary                22.2
Delemont                  22.2
Franches-Mnt              20.2
Moutier                   20.3
Neuveville                20.6
Porrentruy                26.6
```

We use the pairs() command with an inclusion of smooth curves following the "LOWESS" (locally weighted scatter plot smoothing) method (Fig. 6).

```
> pairs(swiss, panel = panel.smooth, main = "Swiss Data in 1888",
+          col = 3)
```

(col = 1 means red; col = 2 means blue; col = 3 means green)

We build a multiple regression model regressing *Fertility* on all the remaining variables. The symbol "~" means we use all other variables as predictors and there is no need to type all of them.

Fertility $= \beta_0 + \beta_1$ *Agriculture $+ \beta_2$ *Examination $+ \beta_3$ *Education $+ \beta_4$ *Catholic $+ \beta_5$ *Infant.Mortality + Error.

```
> SwissReg <- lm(Fertility ~ ., data = swiss)
```

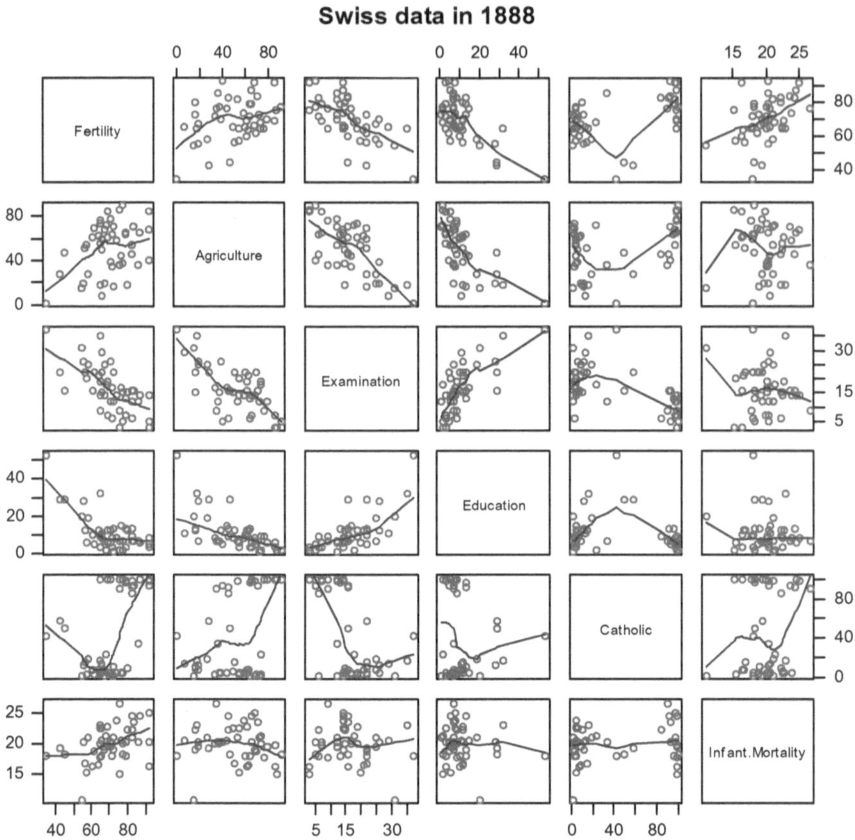

FIGURE 6 Scatter plot matrix of all variables in Swiss data set with nonparametric regression curves.

```
> summary(SwissReg)
Call:
lm(formula = Fertility ~ ., data = swiss)
Residuals:

     Min        1Q    Median        3Q       Max
-15.2743   -5.2617    0.5032    4.1198   15.3213

Coefficients:

               Estimate Std.     Error   t value    Pr(>|t|)
(Intercept)       66.91518   10.70604     6.250    1.91e-07   ***
Agriculture       -0.17211    0.07030    -2.448     0.01873   *
Examination       -0.25801    0.25388    -1.016     0.31546
Education         -0.87094    0.18303    -4.758    2.43e-05   ***
Catholic           0.10412    0.03526     2.953     0.00519   **
Infant.            1.07705    0.38172     2.822     0.00734   **
Mortality
---
```

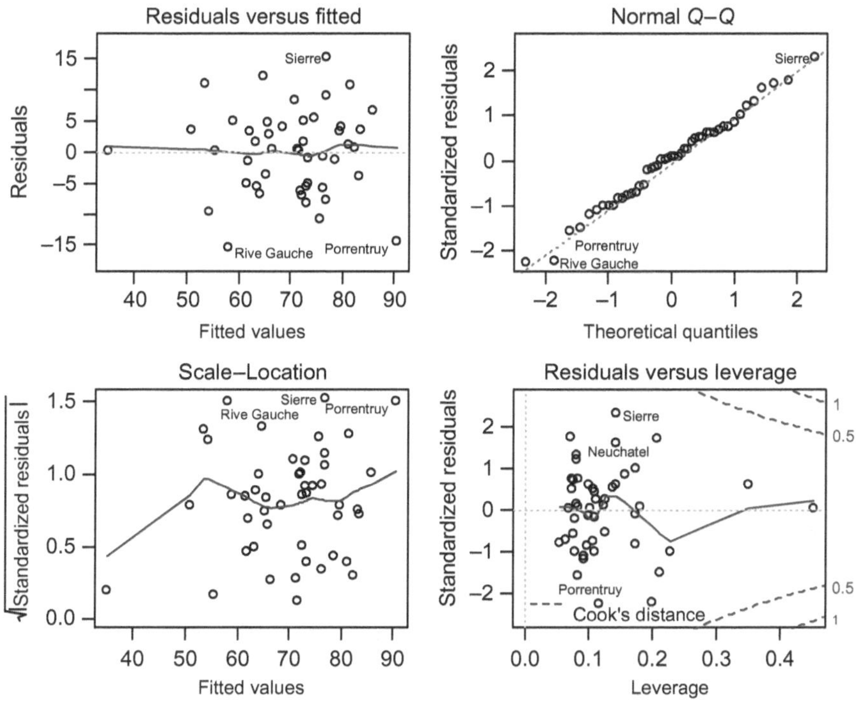

FIGURE 7 Residuals versus fitted plot (homoscedasticity) and *Q–Q* plot (normality) between fertility and all other variables in `Swiss` data.

```
Signif. codes: 0 '***' 0.001 '**' 0.01 '*' 0.05 '.' 0.1 ' ' 1
Residual standard error: 7.165 on 41 degrees of freedom
Multiple R-squared: 0.7067, Adjusted R-squared: 0.671
F-statistic: 19.76 on 5 and 41 DF, p-value: 5.594e-10
```

From the analysis, we have found that R^2 is significant. Are the assumptions about multiple regression model met? We look at the residuals versus fitted plot for homoscedasticity and *Q–Q* plot for normality (Fig. 7).

```
plot(aov(Fertility ~ ., data = swiss))
```

The LOWESS curve in the residuals versus fitted plot seems to be more or less the *x*-axis. The residuals seem to spread evenly on either side of the *x*-axis. There seems to be three outliers. The *Q–Q* plot is giving a clean bill of health on normality.

3.3 Scatterplot Matrices of Data Segregated by a Categorical Variable

Here is another kind of matrix of plots. To demonstrate this, we will use the `iris` data. This is a famous (Fisher's or Anderson's) data set. This data set gives the measurements in centimeters of the variables sepal length and width and petal length and width for 50 flowers from each of 3 species of iris. These species are *Iris setosa*, *I. versicolor*, and *I. virginica*. This data set contains five variables named `Sepal.Length`, `Sepal.Width`, `Petal.Length`, `Petal.Width`, and `Species`. Let us download the data.

```
> data(iris)
> dim(iris)
[1] 150 5
> head(iris)
  Sepal.Length Sepal.Width Petal.Length Petal.Width Species
1          5.1         3.5          1.4         0.2  setosa
2          4.9         3.0          1.4         0.2  setosa
3          4.7         3.2          1.3         0.2  setosa
4          4.6         3.1          1.5         0.2  setosa
5          5.0         3.6          1.4         0.2  setosa
6          5.4         3.9          1.7         0.4  setosa
```

Now, we want scatter plots of every variable versus every other variable species by species (Fig. 8). To do this we will use red for setotsa, green3 for versicolor, and blue for virginica as background color (`bg`).

```
> pairs(iris[1:4], main = "Iris Data - 3 Species",
+         pch = 21, bg = c("red", "green3", "blue")
+         [unclass(iris$Species)])
```

4 TIME SERIES PLOTS

We look at a time series data—`LakeHuron` from the package `datasets`. This gives the levels of Lake Huron in feet from 1875 to 1972, a collection of 98 observations. We want to plot these data so that we can see how the level rises and falls with years. Obviously, the *x*-axis has to be "Year" and the *y*-axis "Level."

```
> data(LakeHuron)
> LakeHuron
Time Series:
Start = 1875
End = 1972
Frequency = 1
 [1] 580.38 581.86 580.97 580.80 579.79 580.39 580.42 580.82 581.40 581.32
[11] 581.44 581.68 581.17 580.53 580.01 579.91 579.14 579.16 579.55 579.67
...
[81] 579.61 578.76 578.18 577.21 577.13 579.10 578.25 577.91 576.89 575.96
[91] 576.80 577.68 578.38 578.52 579.74 579.31 579.89 579.96
```

Note that the data are in time series mode even though the time is not described explicitly in the display of the observations.

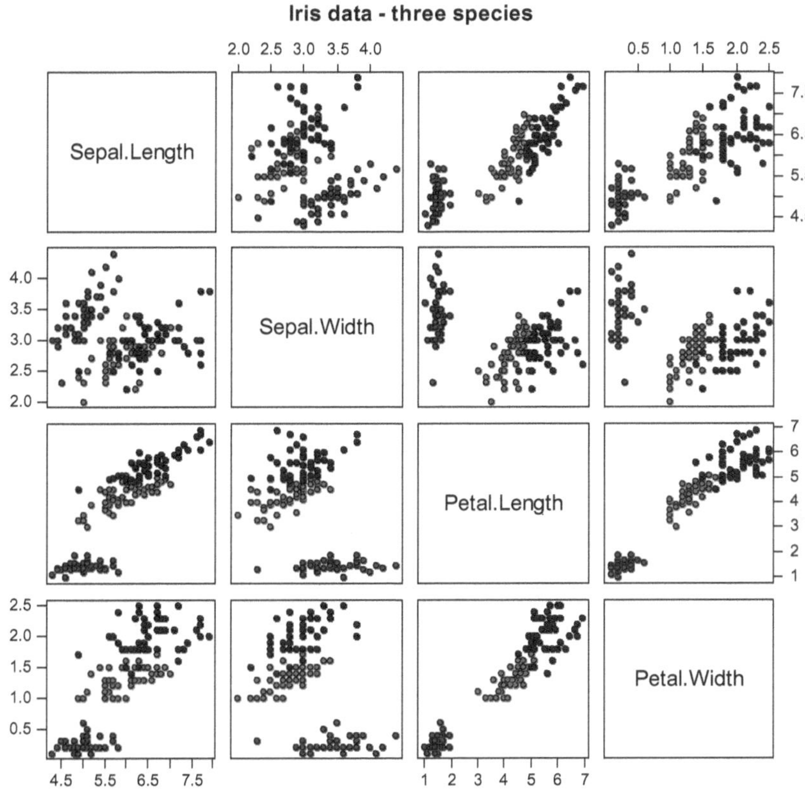

FIGURE 8 Scatter plot matrix of different variables for three species of *Iris* from `Iris` data set.

```
> dim(LakeHuron)
NULL
> length(LakeHuron)
[1] 98
> class(LakeHuron)
[1] "ts"
```

The data is clearly identified as a time series ("ts") data. This simplifies plotting (Fig. 9).

```
> plot(LakeHuron, type = "b", col = "red",
+       main = "Levels of Lake Huron", col.main = "blue", pch = 21,
+       ylab = "Feet")
```

4.1 Three Graphs in a Single Frame

The purpose is now to demonstrate how three graphs can be put in a single frame esthetically pleasing. One solution is to stretch one graph from one to

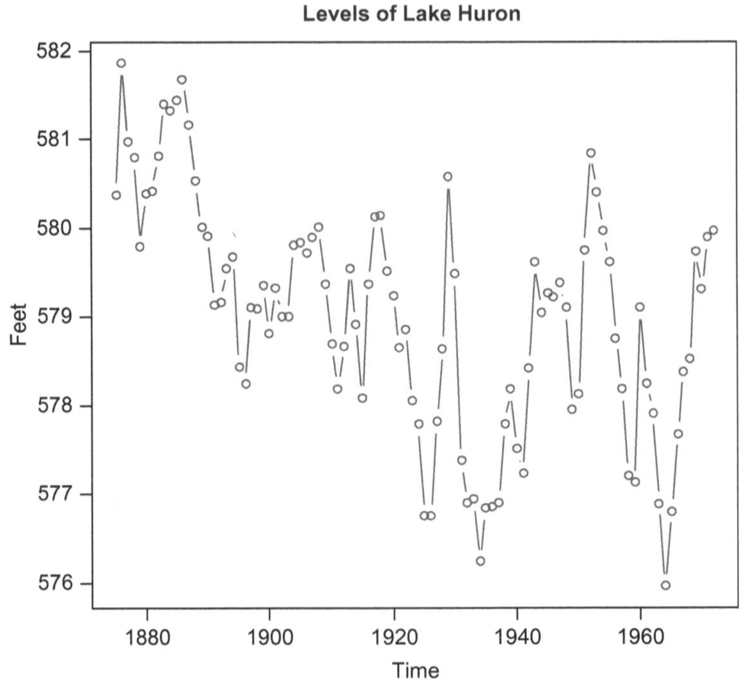

FIGURE 9 Time series plot of Levels of Lake Huron from the year 1875 to 1972 as in `LakeHuron` data set.

the other at the top and the other two occupying their normal positions. This can be achieved by the command `layout`.

```
> layout(matrix(c(1, 1, 2, 3), 2, 2, byrow = TRUE))
```

The interpretation of the input `matrix c(1, 1, 2, 3)` in the `layout` command is that Graph 1 goes into Slots 1 and 2 in the frame, Graph 2 into Slot 3, and Graph 4 into Slot 4. Here is the first graph in Slots 1 and 2.

```
> plot(Year, LakeHuron, type = "b", col = "blue",
+       main = "Levels of Lake Huron", col.main = "red",
+       ylab = "Feet", col.axis = "red")
> text(10, 577, "Year 1 = 1875 \n Year 98 = 1972")
```

Here is the second graph in Slot 3.

```
> plot(Year, LakeHuron, type = "b", col = "blue",
+       main = "Levels of Lake Huron",col.main = "red",
+       ylab = "Feet", col.axis = "red")
> text(10, 577, "Year 1 = 1875 \n Year 98 = 1972")
```

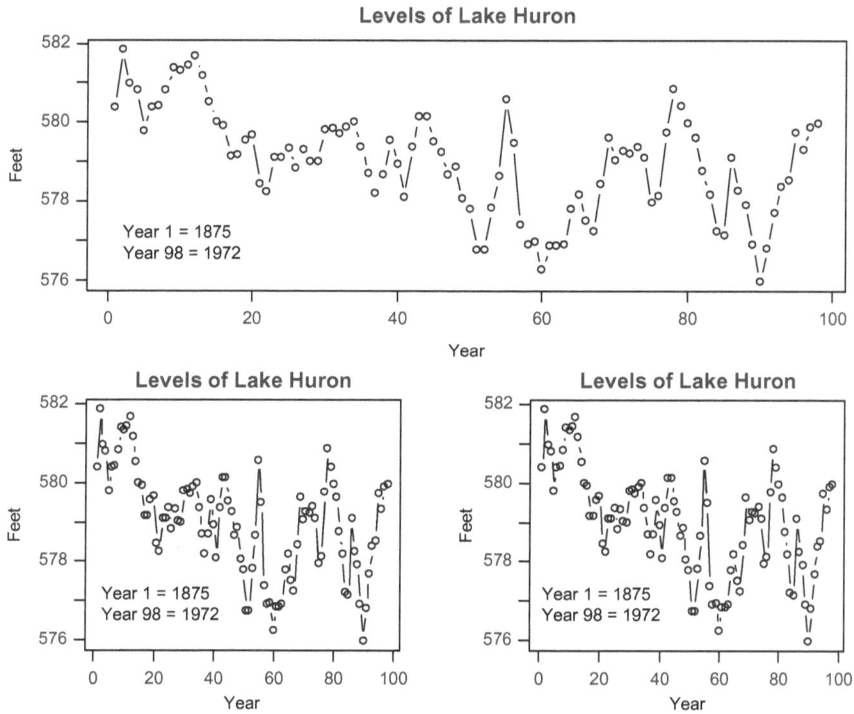

FIGURE 10 Three time series plots of Levels of Lake Huron from the year 1875 to 1972 in one frame.

Here is the third graph in Slot 4.

```
> plot(Year, LakeHuron, type= "b", col= "blue", main= "Levels of
+       LakeHuron",col.main= "red",ylab= "Feet",col.axis= "red")
> text(20, 577, "Year 1 = 1875 \n Year 98 = 1972")
```

The final output is presented in Fig. 10.

4.2 Two Different Time Series Data Sets in a Single Plot

The goal in this section is to plot two time series data over the same period in a single graph. We focus on the following two data sets.

Data Set 1: `AirPassengers`—Monthly totals of international airline passengers from 1949 to 1960.

Data set 2: `airmiles`—The revenue passenger miles flown by commercial airlines in the United States for each year from 1937 to 1960.

Both data sets are available in the `datasets` package. In order to bring both the datasets to a common time unit, we summed air passenger numbers year-by-year.

Download the datasets and determine their classes.

```
> data(AirPassengers)
> class(AirPassengers)
[1] "ts"
> data(airmiles)
> class(airmiles)
[1] "ts"
```

Look at the data airmiles.

```
> airmiles
Time Series:
Start = 1937
End = 1960
Frequency = 1
 [1]  412  480   683  1052  1385  1418  1634  2178  3362  5948  6109  5981
[13] 6753 8003 10566 12528 14760 16769 19819 22362 25340 25343 29269 30514
```

Look at the data AirPassengers.

```
> AirPassengers
      Jan  Feb  Mar  Apr  May  Jun  Jul  Aug  Sep  Oct  Nov  Dec
1949  112  118  132  129  121  135  148  148  136  119  104  118
1950  115  126  141  135  125  149  170  170  158  133  114  140
...
1959  360  342  406  396  420  472  548  559  463  407  362  405
1960  417  391  419  461  472  535  622  606  508  461  390  432
```

We sum the monthly air passenger numbers year-by-year. We use the loop command to achieve the objective. The years under AirPassengers is subsumed by the years under airmiles. Consequently, we build a plot of airmiles first. We stretch the AirPassengers data to 24 years with the first 12 years being codified as "NA."

```
> AP1 <- rep (0, 24)
> j <-0
> for (i in 1:24)
{if (i < 14)
{AP1[i] = NA
}
else
{AP1[i] <- sum(AirPassengers[(12*j+1):(12*j+12)])
j <- j + 1
}
}
```

```
> AP1
 [1]    NA    NA    NA    NA    NA    NA    NA    NA    NA NA NA NA NA 1520 1676
[16] 2042 2364 2700 2867 3408 3939 4421 4572 5140
```

```
> class(AP1)
[1] "numeric"
```

We convert the data AP1 into a time series.

```
> AP1 <- ts(AP1, start = 1937, end = 1960)
> class(AP1)
[1] "ts"
> AP1
Time Series:
Start = 1937
End = 1960
Frequency = 1
 [1]    NA    NA    NA    NA    NA    NA    NA    NA    NA NA NA NA NA1520 1676
[16] 2042 2364 2700 2867 3408 3939 4421 4572 5140
```

We plot `airmiles` data first.

```
> par(mar = c(5, 5, 2, 5))
> plot(airmiles, lwd = 2, col = "red", ann = F, las = 2)
> title(main = "Time series plots")
> mtext("Airmiles", side = 2, line = 3.5)
```

The `par()` command lets you plot another time series in the same graph with the same x-axis and y-axis being depicted on the fourth side of the graph (Fig. 11).

```
> par(new = T)
> plot(AP1, ann = F, axes = F, col = "blue", las = 2)
> axis(4)
> mtext("Air Passengers", side = 4, line = 3)
> legend("topleft", legend = c("Annual air miles",
+          "Annual total no. of passengers"), pch = c(16, 16),
+          col = c("red", "blue"))
```

5 PIE CHARTS

A pie chart is a circular chart having different sectors that represent the proportions of an individual variable in relation to the total. R can create pie charts as well. Here is an example.

Altman (1991) sought information about caffeine consumption and marital status from 3888 pregnant women. Pregnancy is a stressful period. Caffeine is supposed to reduce stress. One can get caffeine from coffee and colas. One

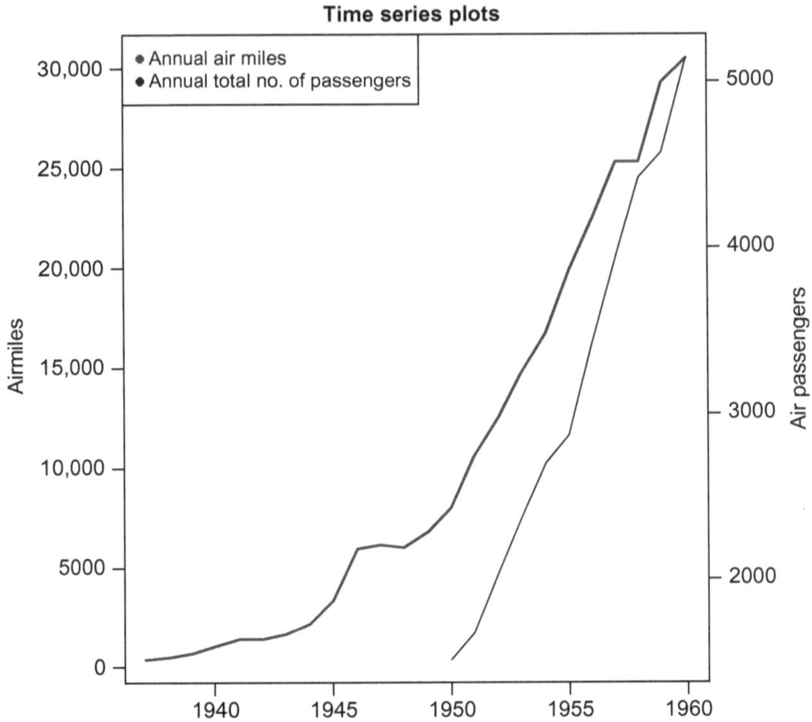

FIGURE 11 Two different time series plots with same *x*-axis.

can also measure how much caffeine is ingested by taking into account all that was drunk in the course of a typical day. The goal of the study is to examine how caffeine consumption is associated with marital status of women. The information on these two categorical variables is cross-classified. Here is the data (*Source*: Dalgaard, 2008).

Marital Status	Caffeine Consumption			
	0	1–150	151–300	>300
Married	652	1537	598	242
Prev. married	36	46	38	21
Single	218	327	106	67

We input the data into an R console as a matrix and label the rows and columns. Since the labels are descriptive, they can be typed into a column vector with each description typed up within double inverted commas.

```
> caff.marital <- matrix(c(652, 1537, 598, 242, 36, 46, 38, 21, 218,
+                          327, 106, 67), nrow = 3, byrow = TRUE)
> colnames(caff.marital) <- c("0", "1-150", "151-300", ">300")
> rownames(caff.marital) <- c("Married", "Prev.married", "Single")
```

The output is given by:

```
> caff.marital
                 0   1-150   151-300   >300
Married        652    1537       598    242
Prev.married    36      46        38     21
Single         218     327       106     67
```

Our goal is that we want to look at how caffeine consumption varies across each marital status. A pie chart of the numbers for each marital status can be drawn. We will thus have three pie charts. We need all the three pies in a single frame. Three pie charts are to be arranged in a 2×2 grid. We use the `par()` command.

A pie will have slices. Each slice has to be described. In our example, we will have four slices in a pie. Each slice is identified with how much caffeine is consumed. The size of the description can be controlled. The input is `mex()`. Also we can spell out the margin size for each pie. The command is:

```
> par(mfrow= c(2, 2), mex= 0.8, mar= c(1, 1, 2, 1), oma= c(1, 1, 4, 1))
```

Some explanation is needed for the inputs. The input "mex" gives us a choice on the size of the description of the slices. The default is mex = 1. The input "mar" lets us spell out the amount of space that we want at the bottom, left, top, and right side of the graph. The default is `mar = c(5, 4, 4, 2) + 0.1`. We have changed the defaults. We choose the colors for the slices. We build the pies one-by-one (Fig. 12).

```
> slices <- c("red", "blue", "green", "white")
```

Pie No. 1:

```
> pie(caff.marital["Married", ], main= "Married", col = slices)
```

Pie No. 2:

```
> pie(caff.marital["Prev.married", ], main = "Prev.married",
+      col = slices)
```

and Pie No. 3:

```
> pie(caff.marital["Single", ], main= "Single", col = slices)
> title (main= "Caffeine consumption by the status of the mother",
+        outer = T)
```

6 SPECIAL BOX PLOTS

A box plot is a combination of a box and two whiskers. The upper edge of the box is the third quartile Q_3 of the data. The bottom edge of the box is the first quartile Q_1 of the data. The median of the data passes through the interior of the box. The interquartile range of the data is $Q_3 - Q_1$, which is a measure of

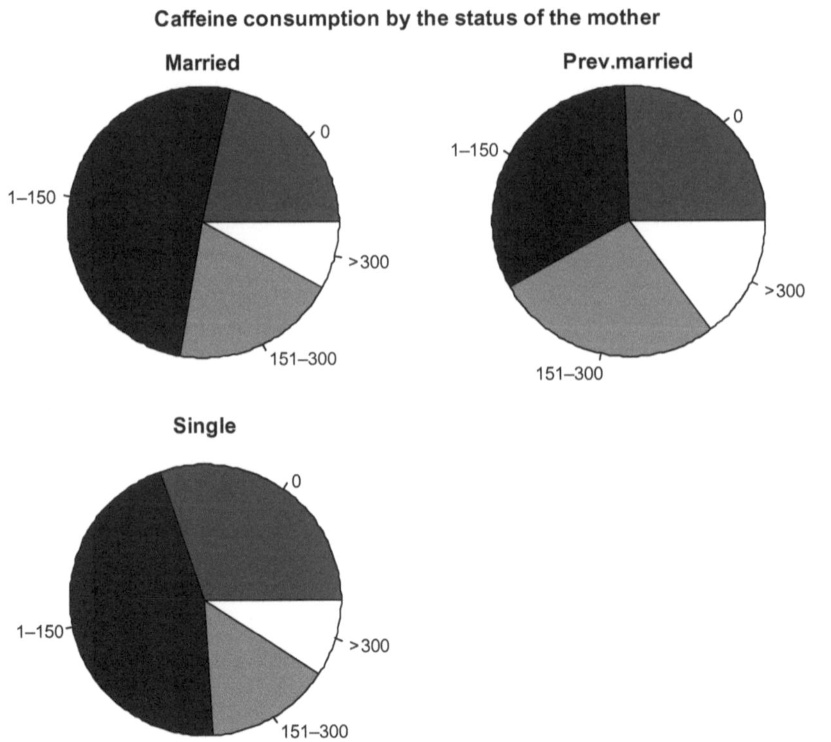

FIGURE 12 Pie charts of caffeine consumption by pregnant women categorized by their marital status.

amount of variation present in the data. The upper whisker is marked either at the maximum of the data or $Q_3+1.5(Q_3-Q_1)$ whichever is lower. The lower whisker is marked either at the minimum or $Q_1+1.5(Q_3-Q_1)$, whichever is higher. Any observation outside the whiskers is deemed as an outlier. Box plots are used to check normality. For a theoretical normal distribution, the median line is at the center of the box and whiskers are of equal length. If the data size is small, box plots can be used to check normality. An important feature of a box plot is that it is built on quantiles of the data only.

We build box plots. The data of interest, `InsectSprays` comes from entomological research and available in the "datasets" package. Six different sprays are experimented (A, B, ..., F) on the survival of a colony of insects. The number of insects survived (count) is the response variable of interest. The ultimate goal is to find the best spray, which we will not discuss.

```
> data(InsectSprays)
> dim(InsectSprays)
[1] 72 2
> head(InsectSprays)
```

```
  Count  spray
1    10      A
2     7      A
3    20      A
4    14      A
5    14      A
6    12      A
```

Here is the traditional box plot of the data by spray type (Fig. 13).

```
> boxplot(count ~ spray, data = InsectSprays, col = "lightgray")
> title(main = "Box Plots of Counts by Type of Spray Used",
+         sub = "Spray Type")
```

Each data set is skewed. One may have to think about transforming the data to achieve a semblance of symmetry before proceeding with analysis. A number of variations of box plots is available one of which is `notch`. The input notch = TRUE allows us to compare medians of pairs of different sprays (Fig. 14). A notch is drawn on each side of the boxes. If the notches of two plots do not overlap it means that there is "strong evidence" that the two medians differ.

```
> boxplot(count ~ spray, data = InsectSprays, notch = T,
+           col = "lightgray")
> title(main = "Box Plots of Counts by Type of Spray Used",
+         sub = "Spray Type")
```

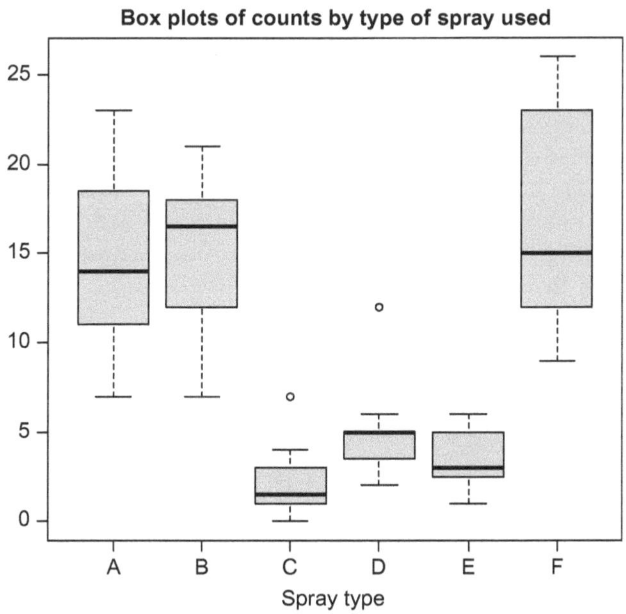

FIGURE 13 Common box plot of spray type versus counts of insects survived from `InsectSprays` data set.

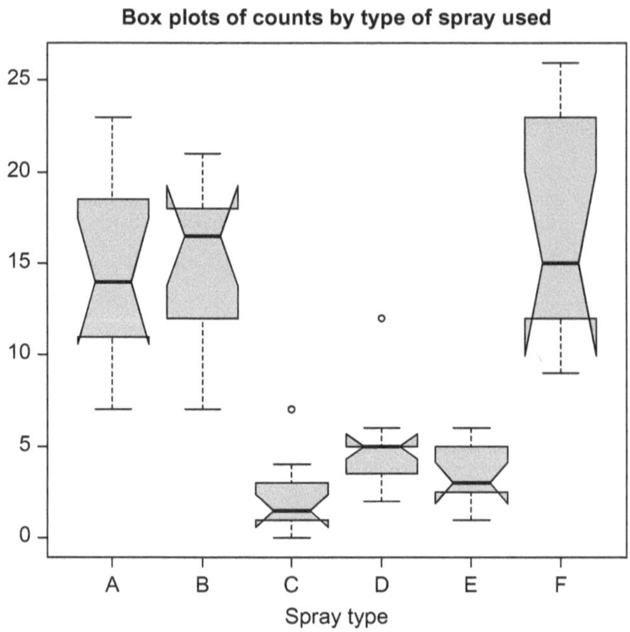

FIGURE 14 Box plot with notch of spray type versus counts of insects survived from `InsectSprays` data set.

7 *xy* PLOTS

For an *xy* plot, we need a response variable *y* and a predictor *x*. There is another variable *z* in the background.

We download the data `quakes` from `datasets` package. This is a data frame about the earthquakes that occurred around the Fiji islands since 1964 and contains 1000 observations on five variables: lat (latitude); long (longitude); depth; mag (magnitude on Richter scale); and stations (number of stations reporting the quake).

```
> data(quakes)
> dim(quakes)
[1] 1000 5
> head(quakes)

      Lat    long  depth  mag  stations
1  -20.42  181.62    562  4.8        41
2  -20.62  181.03    650  4.2        15
3  -26.00  184.10     42  5.4        43
4  -17.97  181.66    626  4.1        19
5  -20.42  181.96    649  4.0        11
6  -19.68  184.31    195  4.0        12
```

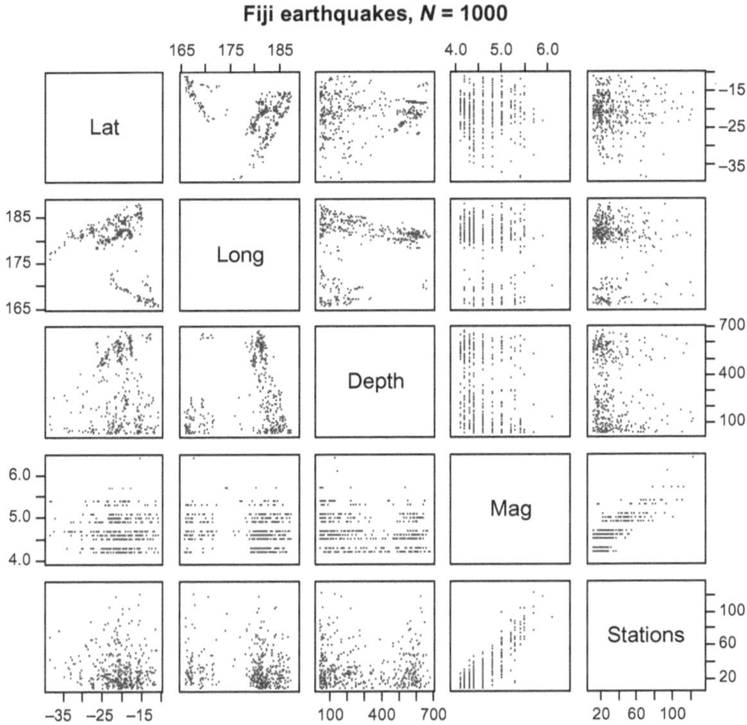

FIGURE 15 Bivariate plot of different variables from `quakes` data set.

We do the bivariate plots (Fig. 15).

```
> pairs(quakes, main= "Fiji Earthquakes, N= 1000",
+         cex.main= 1.2, pch= ".", col = "red")
```

We look at the latitude and longitude graph. The quakes seem to occur in two distinct regions.

We will do an xyplot focusing on "depth." We split the data into several subgroups based on *z*. For example, we can find quintiles of the data on *z*. Look at the *x*–*y* data in the first quintile, second quintile, third quintile, fourth quintile, and, finally, fifth quintile. We want plots of *x* and *y* in each quintile. The command xyplot gives the five plots at the same in a single frame. We can divide the data in any way we like. The command equal.count divides the data in an interesting way.

The depth of the quakes varies from 40 to 680 km. Divide the data into eight equal subgroups based on depth. Each subgroup will have 125 observations. We allow overlap. The subgroups can have 10% of observations common. This means each subgroup can have 125 + 13 = 138 observations. Look at the entire range of depth values. Arrange them in increasing order of magnitude. Pick up the first 138 observations. Take the last 13 observations of this group and then add the next 125 observations. And so on. The command equal.count works out the details.

```
> Depth <- equal.count(quakes$depth, number = 8, overlap = 0.1)
> Depth
Data:
[1]    562 650  42 626 649 195  82 194 211 622 583 249 554 600 139 306  50 590
[19]   570 598 576 211 512 125 431 537 155 498 582 328 553  50 292 349  48 600
[37]   206 574 585 230 263  96 511  94 246  56 329  70 493 129 554 223  46 593

[955] 530 497  63  82 605 234  41  40 137 223 109 595 512 613  60  43 172  54
[973]  68 217 102 178 251  42 575  43 577  42  75  71  60 291 125  69 614 108
[991] 575 409 243 642  45 470 248 244  40 165

Intervals:
    min    max  count
1   39.5   63.5   138
2   60.5  102.5   138
3   97.5  175.5   138
4  161.5  249.5   142
5  242.5  460.5   138
6  421.5  543.5   137
7  537.5  590.5   140
8  586.5  680.5   137
```

Overlap between adjacent intervals:

```
[1]  16  14  19  15  14  15  15
```

We look at the graph of Depth. The intervals are depicted in the graph (Fig. 16).

```
> plot(Depth)
> range(quakes$depth)
[1]  40  680

> 1000/8
[1] 125
```

We download and activate the `lattice` package and then create latitude–longitude plot at each of the depth regions (Fig. 17).

```
xyplot(long ~ lat| Depth, data = quakes)
```

8 CURVES

The `curve()` command can be used to plot any type of function of a single numerical variable. For example, we plot the sine function. We also plot the cosine function and superimpose it on the sine curve (Fig. 18).

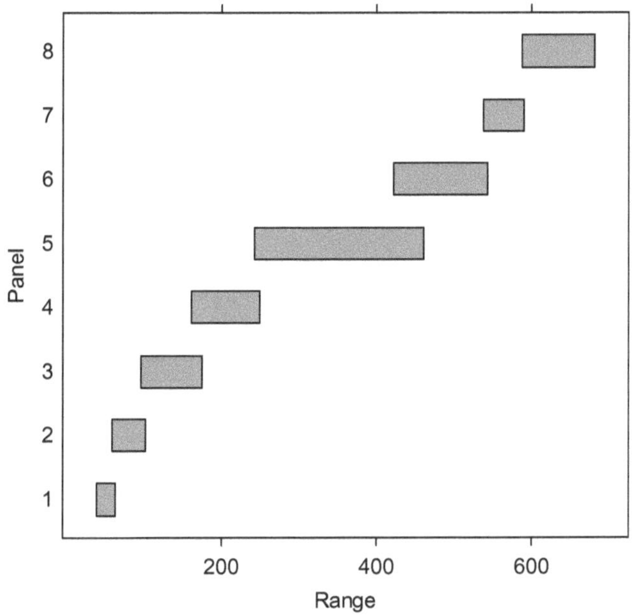

FIGURE 16 Intervals of depth with equal counts and overlap of 10% from `quakes` data set.

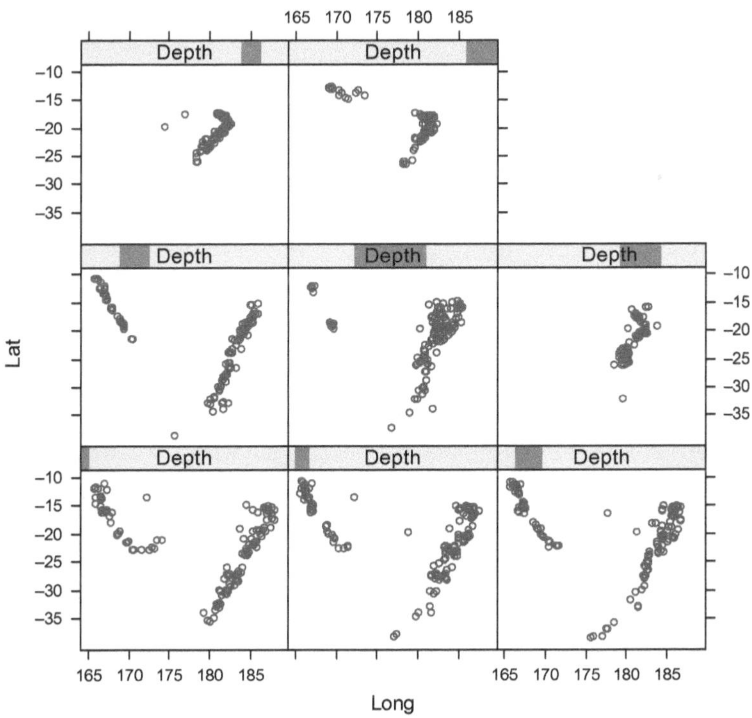

FIGURE 17 *xy* Plot on latitude versus longitude at each of the eight depth regions from `quakes` data set.

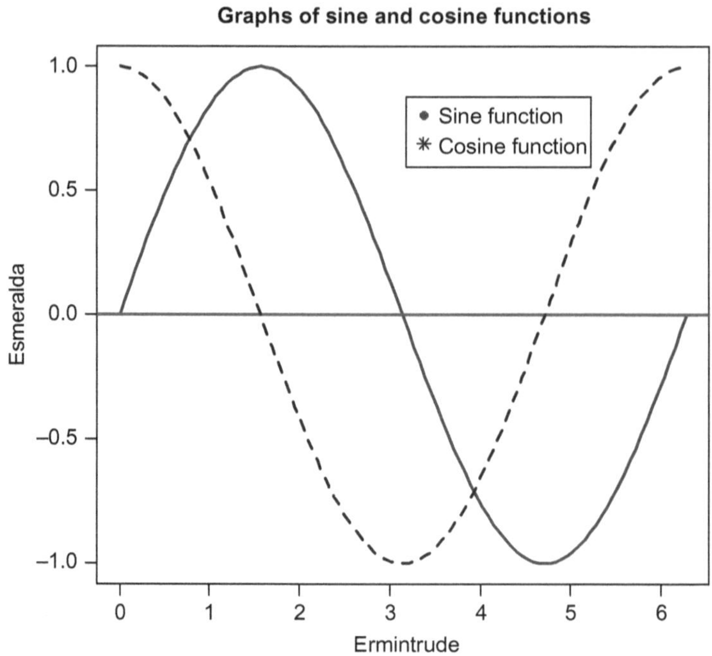

FIGURE 18 Curves for sin(x) and cos(x) functions.

```
> curve(sin(x), from = 0, to = 2*pi, col = "red", lwd = 2, lty = 1,
+         xlab = "Ermintrude", ylab = "Esmeralda",
+         main = "Graphs of sine and cosine functions")
```

The x-axis is created.

```
> abline(a = 0, b = 0, col = "magenta", lwd = 2)
```

In order to superimpose the cosine curve onto the sine curve we include the input `add = T` in the `curve()` command.

```
> curve(cos(x), col = "blue", lwd = 2, lty = 2, add = T)
```

We create a legend and plant our legend anywhere we want using the `locator()` argument. The mouse pointer is dragged down to the graph becomes "+" on the graph. If we click the mouse at the chosen position, the legend appears at that position.

```
> legend(locator(1), legend = c("Sine Function",
+          "Cosine Function"), pch = c(16, 8), col = c("red", "blue"))
```

We try another example. The data are on alligators (*source*: Agresti, 1984). This data came from the Florida Game and Fresh Water Fish Commission. We want to identify the factors influencing the primary food choice of alligators. For 59 alligators sampled in Lake George, Florida, the numbers pertain to the alligator's length (in meters) and primary food type found in the alligator's stomach. Primary food type has three categories: Fish, Invertebrate, and Other. The invertebrates are primarily apple snails, aquatic insects, and crayfish. The "Other" category includes reptiles (primarily turtles, though one stomach contained tags of 23 baby alligators that had been released in the lake during the previous year).

```
Size: 1.24  1.45  1.63  1.78  1.98  2.36  2.79  3.68  1.30
Food: I     I     I     I     I     F     F     0     I
Size: 1.45  1.65  1.78  2.03  2.39  2.84  3.71  1.30  1.47
Food: 0     0     I     F     F     F     F     I     I
Size: 1.65  1.78  2.03  2.41  3.25  3.89  1.32  1.47  1.65
Food: I     0     F     F     0     F     F     F     F
Size: 1.80  2.16  2.44  3.28  1.32  1.50  1.65  1.80  2.26
Food: I     F     F     0     F     I     F     F     F
Size: 2.46  3.33  1.40  1.52  1.68  1.85  2.31  2.56  3.56
Food: F     F     I     F     I     F     F     F     0   F
Size: 1.42  1.55  1.70  1.88  2.31  2.67  3.58  1.42  1.60
Food: I     I     I     I     F     F     F     F     I
Size: 1.73  1.93  2.36  2.72  3.66
Food: 0     I     F     I     F
```

Here, the response variable is "Food Type," which is a nominal categorical variable. The independent variable or predictor is "Size," which is quantitative. We want to investigate how the food choice is dependent on size. In such situations, one entertains a multinomial logistic regression model.

$$\Pr(\text{Food}=F)=\frac{\exp\{\alpha_1+\beta_1{*}\text{Size}\}}{D}$$

$$\Pr(\text{Food}=I)=\frac{\exp\{\alpha_2+\beta_2{*}\text{Size}\}}{D}$$

$$\Pr(\text{Food}=O)=\frac{1}{D}$$

$$D=1+\exp\{\alpha_1+\beta_1{*}\text{Size}\}+\exp\{\alpha_2+\beta_2{*}\text{Size}\}$$

This model is fitted to the data using R. The estimated model is given by:

$$\Pr(\text{Food}=F)=\frac{\exp\{1.62-0.11{*}\text{Size}\}}{D}$$

$$\Pr(\text{Food}=I)=\frac{\exp\{5.70-2.47{*}\text{Size}\}}{D}$$

$$\Pr(\text{Food}=O)=\frac{1}{D}$$

$$D=1+\exp\{1.62-0.11{*}\text{Size}\}+\exp\{5.70-2.47{*}\text{Size}\}.$$

We graph this model to understand the connection between food choice of the alligators and size (Fig. 19). We use the curve command.

```
> curve(exp(1.62 - 0.11*x)/(1 + exp(1.62 - 0.11*x) + exp(5.70 -
2.47*x)), xlim= c(1, 4), xlab= "Size", ylab= "Probability")
```

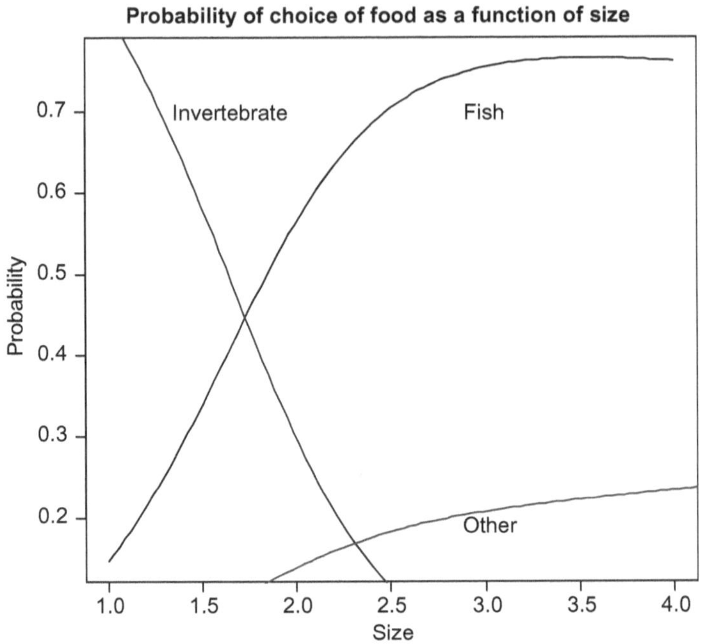

FIGURE 19 Probability of choice of foods as a function of size in alligators sampled in Lake George, Florida.

```
> curve(exp(5.70 - 2.47*x)/(1 + exp(1.62 - 0.11*x) + exp(5.70 -
2.47*x)), add = TRUE, col = "blue")
> curve(1/(1+ exp(1.62 - 0.11*x) + exp(5.70 - 2.47*x)), add = TRUE,
col = "red")
> title(main = "Probability of Choice of Food as a function of Size")
> text(3, 0.7, "FISH")
> text(1.5, 0.7, "Invertebrate")
> text(3, 0.2, "Other")
```

9 LOWESS

LOWESS is the acronym of "locally weighted scatter plot smoother." This falls under the realm of nonparametric regression. Suppose Y is a response variable and X a covariate. A simple linear regression model is given by:

$$Y = \beta_0 + \beta_1 X + \text{Error}$$

with Error being assumed to be normally distributed with mean zero and standard deviation σ. This is a parametric model with three parameters: β_0, β_1, and σ. In nonparametric regression, the model is

$$Y = g(X) + \text{Error}$$

where g is an unknown function of X. If we spell that g is a linear function of X, the model becomes parametric. If we spell that g is a quadratic function of X, the model is again parametric. In nonparametric regression, g is totally arbitrary. The data will be used to find a smooth function f, which approximates g. LOWESS is a method which finds f.

How does this method work? Suppose $(x_1, y_1), (x_2, y_2), \ldots, (x_n, y_n)$ are the data generated on (X, Y). Assume for simplicity, $x_1 < x_2 < \ldots < x_n$.

We need to choose f (smoother span), a number between 0 and 1. Start with the point x_1. Suppose $f = ½$. Choose 50% of x's closest to x_1. If $n = 100$, the points will then be $x_1, x_2, \ldots, x_{50}$. Minimize

$$\sum_{i=1}^{50} w_i (y_i - \beta_0 - \beta_1 x_i)^2$$

with respect to β_0 and β_1, where w's are weights coming from a certain kernel K. The weight w_1 is the largest for x_1 and progressively decrease as x_i's move away from x_1. The local observation x_1 gets the maximum weight (importance). Using the residuals of this fit and estimated σ, the weights are modified and β_0 and β_1 recalculated. Again use the residuals coming from this exercise and the resultant estimate of σ, the weights are modified. One can ask for as many iterations as one wishes to have. If iter = 3, we want three iterations. Using the β_0 and β_1 estimated at the final iteration, predict the value of Y for x_1.

We move to the next point x_2.

Choose 50% of x's closest to x_2. If $n = 100$, the points will then be $x_1, x_2, \ldots, x_{50}$ or $x_2, \ldots, x_{51}$. Minimize

$$\sum_{i=1}^{50} w_i (y_i - \beta_0 - \beta_1 x_i)^2$$

with respect to β_0 and β_1, where w's are weights coming from a certain kernel K. The weight w_2 is the largest for x_2 and progressively decrease as x_i's move away from x_2. The procedure for getting the final estimates is similar to what was described for the point x_1.

Continue this way through all x's. Join the predicted values. This is the LOESS or LOWESS of the data.

We work on some data. The data "cars" is available in the "base." The data give the speed of cars and the distances taken to stop when brakes are applied. It has two variables: X = Speed (mph) and Y = Stopping distance (ft).

```
> data(cars)
> dim(cars)
[1] 50 2
> head(cars)
```

```
   Speed  dist
1      4     2
2      4    10
3      7     4
4      7    22
5      8    16
6      9    10
```

We obtained the scatter plot of the data. The scatter plot indicated a quadratic relationship between the variables. The variables on logarithmic scale indicated a linear relationship. We fitted a LOWESS curve at three different smoothing costants, $f=0.1$, 2/3, 0.9 ($f=2/3$ and iter=3 are the defaults). The input "iter" stands for iterations. The original scatter plot along with LOWESS curves are plotted in the same frame using the `par()` command (Fig. 20).

```
par (mfrow= c(2, 2), oma = c(1, 1, 4, 1))
plot(cars, xlab= "Speed (mph)", ylab= "Stopping Distance (ft)")
title(main= "Car Data: Speed vs Stopping Distance")
plot(cars, xlab= "Speed (mph)", ylab= "Stopping Distance (ft)",
     las = 1, log= "xy")
```

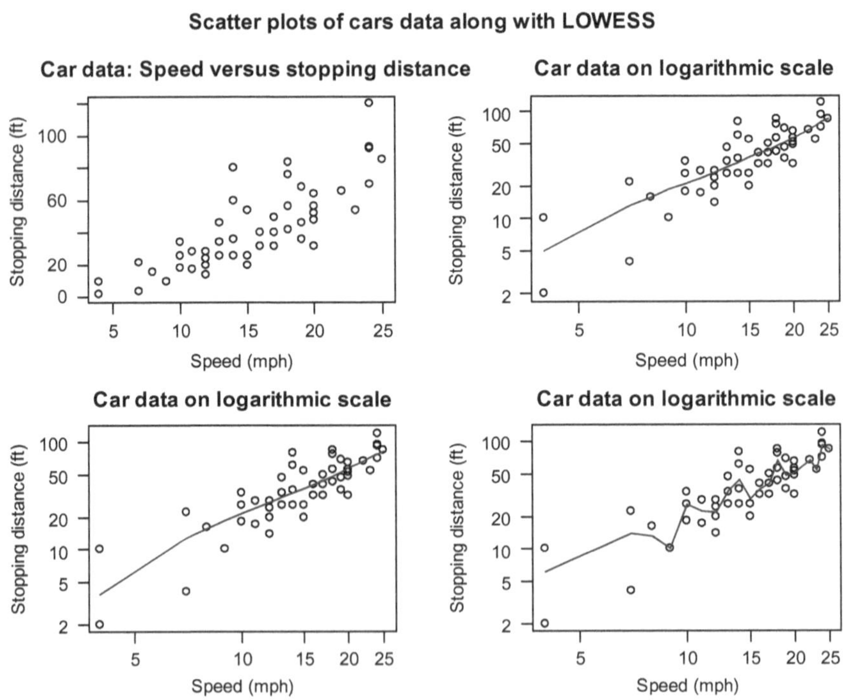

FIGURE 20 Speed versus stopping distance of cars as recorded in "*cars*" data set (normal scale, on a logarithmic scale with $f=2/3$, $f=0.1$, and $f=0.9$).

```
title(main= "Car Data on Logarithmic Scale")
lines(lowess(cars$speed, cars$dist, f=2/3, iter=3), col= "red")
plot(cars, xlab= "Speed (mph)", ylab= "Stopping Distance (ft)",
     las = 1, log= "xy")
title(main= "Car Data on Logarithmic Scale")
lines(lowess(cars$speed, cars$dist, f=0.9, iter=3), col= "red")
plot(cars, xlab= "Speed (mph)", ylab= "Stopping Distance (ft)",
     las = 1, log= "xy")
title(main= "Car Data on Logarithmic Scale")
lines(lowess(cars$speed, cars$dist, f=0.1, iter=3), col= "red")
title(main= "Scatterplots of Cars data along with LOWESS", outer= T)
```

10 SUNFLOWER PLOTS

In the context of bivariate data, several pairs (x, y) of data may be the same. When the `plot()` command is used, all these identical pairs of points show up as a single point possibly denser than the single ones. We want to identify and count all these identical pairs by some means. The `sunflowerplot()` command identifies all the identical pairs by sunflowers and petals. If five of the (x, y)-values are the same, it is identified by a flower with five petals. We use the `iris` data again. This time we want to separate the data on setosa flowers from the rest.

```
> setosa <- subset(iris, iris$Species == "setosa")
> dim(setosa)
[1] 50 5
> head(setosa)
  Sepal.Length Sepal.Width Petal.Length Petal.Width Species
1          5.1         3.5          1.4         0.2  setosa
2          4.9         3.0          1.4         0.2  setosa
3          4.7         3.2          1.3         0.2  setosa
4          4.6         3.1          1.5         0.2  setosa
5          5.0         3.6          1.4         0.2  setosa
6          5.4         3.9          1.7         0.4  setosa
```

Now, plot petal length versus petal width in a scatter plot (Fig. 21a),

```
> plot(setosa$Petal.Length, setosa$Petal.Width, pch = 16,
+       col = "red", main = "Scatter Plot of Petal Length vs Petal
+       Width", sub = "Iris Setosa")
```

There are 50 data points in the subset "`setosa`," but only 22 points show up. This means some of the data points are identical. The sunflower plot identifies all 50 data points.

```
> sunflowerplot(setosa[ , 3:4], cex = 0.5, cex.fact = 1,
+                 size = 0.25, seg.lwd = 0.8, pch = 16)
```

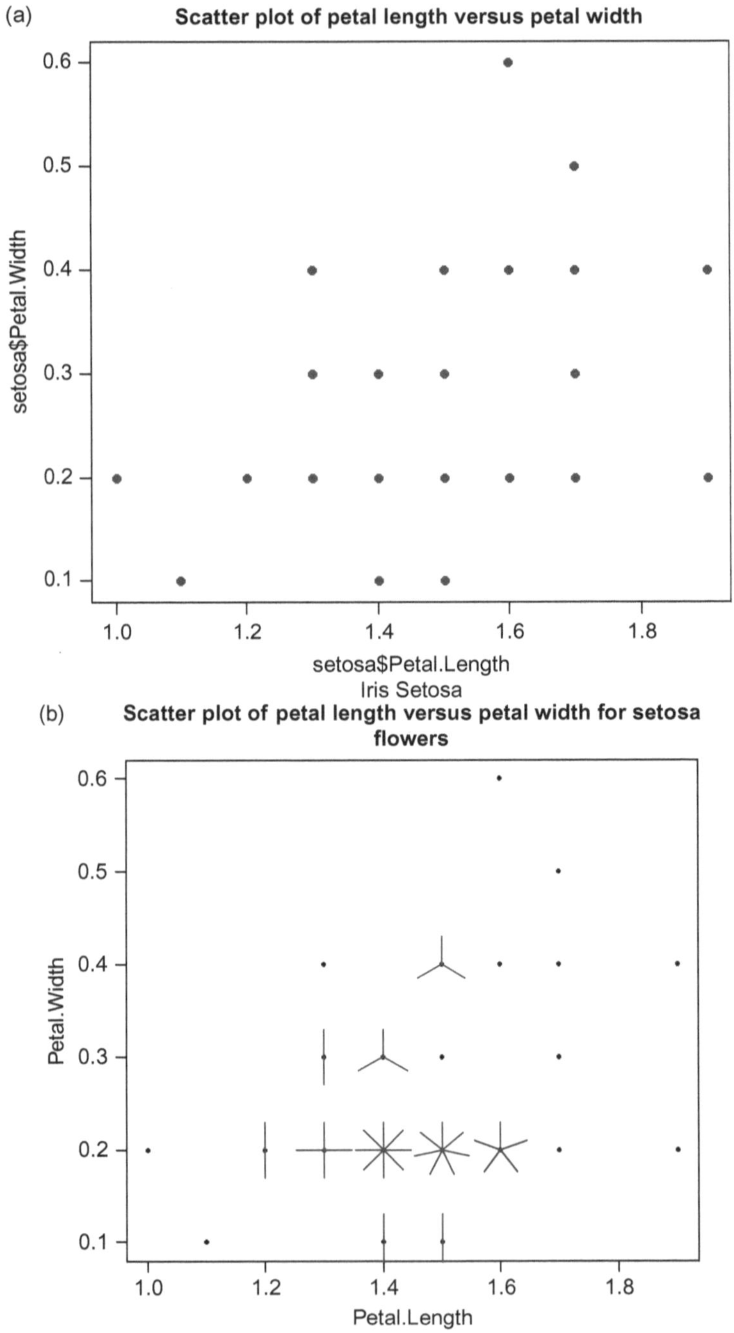

FIGURE 21 (a) Scatter plot of petal width versus petal length of setosa flowers from `iris` data. (b) Sunflower plot of petal width versus petal length of setosa flowers from `iris` data.

```
> title(main = "Scatter Plot of Petal Length versus Petal Width for
+          Setosa Flowers")
```

Explanation of inputs: `seg.lwd`=width of petal leaves; `size`=length of the leaves; `cex.fact`=1 means that the size of the background point in a flower should be the same as that of single point; `cex`=size of the points.

Now, look at the data points. Those overlapped data are showed as petals of flower (Fig. 21b).

11 VIOLIN PLOTS

A violin plot (Hintze and Nelson, 1998) is a combination of a box plot and a density plot. It starts with a box plot and then adds a rotated kernel density plot to each side of the box plot. For this, we download the package `vioplot` and then activate it. We illustrate the command on the data set `USArrests` available in the `datasets` package.

```
> data(USArrests)
> dim(USArrests)
[1] 50 4
> head(USArrests)
            Murder  Assault  UrbanPop  Rape
Alabama       13.2      236        58  21.2
Alaska        10.0      263        48  44.5
Arizona        8.1      294        80  31.0
Arkansas       8.8      190        50  19.5
California     9.0      276        91  40.6
Colorado       7.9      204        78  38.7

> summary(USArrests)
     Murder           Assault          UrbanPop           Rape
 Min.   : 0.800   Min.   : 45.0   Min.   : 32.00   Min.   : 7.30
 1st Qu.: 4.075   1st Qu.: 109.0  1st Qu.: 54.50   1st Qu.: 15.07
 Median : 7.250   Median : 159.0  Median : 66.00   Median: 20.10
 Mean   : 7.788   Mean   : 170.8  Mean   : 65.54   Mean   : 21.23
 3rd Qu.: 11.250  3rd Qu.: 249.0  3rd Qu.: 77.75   3rd Qu.: 26.18
 Max.   : 17.400  Max.   : 337.0  Max.   : 91.00   Max.   : 46.00
```

We create the violin plot of murder data from `USArrests` (Fig. 22).

```
> vioplot(USArrests$Murder, horizontal = T, col = "green")
```

As an another example, we will combine a scatter plot of a bivariate data with individual violin plots (Fig. 23).

```
> plot(USArrests$Murder, USArrests$Rape, xlim = c(-5, 20),
+        ylim = c(-10, 50), xlab = "Murder Rate",
+        ylab = "Rape Rate", main = "Scatter plot with Violin plots",
+        pch = 16, col = "red")
```

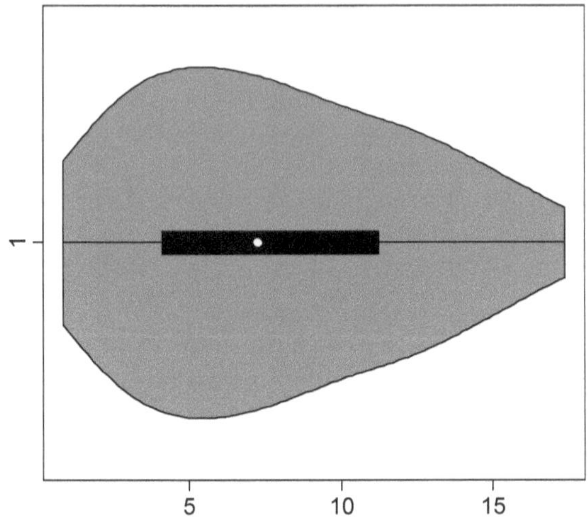

FIGURE 22 Violin plot of murder rate in the United States from `USArrests` data.

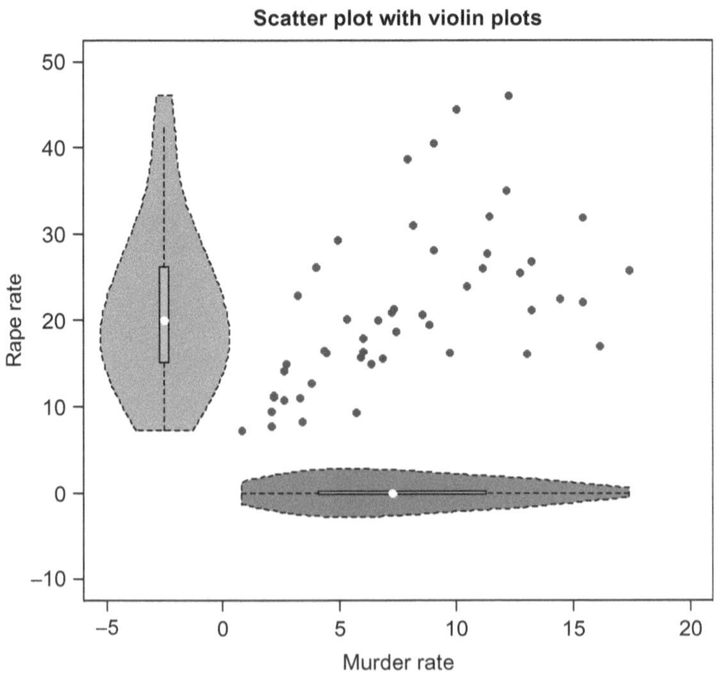

FIGURE 23 Scatter plot and violin plot together.

We created a wider margin for both the variables to accommodate violin plots.

```
vioplot(USArrests$Murder, col = "tomato", horizontal = T, at = 0,
+         add = T, lty = 2, rectCol = "gray", wex = 7)
vioplot(USArrests$Rape, col = "cyan", horizontal = F, at = -2.5,
+         add = T, lty = 2, rectCol = "gray", wex = 7)
```

Explanation of inputs: `wex` controls the width of the violin plot. `lty` means line type and `rectCol` means the color of the rectangle inside the violin plot.

12 BEAN PLOTS

A bean plot (Kampstra, 2008) is a combination of a box plot and a density plot with a rug embedded. To create a bean plots we download the package `beanplot` and then activate it. We will again use the murder data from `USArrests` for an illustration (Fig. 24).

```
> beanplot(USArrests$Murder, col = "cyan", horizontal = T,
          beanlines = "mean")
> title(main = "Murder Rate for 100,000 Residents",
        xlab = "Murder Rate")
```

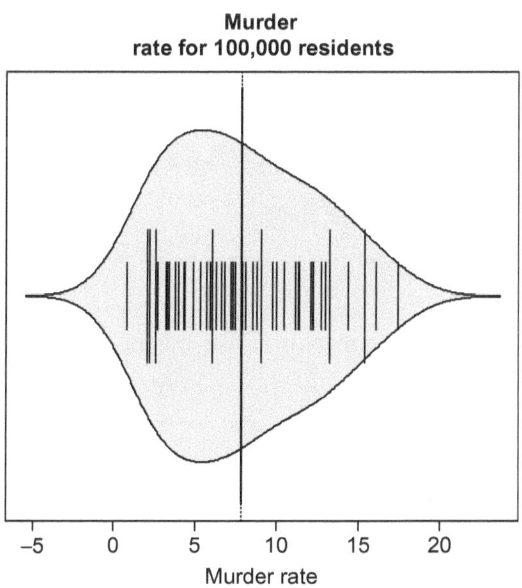

FIGURE 24 Bean plot of murder rate in the United States from `USArrests` data.

Features of the bean plot:

- The long center line is the mean.
- The small lines represent all data points.
- The line on the extreme right is the maximum.
- The line on the extreme left is the minimum.
- The nonparametric density curve is outlined above the lines and its mirror image is given below the lines.

13 BUBBLE CHARTS

The goal is to present a scatter plot of two variables with reference to a third variable. We focus on `USArrests` for an illustration. The variables of interest are: *Murder* and *Rape*. We want to see how these two variables move in tandem when we take into account the size of the urban population. If the size of the urban population is large, we expect these two rates to be high. We want to capture this type of relationship graphically.

We take the x-axis to be the *Murder Rate* and y-axis *Rape Rate*. For each state in the data set, we create a circle whose area is proportional to the size of its urban population. We scale down the numbers in the column `UrbanPop`.

```
> radius <- sqrt(USArrests$UrbanPop/pi)
> head(radius)
[1]  4.296740  3.908820  5.046265  3.989423  5.382026  4.982787
```

The following command gives a bubble chart (Fig. 25).

```
> symbols(USArrests$Murder, USArrests$Rape, circles = radius,
+          inches = 0.25, fg = "white", bg = "red",
+          xlab = "Murder Rate",
+          ylab = "Rape Rate",
+          main = "Bubble Chart by Urban Population Size")
```

We identify the states on the chart.

```
> text(USArrests$Murder, USArrests$Rape, rownames(USArrests),
+      cex = 0.5)
```

Explanation of inputs: `fg`=color of the border of the circles; `bg`=background color of the circles.

14 3D SURFACE PLOT

Three-dimensional surface plot is about plotting any real function $z=f(x, y)$ of two numeric variables x and y. In a statistical context, we deal with data on a pair (x, y) of variables and their distribution. The distribution could be parametric such as bivariate normal or nonparametric such as kernel density. In

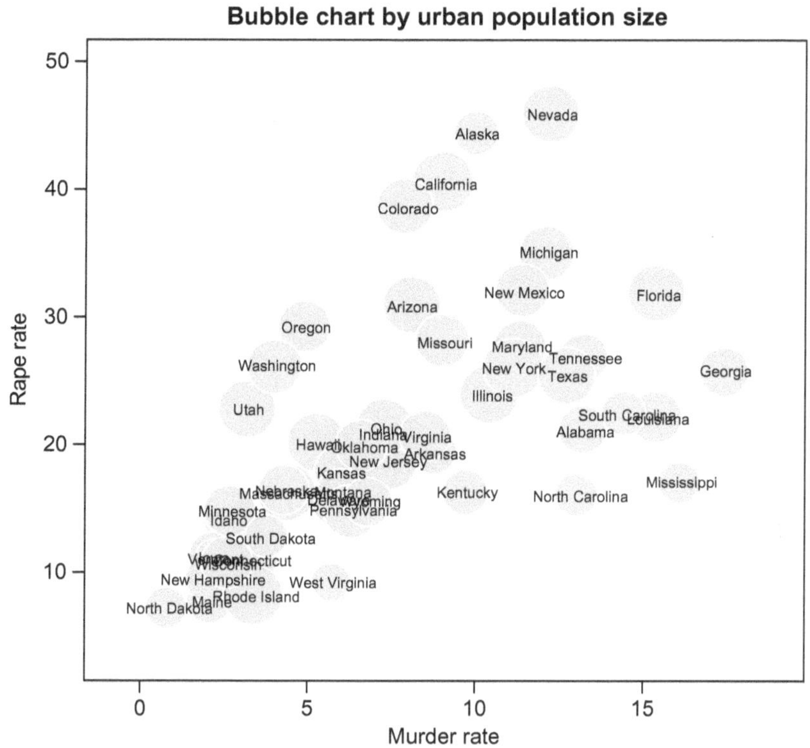

FIGURE 25 Bubble chart of murder rate and rape rate in different urban populations of the United States from `USArrests` data.

this section, we use kernel density to describe the distribution of (x, y) utilizing its data. We will not provide information on kernel density estimation. The reader may refer to Devroye and Gyorfi (1985) or Green and Silverman (1993). The command `kde2d` in the `MASS` package estimates kernel density given data on (x, y). We activate the package `MASS`. For surface plotting, we download and activate the package `rgl`. For an illustration, we will use the `quakes` data again. The x-variable is the longitude and y-variable latitude at which an earthquake occurred.

```
> data(quakes)
```

We create a folder and store the kernel density values z (`kde2d`) at all plausible range of longitude and latitude.

```
> quakeD <- kde2d(quakes$long, quakes$lat)
```

We generate the three-dimensional surface plot using these density values (Fig. 26).

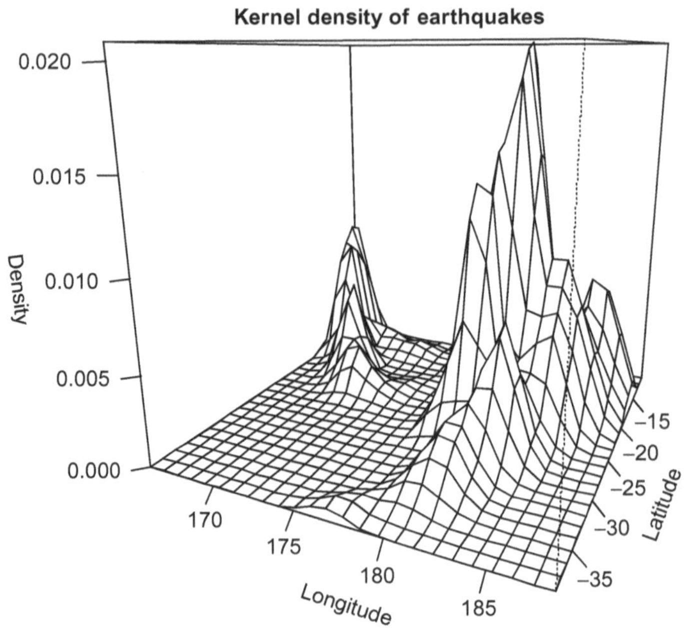

FIGURE 26 A 3D surface showing the 2D density estimate of locations of earthquakes in Fiji island (from `quakes` data).

```
> persp(quakeD, xlab = "Longitude", ylab = "Latitude",
+        main = "Kernel density of earthquakes", zlab = "Density",
+        ticktype = "detailed", theta = 25)
```

It is clear from the graph that a preponderance of earthquakes occurs in the longitude region from 175° to 190° and also in the region covered by 165° to 175° longitude and −25° to −10° latitude.

Explanation of inputs: `ticktype` provides the numbers on the axes; `theta` helps the orientation of the graph.

We modify the density plot to bring into focus only the density features. We also included colors of choice on z-values. We removed the border from the three-dimensional plot using the input "`box = FALSE`" (Fig. 27).

```
numRz <- nrow(quakeD$z)
numCz <- ncol(quakeD$z)
D.colors <- colorRampPalette( c("blue", "green", "yellow",
"orange", "red") )
nbcol <- 100
color <- D.colors(nbcol)
zfacet <- quakeD$z[-1, -1] + quakeD$z[-1, -numCz] + quakeD$z
[-numRz, -1]+ quakeD$z[-numRz, -numCz]
```

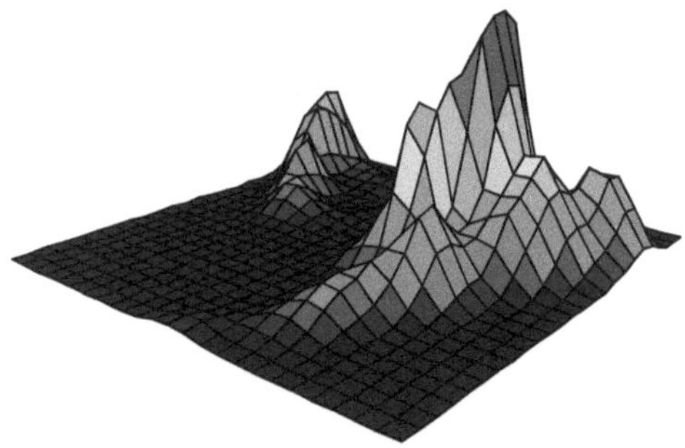

FIGURE 27 Modified 3D surface plot on `quakes` data.

```
facetcol <- cut(zfacet, nbcol)
persp(quakeD, expand = 0.5, theta = 40, col = color[facetcol],
box = FALSE)
```

15 CHERNOFF FACES—GRAPHICAL PRESENTATION OF MULTIVARIATE DATA

Plotting multivariate quantitative data is fraught with many difficulties. Bivariate data can be plotted in two-dimensions. The command `plot()` takes care of the plotting. Three-dimensional data can be plotted using two-dimensions (for example, a bubble plot). How do we plot data that is more than three-dimensional? "Chernoff faces" (Chernoff, 1973) is one solution offered in the literature. The R command `faces()` and `faces2()` inside the package `TeachingDemos` facilitates drawing Chernoff faces. This command accommodates up to 18 variables. The data has to be in matrix form.

A face is identified with 18 features. The R command matches the variables in the data with the following features of the face in the order the variables are present in the data.

- Width of center
- Top versus bottom width (height of split)
- Height of Face
- Width of the top half of the face
- Width of the bottom half of the face
- Length of nose
- Height of mouth
- Curvature of mouth (abs <9)
- Width of mouth
- Height of eyes
- Distance between eyes (0.5–0.9)
- Angle of eyes/eyebrows
- Circle/ellipse of eyes

- Size of eyes
- Position of left/right eyeballs/eyebrows
- Height of eyebrows
- Angle of eyebrows
- Width of eyebrows

All the variables have to be scaled to be in between 0 and 1. One of the inputs in the `faces2()` command is `scale` with four options.

`Columns`: Each column is scaled on its own to range from 0 to 1.

`All`: The entire data set is scaled to range from 0 to 1.

`Center`: Each column is scaled so that the mean of the column is 0.5 and every other entry is between 0 and 1.

`None`: No scaling is done because the data are already scaled.

The default is "`columns`."

Esthetically, the face is divided into two parts: the top part and bottom part. The center of the face is an imaginary line separating the top and bottom parts of the face. The width of the imaginary line is the first feature of the face. Subjects with a small value for the first variable will have a narrow middle of the face. The nose (sixth feature of the face) runs across the top and bottom parts of the face. The top part has eyes and eyebrows. The bottom part has mouth. We work with `USArrests` data. The data has only four variables and the features of the face are recycled over the same variables (Fig. 28).

```
> faces2(USArrests)
```

Another package that gives faces is `aplpack` with better control. The command `faces()` does the job. Here is the output on `USArrests` data. One nice feature of this package is that it tells you which variable corresponds to which feature of the data (Fig. 29).

```
> faces(USArrests, main = "Crime data of States")
effect of variables:
modified item           Var
"height of face"        "Murder"
"width of face"         "Assault"
"structure of face"     "UrbanPop"
"height of mouth"       "Rape"
"width of mouth"        "Murder"
"smiling"               "Assault"
"height of eyes"        "UrbanPop"
"width of eyes"         "Rape"
"height of hair"        "Murder"
"width of hair"         "Assault"
"style of hair"         "UrbanPop"
"height of nose"        "Rape"
"width of nose"         "Murder"
"width of ear"          "Assault"
"height of ear"         "UrbanPop"
```

FIGURE 28 Chernoff faces on `USArrests` data.

How does one sort the faces? One has to choose a variable in the data by which one would like to sort the faces. We choose `Murder`. We sort this variable first.

Download the package `doBy`. Use the command `orderBy` on Murder. Create a new folder for the sorted data. Apply the `faces()` command on the new data (Fig. 30).

```
> orderBy(~Murder, data = USArrests)
              Murder  Assault  UrbanPop  Rape
North Dakota     0.8       45        44   7.3
Maine            2.1       83        51   7.8
New Hampshire    2.1       57        56   9.5
...
Louisiana       15.4      249        66  22.2
Mississippi     16.1      259        44  17.1
Georgia         17.4      211        60  25.8
```

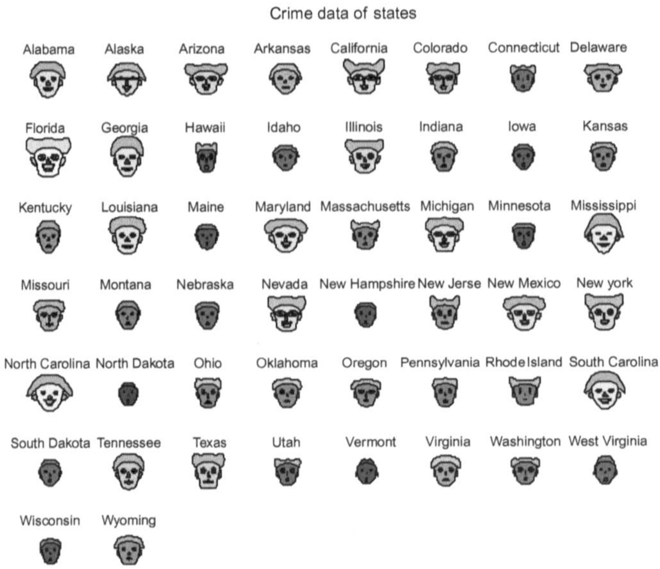

FIGURE 29 Chernoff faces on `USArrests` data using `aplpack`.

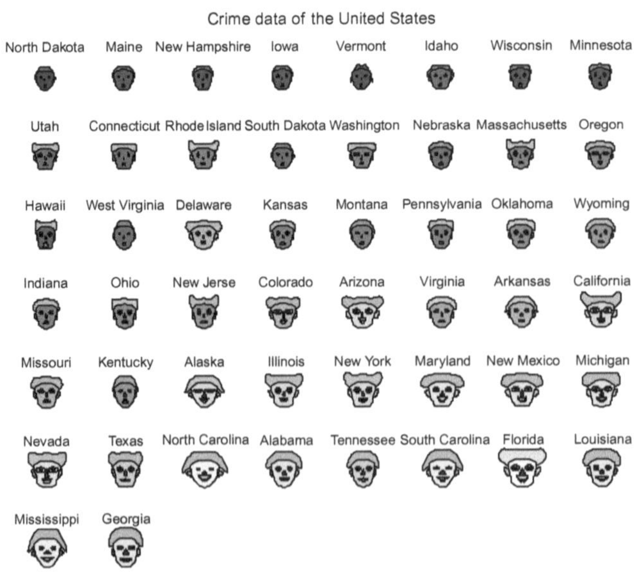

FIGURE 30 Chernoff faces on `USArrests` data sorted according to murder rate using `aplpack`.

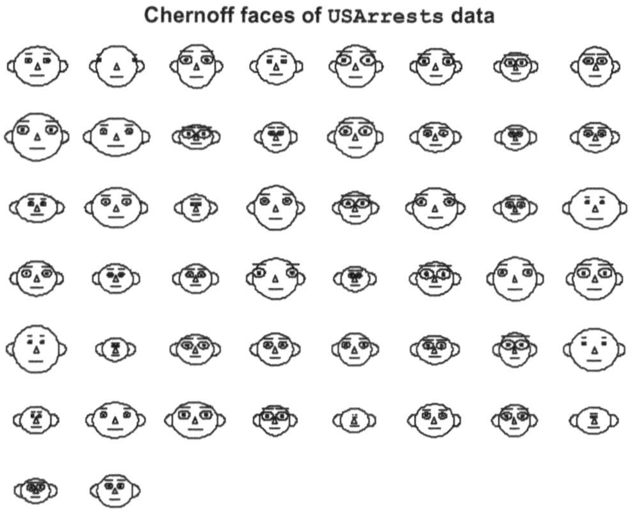

FIGURE 31 Chernoff faces on USArrests data using symbols.

```
> MB1 <- orderBy(~Murder, data = USArrests)
> faces(MB1, main = "Crime Data of United States")
```

We use the symbols package to create Chernoff faces on USArrests data (Fig. 31).

```
> symbol(USArrests, type = "face", main = "Chernoff faces of
USArrests Data")
```

16 MAPS

16.1 Drawing Common Maps

Geographical maps can be created using R. We will download and activate two packages—ggplot2 and maps.

We want to create USA map with all contiguous states (Fig. 32).

```
> map("state")
```

We create USA map without states outlined (Fig. 33).

```
> map("state", interior = FALSE)
```

We draw a map of Ohio, Indiana, and Kentucky using the input region (Fig. 34).

```
> map('state', region = c('indiana', 'kentucky', 'ohio'), fill =
TRUE, col = c("red", "yellow", "green"), bg = "mistyrose")
```

FIGURE 32 U.S. map with states.

FIGURE 33 U.S. map without states.

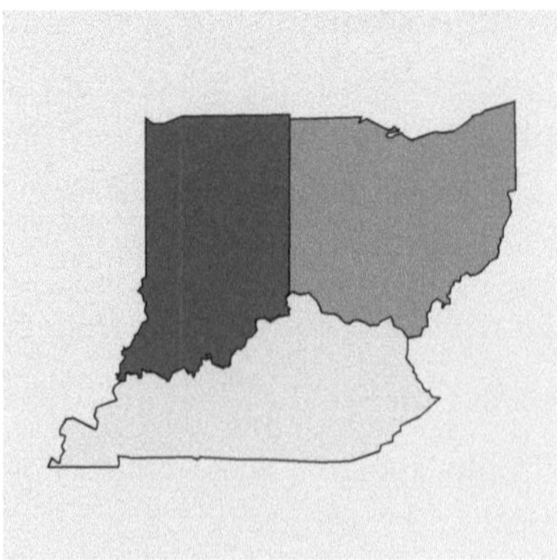

FIGURE 34 A region of the United States containing three states—Ohio, Indiana, and Kentucky.

FIGURE 35 Map of Illinois state with all counties.

We draw the map of Illinois with all its counties (Fig. 35).

```
> map("county", "illinois")
```

Note: Not every countries map is available in R yet. It is better to check with R about the availability of the maps a region of interest. To get the map information about countries use the commands:

```
> map_data("world")
> sort(unique(map_data("world")$region))
```

16.2 Creating a Choropleth Map

We create a map of all the states of USA that are color shaded depending on the values of the variable of interest. As an example, we will again use the `USArrests` data. But, to draw the map first we need to arrange the `USArrests` data into correct format. For the map of our interest, we create a new folder that contains `USArrests` plus a new column christened as `state` with the names of the states in lower case letters.

```
> arrestData <- data.frame(state = tolower(rownames(USArrests)),
USArrests)
> head(arrestData)
```

	State	Murder	Assault	UrbanPop	Rape
Alabama	alabama	13.2	236	58	21.2
Alaska	alaska	10.0	263	48	44.5
Arizona	arizona	8.1	294	80	31.0
Arkansas	arkansas	8.8	190	50	19.5
California	california	9.0	276	91	40.6
Colorado	colorado	7.9	204	78	38.7

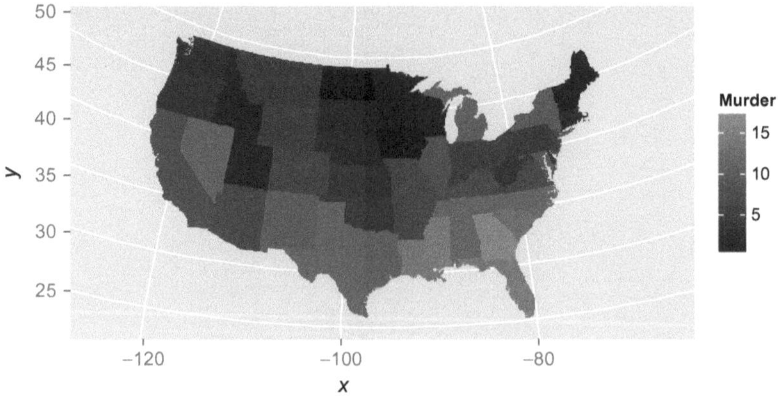

FIGURE 36 Choropleth map of the United States shaded according to the murder rate in `USArrests`.

We create a folder storing map data of the states.

```
> stMap <- map_data("state")
```

To generate a choropleth map of U.S. states with shading based on the murder rate, we will use the following command (Fig. 36). (For an excellent treatment in this connection, see Chang, 2012).

```
> ggplot(arrestData, aes(map_id = state, fill = Murder)) +
geom_map(map = stMap) + expand_limits(x = stMap$long, y = stMap$lat)
+ coord_map("polyconic")
```

The gradation of blue color is default. We can change the color according to our choice (Fig. 37).

```
ggplot(arrestData, aes(map_id = state, fill = Murder)) +
geom_map(map = stMap, col = "black") + scale_fill_gradient2(low =
"chartreuse2", mid = "yellow1", high = "darkred", midpoint =
median(arrestData$Murder)) + expand_limits(x = stMap$long, y = stMap
$lat) + coord_map("polyconic") panel.background = element_blank()
```

Note: You can use R color chart to pick the colors.

16.2.1 Creating Maps with Custom Colors Depending on Values

We create a map of Ohio coloring different counties with desired colors depending on the population size. We have population count for each of 88 counties in OH according to 2010 census in a ".csv" file. We import the data in R. Then we create color for each county depending on their population size and draw the map (Fig. 38).

```
library(maps)
OHdata <- read.csv("CountyPop.csv", header = T, sep = ",")
```

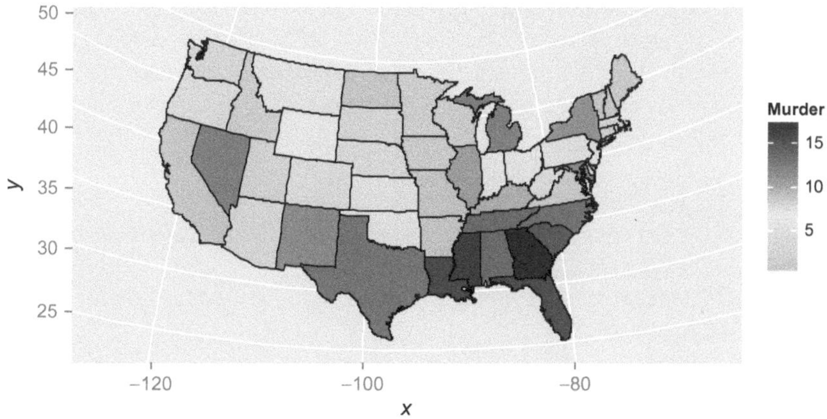

FIGURE 37 Choropleth map of the United States shaded according to the murder rate in `USArrests` using a palette of colors (different gray shades in the print version).

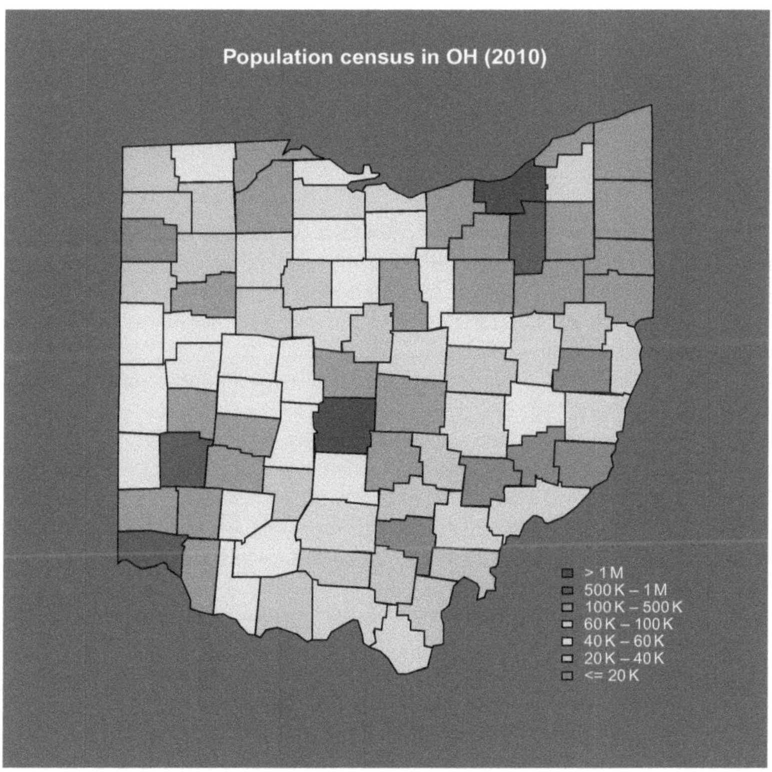

FIGURE 38 Choropleth map of OH counties depending on the population size using different colors (different gray shades in the print version).

```
OHdata.color <- array(NA, 88)
for (i in 1:88)
{ if(OHdata[i,2] > 1000000) OHdata.color[i] <- "brown"
else if(OHdata[i, 2] > 500000 & OHdata[i, 2] <= 1000000) OHdata.
color[i] <- "red"
else if(OHdata[i, 2] > 100000 & OHdata[i, 2] <= 500000) OHdata.
color[i] <- "orange"
else if(OHdata[i, 2] > 60000 & OHdata[i, 2] <= 100000) OHdata.color
[i] <- "peachpuff2"
else if(OHdata[i, 2] > 40000 & OHdata[i, 2] <= 60000) OHdata.color
[i] <- "yellow2"
else if(OHdata[i, 2] > 20000 & OHdata[i, 2] <= 40000) OHdata.color
[i] <- "lightgreen"
else if(OHdata[i, 2] <= 20000) OHdata.color[i] <- "steelblue" }
map("county", "Ohio", fill = TRUE, col = OHdata.color[], bg =
"wheat4")
title(main = "Population census in OH (2010)", col.main = "white",
font = 4)
legend("bottomright", c("> 1M", "500K - 1M", "100K - 500K", "60K -
100K", "40K - 60K", "20K - 40K", "<= 20K"), fill = c("brown", "red",
"orange", "peachpuff2", "yellow2", "lightgreen", "steelblue"),
bty = "n", cex = 0.7, ncol = 1, horiz = F, text.col = "white")
```

REFERENCES

Agresti, A., 1984. Analysis of Ordinal Categorical Data, second ed. Wiley, New York, NY.

Altman, D.G., 1991. Practical Statistics for Medical Research. Chapman & Hall, London.

Chang, W., 2012. R Graphics Cookbook. O'Reilly Media, Inc., Sebastopol, CA.

Chernoff, H., 1973. The use of faces to represent points in K-dimensional space graphically. J. Am. Stat. Assoc. 68, 361–368. http://dx.doi.org/10.2307/2284077, JSTOR 2284077.

Cook, D., Swayne, D.F., 2007. Interactive and Dynamic Graphics for Data Analysis: With R and GGobi. Springer, New York, NY.

Dalgaard, P., 2008. Introductory Statistics with R. Springer, New York, NY.

Devroye, L., Gyorfi, L., 1985. Nonparametric Density Estimation: The L1 View. Wiley Series in Probability and Statistics, Wiley, New York, NY.

Green, P.J., Silverman, B.W., 1993. Nonparametric Regression and Generalized Linear Models: A Roughness Penalty Approach. Chapman & Hall/CRC Monographs on Statistics & Applied Probability, New York, NY.

Hintze, J.L., Nelson, R.D., 1998. Violin plots: a box plot-density trace synergism. Am. Stat. 52 (2), 181–184.

Kampstra, P., 2008. Beanplot: a boxplot alternative for visual comparison of distributions. J. Stat. Softw. 28 (1), 1–9. http:/www.jstatsoft.org/v28/c01/, Code Snippets.

Murrell, P., 2011. R Graphics. CRC Press, Boca Raton, FL.

Sarkar, D., 2008. Lattice: Multivariate Data Visualization with R. Springer, New York, NY.

Wickham, H., 2009. ggplot2: Elegant Graphics for Data Analysis. Springer, New York, NY.

Chapter 4

Matrix Algebra Topics in Statistics and Economics Using R

Hrishikesh D. Vinod[1]
Professor of Economics, Fordham University, Bronx, New York, USA
[1]Corresponding author: e-mail: vinod@fordham.edu

ABSTRACT
This chapter provides a review of certain matrix algebra topics useful in Economics and Statistics which can be implemented by using the R software and graphics system. We illustrate uses of newer versions of R packages including graphics for visualization of relations. In addition to basics, we include newer innovative topics having practical relevance including decision payoff matrices, new generalized asymmetric correlations, sparse matrices, heteroscedasticity and autocorrelation consistent covariance matrices, projection matrices, robust multivariate outlier detection tools, generalized canonical correlations, demographic transitions, and others.

Keywords: Affine transformation, Invariance, Equivariance, Standardization, Multivariate normality test, Invariant coordinates, Sparse matrices, Multivariate components, Projection matrices, Population dynamics, Generalized correlations, Heteroscedasticity correction, Autocorrelation, Principal components, Ridge regression, Singular value decomposition, QR decomposition, Payoff matrices, Decision analysis, Econometrics, Regression, Generalized measures of correlation, Robust methods, Outlier detection, Nonparametric regression

1 INTRODUCTION

This chapter discusses some results from matrix algebra used in Statistics and Econometrics, where the use of the free R software and graphics system is feasible. An aim of this chapter is to facilitate learning of the matrix theory and its applications using R software tools. There are many books for matrix algebra including Gantmacher (1959) and Rao (1973). With the wide availability of the Internet, the open source and free R software has attracted researchers from around the world allowing for exponential growth in the

Handbook of Statistics, Vol. 32. http://dx.doi.org/10.1016/B978-0-444-63431-3.00004-8

number of users of R. To the best of my knowledge, the first matrix algebra book using R is Vinod (2011).

It is impossible to review thousands of matrix algebra tools scattered over thousands of R packages in a single chapter. We focus on those tools from only a few packages which are important in Statistics and Economics and which deserve attention. We assume that the reader has some familiarity with basics of R including downloading of packages from CRAN and bringing them into current memory with the "require" command.

Official R uses as the assignment symbol " <- " which requires four strokes, since it generally needs to be surrounded by spaces to avoid confusion. Fortunately, R (similar to FORTRAN) allows us to use an alternate symbol "=" for assignment, which requires only one stroke and no surrounding spaces. Hence, I will use the shorter assignment symbol to save printing space throughout, even though my code may not be portable to S-plus. After all, a global search for "=" and replacement with " <- " except in function defaults is not hard.

An R session should begin, in my opinion, with the following code which removes the R prompt ">" completely and replaces the "+" used by R for "continuation of command" symbol with two blank spaces to permit direct copy and paste from any electronic copy of this chapter. Since R ignores everything after the # symbol in a command line, we often use it throughout this chapter to explain the meaning of individual R commands.

```
#snippet cleanup
rm(list=ls()) #clean out memory
options(prompt = " ", continue = " ", width = 68,
useFancyQuotes = FALSE) #my recommended options
print(date()) #date stamp
```

The outline of the remaining section in this chapter is as follows. Section 2 begins with basic matrix manipulations in R, followed by Section 3 for descriptive statistics. Section 4 reviews invariance and equivariance under affine tranformations. Section 5 reviews my chapter 5 dealing with decision analysis from my book, Vinod (2011). By providing a link to my R software we are able to save space in this section. Section 6 considers matrix algebra in regression models. Section 7 focuses on the correlation matrix and describes a new generalized correlation matrix for assessment of potential causality, while relaxing the linearity assumption as explained in my paper, Vinod (2013). Section 8 discusses matrices in study of population dynamics, while Section 9 describes multivariate component analysis. Section 10 deals with R tools for working with sparse matrices.

2 BASIC MATRIX MANIPULATIONS IN R

We illustrate some matrix algebra concepts in R by using *Motor Trend* magazine's automobile engineering specifications data for 32 automobiles discussed in Henderson and Velleman (1981) and elsewhere. The data are called "mtcars" and are always available in R by that name. The command "`names(mtcars)`" will list the names of all 11 variables. The command "`attach(mtcars)`" allows us to treat each name as a 32×1 vector representing the variable. The R function "`cbind`" stands for "column binding" of vectors into a matrix.

```
names(mtcars)
attach(mtcars)
A=cbind(mpg,disp,hp);A #32 by 3 matrix
B=cbind(drat,wt,qsec);B#32 by 3 matrix
```

The reader should type "?mtcars" to obtain detailed information about the cars data and meaning of all symbols such as miles per gallon (mpg), horsepower (hp), and weight (wt). The last two lines from the above code have the "`cbind`" function which "column binds" the indicated three variables into our A and B matrices. The names are accessible, thanks to the "attach" command.

Additional information about any R function is found by typing `?anyfn` without spaces after the "?" at the R console.

```
Bd=t(B);Bd #B dash=transpose of B, so Bd is 3 by 32
Bd %*% A #row column matrix multiplication
crossprod(B,A) #cross product B'A of two matrices
```

The transpose of B is defined by the code "`Bd=t(B)`," which is a 3×32 matrix. The row–column matrix multiplication of Bd with A is given by the command (`Bd %*% A`) to yield a 3×3 matrix saved by us in the R object called "BdA" in the following code: Note that typing `?"%*%"` without any spaces and with quotes gives details about the set of three symbols in R for matrix multiplication. Matrix multiplication of two vectors gives their inner product.

Verify that the R function "crossprod" is a convenient replacement of the code "(`Bd %*% A`)," since both commands give same answer. In fact, "crossprod" is numerically more reliable and more efficiently executed in R than the alternative. "`crossprod(mpg,wt)`" gives the inner product of two vectors.

R has many more functions similar to "crossprod" for matrix manipulations. An *outer product* of two $n \times 1$ vectors (n=32 here) will be an $n \times n$ matrix. It is given by the following code.

```
mpg %o% wt #32 by 32 outer product of two 32 by 1 vectors
```

Matrix inverse of a square matrix is computed by the function "solve" in R.

```
cpAB=crossprod(A,B)#define crossprod 3 by 3 matrix
cpABInv=solve(cpAB) #create inverse matrix
round(cpABInv %*% cpAB,12) #this should be identity
```

We report only the last line verifying that the inverse matrix times the original square matrix is the identity matrix. We round the answer to 12 decimal spaces by the "round" function on the last line of the code above.

```
        drat  wt  qsec
drat       1   0     0
wt         0   1     0
qsec       0   0     1
```

If the original matrix is not square, similar to our "A," its usual inverse does not exist. R has thousands of "packages" which can be accessed free by the command "library" or "require." The Moore–Penrose *generalized (left) inverse* of a matrix is readily computed by the function "ginv" available in the package "MASS" (Venables and Ripley, 2002).

```
library(MASS) #access the package ''MASS''
ginv(A) #3 by 32 Moore Penrose generalized inverse
round(ginv(A) %*% A, 12) #This is 3 by 3 identity matrix
```

The last line of the above code verifies that the premultiplication by the left inverse yields the identity matrix. That is, the function "ginv" is working correctly.

3 DESCRIPTIVE STATISTICS

The standard summary of data includes vectors of means, variances, and other descriptive statistics obtained by the command "`summary(A)`." I personally like to use the function "basicStats" from the package "fBasics."

```
attach(mtcars)
A=cbind(mpg,disp,hp);A #32 by 3 matrix
head(A, 3)# view first 3 lines of A
summary(A)
require(fBasics)
b1=basicStats(A);b1
```

While we are suppressing some R outputs for brevity, we do include descriptive statistics provided by the function "basicStats" reporting most descriptive characteristics of each column separately. It includes the sample size n along the row nobs, the number of missing data (NAs), lower and upper 95% confidence limits (LCL, UCL) of the mean, etc., with self-explanatory row names.

```
head(A, 3)# view first 3 lines of A
      mpg  disp   hp
[1,]  21.0  160  110
[2,]  21.0  160  110
[3,]  22.8  108   93
b1=basicStats(A);b1
                 mpg        disp         hp
nobs         32.0000     32.0000    32.0000
NAs           0.0000      0.0000     0.0000
```

```
Minimum        10.4000      71.1000     52.0000
Maximum        33.9000     472.0000    335.0000
1. Quartile    15.4250     120.8250     96.5000
3. Quartile    22.8000     326.0000    180.0000
Mean           20.0906     230.7219    146.6875
Median         19.2000     196.3000    123.0000
Sum           642.9000    7383.1000   4694.0000
SE Mean         1.0654      21.9095     12.1203
LCL Mean       17.9177     186.0372    121.9680
UCL Mean       22.2636     275.4065    171.4070
Variance       36.3241   15360.7998   4700.8669
Stdev           6.0269     123.9387     68.5629
Skewness        0.6107       0.3817      0.7260
Kurtosis       -0.3728      -1.2072     -0.1356
```

Sample variance and covariance computation requires deviations from the mean. The following R code begins with an *ad hoc* R function called "dev" to computes deviations of arbitrary vector x from the mean. Next, we use the "apply" function in R designed to avoid time-consuming loops. For example, "apply(A, 2, dev)" applies the *ad hoc* function "dev" to the matrix object "A" where the second argument 2 means the second dimension (i.e., columns) of the matrix A. If the second argument were 1, it would operate on rows of A. Our code defines a new matrix object called "demeanA" which contains deviations from the mean for each column of A.

```
dev=function(x){x-mean(x)}
demeanA=apply(A, 2, dev)
head(demeanA, 3)
```

Top three lines of "demeanA" are reported next.

```
            mpg         disp         hp
[1,]   0.909375    -70.72188   -36.6875
[2,]   0.909375    -70.72188   -36.6875
[3,]   2.709375   -122.72188   -53.6875
```

The variance–covariance matrix of the three variables is computed by the code "`cov(A)`." It can also be computed directly from our "demeanA" matrix by using the "crossprod" function and dividing by $(n - 1) = 31$ here.

```
cov1=cov(A)
n=NROW(A)
cov2=crossprod(demeanA)/(n-1)
diff=round((cov1-cov2),12)
max(diff)
```

The output of the above code is zero, showing that the two matrices cov1 and cov2 are identical.

3.1 Outlier Detection and Normality Tests

Outliers are extreme observations which may not "belong" to the same set as most of the remaining observations. Let IQR =Q3−Q1, denote the interquartile range. All observations below (Q1−1.5*IQR) and above (Q3+ 1.5*IQR) are popularly called outliers. Outlier detection is an important aspect of descriptive statistics readily accomplished by the R command:

```
apply(A,2,boxplot.stats)
```

The Jarque–Bera test for normality applied to each column of "A" is computed by the code:

```
require(tseries)
apply(A,2,jarque.bera.test)#normality test
```

The R output is omitted for brevity, but it shows that the normality is not rejected, despite nonzero skewness and kurtosis in all columns.

3.2 Multivariate Normality Tests

An interesting issue is whether the three columns of *A* are jointly normal. The R package "ICS" by Nordhausen et al. (2008) has two functions for testing multivariate normality using skewness and kurtosis. They are implemented by the code:

```
require(ICS) #needs package in memory
#assume A is in memory
mvnorm.skew.test(A)
require(CompQuadForm)
mvnorm.kur.test(A)
```

The self-explanatory output is next

```
        Multivariate Normality Test Based on Skewness
data: A
U = 12.73, df = 3, p-value = 0.005256

        Multivariate Normality Test Based on Kurtosis
data: A
W = 18.4, w1 = 1.12, df1 = 5.00, w2 = 1.60, df2 = 1.00,
p-value = 0.01886
```

Since the *p*-values are smaller than the usual 0.05, we reject joint Normality and note that joint skewness and excess kurtosis are not likely to be zero.

4 MATRIX TRANSFORMATIONS, INVARIANCE, AND EQUIVARIANCE

Matrix algebra plays an important role in studying the concepts of invariance and equivariance discussed in Vinod (1978) in the context of ridge regression and more recently in Serfling (2009) in the context of standardization and outlier detection.

Affine Transformations Defined

The notion of affine transformations is important in linear algebra. Given a $p \times 1$ vector x_j representing jth row of a typical $n \times p$ matrix $X = \{x_{ij}\}$. Let W denote a $p \times p$ nonsingular matrix.

The affine transformation of the jth observation is: $x_j \to x_j W' + c$, where c is a $p \times 1$ vetor of constants. More generally,

$$X \to XW' + 1_{n \times p} \text{ diag}(c), \tag{1}$$

where $1_{n \times p}$ is a matrix of ones and diag(c) is a diagonal matrix.

Desirable Invariance and Equivariance

What is the desirable effect of affine transformations on decisions and estimators? Although controversies remain, it is generally agreed that under affine transformations: (a) statistical inference should be invariant, and (b) estimated values should be equivariant. We illustrate these concepts with the help of a well-known affine transformation called standardization.

4.1 Data Standardization

If x is an $n \times 1$ vector with mean $\bar{x}$ and standard deviation, sd(x), then standardized version of x is defined as:

$$x_{\text{std}} = \frac{(x - \bar{x})}{\text{sd}(x)}. \tag{2}$$

More generally, if X is an $n \times p$ matrix with the column mean vector $\bar{x}$ the sample covariance matrix is given by the $p \times p$ matrix:

$$\text{cov}(X) = \frac{(X - 1_n'\bar{x})'(X - 1_n'\bar{x})}{(n-1)}, \tag{3}$$

where 1_n denotes an $n \times 1$ column vector of ones, and its transpose is denoted by the prime. The covariance matrix usually has a well-defined square root matrix.

The more general standardization for matrices is:

$$X_{\text{std}} = (X - 1_n'\bar{x})[\text{cov}(\text{X})]^{-1/2}, \tag{4}$$

which involves the inverse of the square root of the variance–covariance matrix.

Computation of standardized matrix using an *ad hoc* R function called "stdze" defined below in conjunction with "apply" is fairly simple. The following code uses the columnwise standard deviations (sd) given by the R function sd, without explicitly involving any square root matrix. Note that our "stdze" uses square roots of variances along the diagonal but ignores the off-diagonal elements of the covariance matrix altogether.

```
stdze=function(x){(x-mean(x))/sd(x)}
stdzA=apply(A, 2, stdze)
head(stdzA, 3)
```

Top three lines of standardized matrix A named "stdzA" are reported next.

```
           mpg         disp          hp
[1,]  0.1508848  -0.5706198  -0.5350928
[2,]  0.1508848  -0.5706198  -0.5350928
[3,]  0.4495434  -0.9901821  -0.7830405
```

The standardized matrix X_{std} for our A is denoted as "stdzA." Now we turn to the notion of invariance and equivariance by comparing side-by-side the descriptive statistics of "mpg" data and its standardized values in the first column of stdzA.

```
require(fBasics)
basicStats(cbind(mpg, stdzA[,1]))
```

Its output shows that the estimates of mean and variance are equivariant while conclusions of hypothesis $\mu = 0$ are invariant under standardization, since the "Mean" values under columns "mpg" and "V2" (for standardized mpg) remain within the respective confidence limits LCL and UCL. The skewness and kurtosis measures are invariant.

```
                    mpg         V2
nobs          32.000000  32.000000
NAs            0.000000   0.000000
Minimum       10.400000  -1.607883
Maximum       33.900000   2.291272
1. Quartile   15.425000  -0.774127
3. Quartile   22.800000   0.449543
Mean          20.090625   0.000000
Median        19.200000  -0.147774
Sum          642.900000   0.000000
SE Mean        1.065424   0.176777
LCL Mean      17.917679  -0.360538
UCL Mean      22.263571   0.360538
Variance      36.324103   1.000000
Stdev          6.026948   1.000000
Skewness       0.610655   0.610655
Kurtosis      -0.372766  -0.372766
```

The covariance matrix upon standardization equals the invariant correlation matrix directly given by the R function "cor."

```
cor1=cor(A) #correlation matrix of A
n=NROW(A)
cor2=crossprod(stdzA)/(n-1) #second corr. matrix
cor3=cov(stdzA) #third correlation matrix
diff1=round((cor1-cor2),12)
diff2=round((cor1-cor3),12)
max(c(diff1,diff2)) #should be zero
```

The output is omitted for brevity.

4.2 Limitations of the Usual Standardization

The standardization above is a very old and established tool. However, it has at least three limitations: (a) nonuniqueness of the square root matrix involved in the definition of Eq. (4), (b) nonroubstness with respect to outliers, and (c) nonequivariance with respect to certain affine transformations.

The covariance matrix V of our data matrix A is computed as the R object "cov1" above. Now, we mention some alternative versions of the square root of cov1, making it nonunique. The Cholesky decomposition writes

$$V = C'C, \quad \text{where } C \text{ is upper triangular}, \tag{5}$$

where we can think of C' as the (left) square root matrix. By the way, C' is lower triangular and its inverse is also lower triangular. The R code to compute it is as follows:

```
#asume cov1 is in R memory
C=chol(cov1)
t(C)
CdC=t(C) %*% C #check that decomposition works
diff=CdC-cov1; max(diff)
```

Our Cholesky-type square root C' for this illustration is as follows:

```
t(C)
```

	mpg	disp	hp
mpg	6.026948	0.00000	0.00000
disp	−105.044411	65.77592	0.00000
hp	−53.216330	17.19597	39.66343

A second and most common definition of the square root matrix is obtained by using the square roots of the middle matrix in an eigenvalue–eigenvector decomposition of V defined as

$$V = G\Lambda G', \quad \text{where } G \text{ is orthogonal and } \Lambda \text{ is diagonal.} \tag{6}$$

Now the square root of V is a matrix defined as:

$$V^{1/2} = G\Lambda^{1/2}G'. \tag{7}$$

This can be computed by an *ad hoc* R function named "mtx.sqrt" created as follows.

```
#cov1 should be in R memory
mtx.sqrt=function(V){
    ei=eigen(V)
    d=ei$values
    d=(d+abs(d))/2
    d2=sqrt(d)
    ans<-ei$vectors %*% diag(d2) %*% t(ei$vectors)
    return(ans)}
rootv=mtx.sqrt(cov1); rootv
rootv %*% rootv #equals cov1
```

The output of the above code is:

```
rootv=mtx.sqrt(cov1); rootv

            [,1]        [,2]        [,3]
[1,]    3.218718   -4.437996   -2.503626
[2,]   -4.437996  117.734231   38.467583
[3,]   -2.503626   38.467583   56.699593

rootv %*% rootv #equals cov1

            [,1]        [,2]        [,3]
[1,]     36.3241   -633.0972   -320.7321
[2,]   -633.0972  15360.7998   6721.1587
[3,]   -320.7321   6721.1587   4700.8669
```

Both "rootv" and "−rootv" are square root matrices, providing yet another reason why square root matrices are not unique.

The second problem with the usual standardization is its nonrobustness. If we define x as a vector of five numbers and define y after replacing the fourth value with an outlier, we find that the mean and standard deviation of y are very sensitive to the outlier. Consider the R code

```
x=c(3,5,7,11,6)#define x vector
y=x; y[4]=101#fourth item is 101 an outlier
apply(cbind(x,y),2,mean)#only y vector has an outlier
apply(cbind(x,y),2,sd)
```

The following output shows that upon inserting only one outlier, the mean of y is some 4 times larger than the mean of x and standard deviation of y is about 14 times larger. Such sensitivity to outliers is called nonrobustness.

```
apply(cbind(x,y),2,mean)

  x      y
6.4   24.4

apply(cbind(x,y),2,sd)

       x           y
2.966479   42.846237
```

This small example shows why standardization which depends on means and standard deviations is not robust with respect to outliers. An obvious solution to the nonrobustness problem is to down-weight the offending outlier observations. There is considerable literature on the choice of weight functions under the robust statistics theme.
The third problem with the usual standardization is not well known. Nordhausen et al. (2008) discuss it by stating that the standardized matrix is well defined only up to postmultiplication by an orthogonal matrix.

4.3 Mahalanobis Distance and Outlier Detection

R installation comes with a function "mahalanobis" which returns the squared Mahalanobis distance D^2 of all rows in a matrix from the "center" vector μ, with respect to (wrt) the covariance matrix Σ, defined for a single column vector x as

$$D^2 = (x-\mu)'\Sigma^{-1}(x-\mu). \tag{8}$$

For our matrix "A," the squared Mahalanobis distance of each observation along a row from the vector of column means wrt the covariance matrix is computed by the R code:

```
D2=mahalanobis(A, center=colMeans(A), cov=cov(A))
head(sqrt(D2), 6)
```

Mahalanobis distance is the squared root. Top six distances of each observation from its mean are reported next for our A matrix. They are plotted as a solid line in Fig. 1 in the sequel.

```
[1] 0.9213683 0.9213683 1.2583771 1.2474196 1.6390516 1.4121237
```

Mahalanobis distance has many applications in diverse fields including detection of outliers. For example, a large Mahalanobis distance from the rest of the sample of points is said to have higher leverage since it has a greater "influence" on coefficients of the regression equation.
It is well known that the mean and standard deviation are very sensitive to outliers. Since Mahalanobis distance uses these nonrobust measures, recently researchers have replaced the center and covariance by more robust measures.

```
require(ICS) #library for robust center, cov
robcov=cov.rob(A)
D2=mahalanobis(A, center=colMeans(A), cov=cov(A))
D2rob=mahalanobis(A, center=robcov$center,
    cov=robcov$cov)
plot(sqrt(D2rob),col="red", typ="l", ylab=
    "Mahalanobis Distance", xlab="Observation Number",lty=2)
lines(sqrt(D2), typ="l")
title("Outlier detection using robust Mahalanobis distances")
```

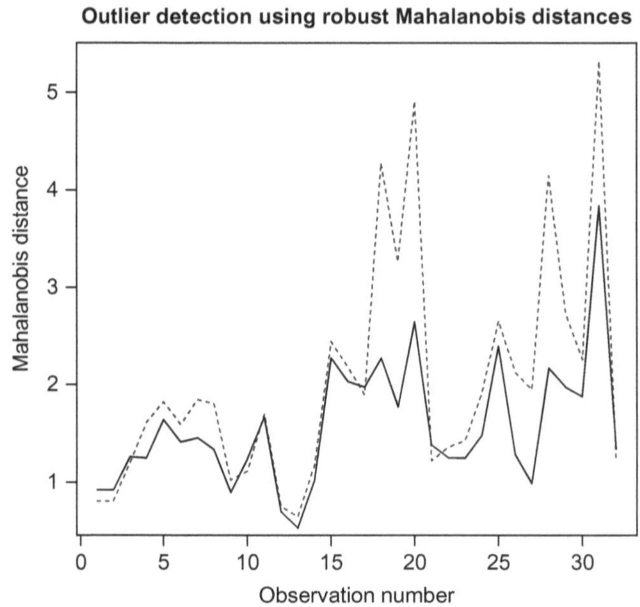

FIGURE 1 Matrix A from cars data Mahalanobis distances (solid line) robust Mahalanobis distances (dashed line).

Figure 1 plots two lines. The solid line is for the Mahalanobis distance √D2, and the dashed line is for the robust Mahalanobis distance √D2rob based on the robust measures of mean and covariance for the matrix A using cars data. It is not surprising that the solid line is less effective in identifying outliers than the dashed line based on robust measures.
Serfling (2009) discusses the use of D2 in "outlyingness functions," proves that it is affine invariant, and indicates applications for spatial distances.

5 PAYOFF MATRICES IN DECISION ANALYSIS

In elementary Business Statistics texts, it is recommended that practical decision makers create an $n \times m$ payoff matrix, usually having human decisions $D_1, D_2, \ldots, D_n$ listed along the matrix rows and external reality outside the control of the decision maker (U for uncontrolled) listed along the columns $U_1, U_2, \ldots, U_m$.

For example, a businessman may decide to build a large factory (D1), medium factory (D2), or a small factory (D3). His net profits might depend on overall economy which can be boom (U1), medium (U2), or a recession (U3). Since the decision maker has no control over the macro economy, he can only choose among D1–D3. The following payoff gives net profits for each situation. For example, the last row shows that if he builds a small factory and the recession hits, his net profits would be 25 million dollars, but if the economy booms, his small factory may have disappointed customers, leading to a smaller profit of 20 million dollars.

```
> payoff

     U1   U2   U3
D1  100   65   30
D2   70   70   40
D3   20   45   25
```

See (Vinod, 2013, chapter 5) for details and general R software for finding solution for large problems.

The *maximax* principle is for optimists (best will always happen). Given the payoff matrix, the decision maker must choose the best decision D. The command "apply(payoff, 2, max)" computes an R object called "rowmax" containing the maximum along each row decision. The optimist maximax decision maker then chooses D1, the highest rowmax.

The *maximin* principle is for the pessimist who focuses on the worst outcome for each decision obtained by the command "apply(payoff, 2, min)." One chooses the largest among these.

Opportunity loss matrix or "regret" matrix measures along each column what "might have been" in terms of maximum profit based on the column maximum. This section focuses on decision makers who are known to always look back and focus on what they "might have earned," had they decided differently. The command "colmax=apply(payoff,2,max) creates an object containing the maximum in each column. The regret in first column is defined by the command "U1r=colmax[1]-U1." Similarly, the second column of the regret matrix has "U2r=colmax[2]-U2," and so forth for other columns. The maximum regret for each decision row is computed by the command "rrowmax=apply(regret,1, max)." For our example, these are 10, 30, and 80 for D1–D3, respectively. Finally, the minimax regret principle focuses on *minimizing the maximum regret* by the command "min(rrowmax)" leading to decision D1 here.

If probabilities associated with each column are known as a vector "prob," in addition to the payoff table, we can use matrix multiplication command: "expected.value=payoff %*% prob." *Maximum of the expected value* "max(expected.value)" command then gives a good solution.

Expected regret is similarly defined by post-multiplication of the regret matrix by the "prob" vector. *Minimizing the expected regret* is then a possible principle for choosing the decision. The reader can download an R function to compute solution by all such decision principles at my Web site: http://www.fordham.edu/economics/vinod/payoff.all.txt.

The principles used in business decisions are worthy of consideration in many social, economic, political, and scientific fields. For example, doctors can use them in the choice of therapy. School administrators, police commanders, politicians, and many practical people can use these tools. Of course, a reliable net payoff matrix needs to be formulated for an exhaustive set of decisions. Construction of the payoff matrix itself is a valuable tool, allowing the decision maker to objectively think about the problem at hand.

6 MATRIX ALGEBRA IN REGRESSION MODELS

Consider the familiar regression model

$$y = X\beta + \epsilon, \tag{9}$$

in matrix notation, where y is a $T \times 1$ vector, X is $T \times p$ matrix, β is a $p \times 1$ vector, and ε is $T \times 1$ vector. In statistics, it is well known that

$$b = (X'X)^{-1}X'y \tag{10}$$

is the ordinary least squares (OLS) regression coefficient vector minimizing $\varepsilon'\varepsilon$, the error sum of squares.

For the cars example, consider the regression of fuel economy measured by mpg on weight and horsepower by the R commands:

```
reg1=lm(mpg~wt+hp); summary(reg1)
require(xtable) #create a Latex table of regression results
xt=xtable(summary(reg1));caption(xt)="Table for Regression"
label(xt)= "tab.mpgwthp"# xt is the table object
print(xt)#code for latex table output to R console
```

The output is tabulated in Table 1.
Additional output is as follows:

```
Residual standard error: 2.593 on 29 degrees of freedom
Multiple R-squared: 0.8268,  Adjusted R-squared: 0.8148
F-statistic: 69.21 on 2 and 29 DF, p-value: 9.109e-12
```

The fitted values of the regression model are given in matrix notation by Xb. If we replace b by the expression in Eq. (10), we have

$$\hat{y} = \text{fitted}(y) = Xb = X(X'X)^{-1}X'y = Hy, \tag{11}$$

which defines the *hat matrix* $H = X(X'X)^{-1}X'$. Note that H is symmetric (transpose of H equals H) and idempotent (squaring of H gives back H). The matrix

TABLE 1 Table for Regression of mpg on wt and hp

	Estimate	Std. Error	*t* value	Pr(>|t|)
(Intercept)	37.2273	1.5988	23.28	0.0000
wt	−3.8778	0.6327	−6.13	0.0000
hp	−0.0318	0.0090	−3.52	0.0015

expression Hy is imagined as the linear map H operating on a vector y. The diagonal elements of H or h_{tt} are very important in studying the properties of the regression fit. The h_{tt} is said to represent leverage, which shows the sensitivity of the regression line to tth individual observation. If observable regression residuals are denoted by $r = y - Hy = (I - H)y$, it can be proved that the standard error (standard deviation of the sampling distribution) is given by

$$\mathrm{SE}(r_t) = s\sqrt{(1 - h_{tt})}, \tag{12}$$

where s is the standard error of the entire regression (standard output of the R command "lm").

6.1 Matrix QR Decomposition

A numerically stable option for computing the regression coefficients b without using matrix inversion is possible by using the QR decomposition of the X matrix.

This topic is described in Vinod, (2011, section 17.3). Given an $n \times p$ matrix X, its QR decomposition rewrites it as:

$$X = QR, \quad \text{where } Q \text{ is orthonormal and } R \text{ is upper triangular.} \tag{13}$$

If the columns of Q are denoted by Q_j, orthonormality of Q means that each Q_j is a linear combination of previous columns $(x_1,x_2,\ldots,x_{j-1})$ of X. Also, Q is orthogonal and satisfies $Q'Q = I$.

The R function for regressions "`lm`" uses "Householder reflections and Givens rotations." When we apply the QR to Eq. (9), we have:

$$y = QR\beta + \epsilon = Q\gamma + \epsilon, \tag{14}$$

where the notation now is $\gamma = R\beta$. In the new notation, $\hat{\gamma} = (Q'Q)^{-1}Q'y$ similar to Eq. (10). The estimate of original regression coefficients can be obtained by $\hat{\beta} = (R)^{-1}Q'y$, where the inverse of upper triangular matrix R is efficiently computed by the backsolve operation. The key point is that using the QR decomposition avoids the matrix inversion of possibly near-singular matrix $(X'X)$ needed for OLS estimation in Eq. (10).

One can apply the QR decomposition to the matrix of regressors in the cars regression reg1 above. The R function "qr" is used for this purpose.

```
q = qr(cbind(wt,hp)) #q contains QR decomposition of X
q$qr #print compact decomposition
#its upper triangle contains the decomposition R
#and its lower triangle that contains information on the Q.
q$rank #prints the rank of A.
```

It is possible to numerically verify that the R code "`lm`" does use the QR decomposition by following a fully worked out example in a free lecture at Penn State University available on the Internet at: http://sites.stat.psu.edu/~jls/stat511/lectures/lec17.pdf.

6.2 Collinearity and Singular Value Decomposition

While the QR decomposition of Eq. (14) is used in the regression context for dealing with computational (rounding-type) errors, a more pernicious error in regression context arises from near-collinearity, which means that the inverse of $X'X$ can be computed by using the QR decomposition, but it is unreliable. Recall from Eq. (7) an eigenvalue–eigenvector decomposition defined by the relation:

$$X'X = G\Lambda G', \tag{15}$$

where $G = \{g_1, g_2, \ldots, g_p\}$ is a $p \times p$ orthogonal matrix of eigenvectors satisfying the (orthogonality) property that its inverse equals its transpose: $G' = G^{-1}$, and where Λ is a diagonal matrix of eigenvalues.

The mean squared error defined as the "expected value" of the "distance" between the estimated coefficients and true values satisfies

$$E||(b-\beta)|| \propto 1/\lambda_{\min}, \tag{16}$$

where $\propto$ denotes "proportional to" and where $\lambda_{\min}$ denotes the smallest eigenvalue: $\min(\Lambda)$.

Near collinearity is precisely defined by: ($\lambda_{\min} \rightarrow 0$), that is, the relevant matrix is almost singular. Since the reciprocal of a near-zero value is infinitely large, it is obvious that collinearity can cause the estimated OLS b coefficient vector to be very "distant" from the vector of true values, β. This is pernicious because it may go unnoticed. See Vinod, (2008, section 1.9) for econometric examples and details.

The singular value decomposition (SVD) provides a deeper understanding of X by splitting it into three matrices:

$$X = U\Lambda^{1/2}G', \tag{17}$$

where U is a $T \times p$ matrix, similar to X itself. It satisfies: $U'U = I$. Note, however, that UU' does NOT equal identity.

The geometric interpretation of U is that it contains standardized sample principal coordinates of X. Given the multidimensional scatter of all data, one places an ellipsoid around it. The first *principal axis* has the greatest spread. The matrices Λ and G are the same as from Eq. (6).

6.2.1 PCR and SVD

Substituting SVD in Eq. (9), we have:

$$y = X\beta + \varepsilon = U\Lambda^{1/2}G'\beta + \varepsilon = U\Lambda^{1/2}\gamma + \varepsilon, \tag{18}$$

where we have used notation γ for $G'\beta$. The OLS estimate of γ is $c = G'b$, where the vector c has as many rows (components) as b.

In principal components regression (PCR), some components of c are simply deleted (weight = 0). Thus, the key decisions in using PCR are to decide how many components to delete and which one(s) to delete. Vinod (2008) explains why one should not delete principal components (eigenvectors) willy-nilly but focus on deleting the relatively "unreliably estimated" components having high sampling variances. These are precisely the components associated with smallest eigenvalues or in order of preference set by: c_p, c_{p-1}, c_{p-2}.

6.2.2 Ridge Regression

Ridge estimator represents a family of estimators parameterized by the biasing parameter $k > 0$. It can be recommended as a tool to solve near-collinearity. The ridge estimator and the corresponding variance–covariance matrix are as follows:

$$b_k = (X'X + kI)^{-1}X'y, \quad V(b_k) = \sigma^2(X'X + kI)^{-1}X'X(X'X + kI)^{-1}. \tag{19}$$

A large number of choices of $k \in [0, \infty)$ are possible. Hoerl and Kennard proved that some k exists which will reduce MSE(b), the mean squared error of the OLS estimator b. All choices lead to shrinkage of $|b|$ by a form of down-weighting of the trailing components $c = G'b$ associated with smaller eigenvalues. See Vinod (2008) for details and R tools for choosing the biasing parameter k.

6.3 Heteroscedastic and Autocorrelated Errors

If regression errors ϵ have nonconstant variances they are said to have heteroscedasticity. Sometimes a researcher may not want to correct this problem by using generalized least squares which changes the model. Then, all one can do is to test whether the presence of this problem makes regression coefficients statistically insignificant.

The usual variance–covariance matrix of regression coefficients used for testing the hypothesis that any component of β equals zero is given by $\text{cov}(b) = s^2(X'X)^{-1}$, where s^2 is the residual variance. Prof. Hal White, an econometrician, and others have proposed to replace it by

$$\text{vcov}HC(b) = s^2(X'X)^{-1}\, X'\Omega X\, (X'X)^{-1}, \tag{20}$$

where Ω is some known function representing the assumed pattern of changing variances. This is easily implemented in R by using the "vcovHC" function in the R package "sandwich" by Zeileis (2004), who also provides examples.

The three matrices in Eq. (20) have two outside matrices identical. Zeileis (2004) calls them "bread," "meat," and "bread" matrices. Statisticians working in robust estimators call them a Huber sandwich. Similar matrices arise in several estimation problems, including when regression errors have autocorrelation. Heteroskedasticity and autocorrelation consistent (HAC) estimation of variance–covariance matrix is given by his R function "vcovHAC" with similar syntax. If one wants to go beyond testing, Vinod (2010) provides new general approaches and R software tools to simultaneously correct for both of these problems, available at: http://www.fordham.edu/economics/vinod/autohetero.txt. This completes our discussion of matrix algebra R software tools used in the context of the regression model.

7 CORRELATION MATRICES AND GENERALIZATIONS

Statisticians have long ago developed alternative standardization procedures with desirable properties by extending Eq. (2) to p vectors to handle multivariate situations. The correlation matrix $\{r_{ij}\}$ is a multivariate descriptive statistics between two or more variables which is free from units of measurement. That is, it is invariant under any linear transformation.

Computation of the correlation matrix is accomplished by the command "cor" as:

```
ca=cor(A);ca
```

It is illustrated for our selected variables as:

```
            mpg         disp         hp
mpg     1.0000000  -0.8475514  -0.7761684
disp   -0.8475514   1.0000000   0.7909486
hp     -0.7761684   0.7909486   1.0000000
```

Note that the correlation matrix is symmetric, $r_{ij} = r_{ji}$, because it measures only a linear dependence between the pairs of variables. The R function "cor.test" allows formal testing of the null hypothesis that the population correlation is zero against various alternatives. I prefer the "rcorr" function of the "Hmisc" package over "cor" because "rcorr" reports three matrices: (i) Pearson or Spearman correlation matrix with pairwise deletion of missing data, (ii) the largest number of data points available for each pair, and (iii) matrix of p-values.

Bounds on the Cross-Correlation

If one knows the correlations r1=r(X,Y) and r2=r(X,Z), is it possible to write bounds for r3=r(Y,Z). Mr. Arthur Charpentier has posted a solution to this problem in terms of the following R function:

```
corrminmax=function(r1,r2,r3){
h=function(r3){
R=matrix(c(1,r1,r2,r1,1,r3,r2,r3,1),3,3)
return(min(eigen(R)$values)>0)}
vc=seq(-1,+1,length=1e4+1)
vr=Vectorize(h)(vc)
indx=which(vr==TRUE)
return(vc[range(indx)])}
```

We illustrate the use of this function for our cars data. We let r1 = r(mpg, disp), r2=r(mpg, hp) and r3=r(disp, hp). The function provides bounds on r3.

```
ca=cor(A)
corrminmax(ca[2,1], ca[3,1], ca[3,2])
```

Even though r1 and r2 are negative, the R function bounds correctly state that r3 must be positive. The function "corrminmax" returns the min and max limits, or bounds on the third correlation coefficient r2 as:

```
#min r3, max r3
[1] 0.3234 0.9924
```

Now we turn to graphics for correlation matrices. In our illustration, the R object "ca" contains the correlation matrix for *A*. There are several tools for plotting them available in R. We illustrate the code of few but omit all outputs to save space.

```
#Need ca containing correlation matrix in R memory
require(sjPlot)
sj1=sjp.corr(ca)
sj1$df #data frame with correlations ordered by col. 1
require(psych)
cor.plot(ca)
```

7.1 New Asymmetric Generalized Correlation Matrix

Zheng et al. (2012) have recently developed generalized measures of correlation (GMC) by using Nadaraya–Watson nonparametric Kernel regressions designed to overcome the linearity assumption of Pearson's standard correlations, $\rho_{X,Y}$. Vinod (2013) developed a weak but potentially useful "kernel-causality" defined by using computer intensive data-driven kernel-based conditional densities: $f(Y|X)$ and $f(X|Y)$. He defines $\delta = \text{GMC}(X|Y) - \text{GMC}(Y|X)$. When $\delta < 0$ we know from the properties of GMCs that *X better predicts Y* than vice versa.

Using better prediction as an indicator of causation (by analogy with Granger causality), I define that "*X* kernel causes *Y*" if $\delta < 0$. The qualifier "kernel" in kernel causality should remind us that this causality is subject to fallacies, such as when the models are misspecified.

Ordinary correlations among *p* variables implies a symmetric $p \times p$ correlation matrix, where positive correlation means that the two variables move in the same direction. Since the GMC's are always positive, they lose important information regarding the direction of the relation. Hence, let us consider new generalized correlation coefficients $\rho^*(Y|X)$ based on signed square roots of the GMC's. We simply assign the sign of the simple correlation to that of the generalized one by defining a sign function, $\text{sign}(\rho_{XY})$, equaling -1 if $(\rho_{XY} < 0)$, and 1 if $(\rho_{XY} \geq 0)$.

Now the off-diagonal elements of our new correlation matrix are as follows:

$$\rho^*(Y|X) = \text{sign}(\rho_{XY})\sqrt{[\text{GMC}(Y|X)]}, \tag{21}$$

Unlike the usual correlation matrix, this one is not symmetric. That is, $\rho^*(Y|X) \neq \rho^*(X|Y)$. When we have p variables, letting $i,j = 1,2,\ldots,p$, the (i,j) location of the $p \times p$ matrix of generalized correlation coefficients (in population) contains $\rho^*(X_i|X_j) = \rho^*_{ij}$, where the row variable X_i is the "effect" and the column variable X_j is the predictor or the "cause." The $p \times p$ matrix of generalized sample correlation coefficients is denoted as: $\{r^*_{ij}\}$.

Let us consider a larger set of six ratio scale variables from the cars data. The following R code selects them, computes the matrix of simple correlations, and plots them is a color scheme to show ellipses as well as numbers using the R package "corrplot" by Wei (2013).

In the following code, the object "ca" contains the correlation matrix.

```
names(mtcars)
attach(mtcars)
mtx=cbind(mpg, disp, hp, drat, wt, qsec)
ca=cor(mtx)
require(corrplot)
corrplot.mixed(ca, upper="number", lower="ellipse")
```

These correlations (subject to the strong assumption of a linear relation) range from a high of 0.89 for "disp" and "wt" to a low of 0.091 between "qsec" and "drat." Figure 2 plots these symmetric correlations with color coding. We display r_{ij} numbers above the diagonal. Instead of reporting the symmetric numbers below the the diagonal, the command "corrplot.mixed" displays appropriate ellipses representing r_{ij} numbers below the diagonal, where negatively sloped ellipses indicate negative correlations.

A table of associated p-values (omitted for brevity) suggests very few p-values exceeding 0.05, implying that most coefficients are statistically significant. However, we do have nonrejection of $\rho_{\text{qsec,drat}} = 0$ and $\rho_{\text{qsec,wt}} = 0$, implying that these few relationships are statistically insignificant at the usual 5% level. Is this all we can learn from a bivariate analysis?

Since the new 6×6 matrix containing all new generalized correlation pairs will obviously have ones along the diagonal, we need to compute the *GMC*($Y|X$) as R^2 of 30 nonparametric nonlinear Kernel regressions giving proper attention to bandwidths (Hayfield and Racine, 2008). The R code for computing my new correlations is as follows:

```
gmcmtx=function(mym){
# mym is a data matrix with n rows and p columns
# some NAs may be present in the matrix
p=NCOL(mym)
#print(c("p=",p))
out1=matrix(1,p,p)# out1 has asymmetric correlations
```

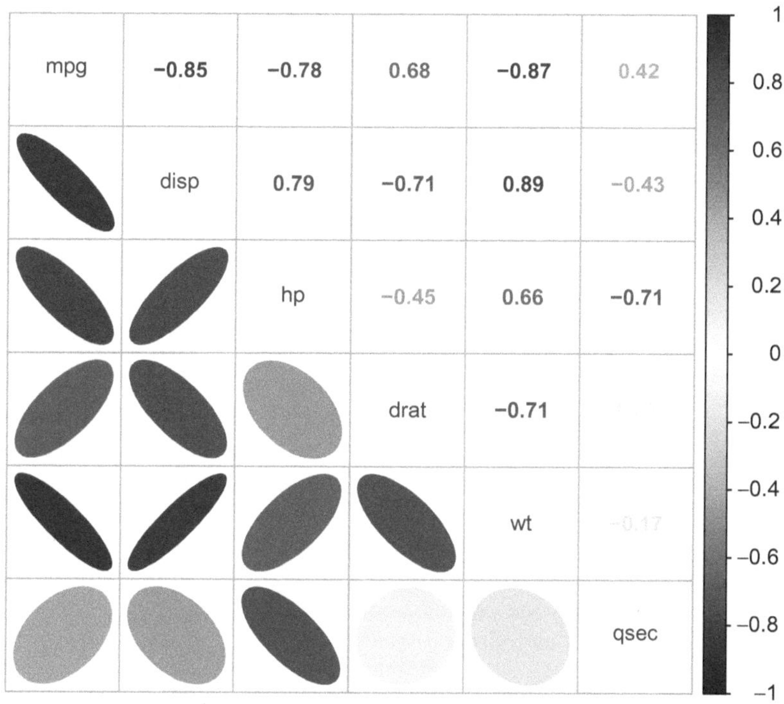

FIGURE 2 Color-coded simple correlations for cars data.

```
for (i in 1:p){
x=mym[,i]
for (j in 1:p){
if (j>i){ y=mym[,j]
ava.x=which(!is.na(x))#ava means available
ava.y=which(!is.na(y))#ava means non-missing
ava.both=intersect(ava.x,ava.y)
newx=x[ava.both]#delete NAs from x
newy=y[ava.both]#delete NAs from y
c1=cor(newx,newy)
sig=sign(c1) #get sign of r(x,y)

#bandwidths for non parametric regressions
bw=npregbw(formula=newx~newy,tol=0.1, ftol=0.1)
mod.1=npreg(bws=bw, gradients=FALSE, residuals=TRUE)
corxy= sqrt(mod.1$R2)*sig #sign times r*(x|y)
out1[i,j]=corxy # r(i,j) has xi given xj as the cause
  bw2=npregbw(formula=newy~newx,tol=0.1, ftol=0.1)
mod.2=npreg(bws=bw2, gradients=FALSE, residuals=TRUE)
coryx= sqrt(mod.2$R2)*sig #sign times r*(y|x)
out1[j,i]=coryx
```

```
}#end i loop
}#end j loop
}#endif
return(out1)}
```

We need to supply this function with the data matrix of six variables and then plot them by the following code.

```
require(np)
cg=gmcmtx(mtx)
colnames(cg)=colnames(mtx)
rownames(cg)=colnames(mtx)
require(xtable)
print(xtable(cg, digits=3))
```

The interpretation of new generalized correlations is straightforward. If $|r^*_{ij}| > |r^*_{ji}|$, it is more likely that the row variable X_i is the "effect" and the column variable X_j is the "cause," or at least X_j is the better predictor of X_i, than vice versa. For example, letting $i = 1, j = 2$ the entries in Table 2 show $0.951 = |r^*_{12}| > |r^*_{21}| = 0.894$. This suggests that "disp" better predicts "mpg" than vice versa.

Now, the following code creates a color-coded asymmetric plot by calling the "corrplot" function as follows.

```
require(corrplot)
corrplot(cg, method="ellipse")
```

Figure 3 plots our new generalized asymmetric correlation coefficients defined in Eq. (21) with color coding similar to Fig. 2.

Finally, we claim that asymmetric correlations in Table 2 contain useful causation information in their asymmetry itself. We have noted that when the asymmetry satisfies $|r^*_{ij}| > |r^*_{ji}|$, the variable X_j is more likely to be the

TABLE 2 Table of Asymmetric Generalized Correlations Among Car Variables

	mpg	disp	hp	drat	wt	qsec
mpg	1.000	−0.951	−0.938	0.685	−0.916	0.738
disp	−0.894	1.000	0.931	−0.770	0.901	−0.761
hp	−0.853	0.817	1.000	−0.554	0.693	−0.927
drat	0.688	−0.946	−0.744	1.000	−0.750	0.549
wt	−0.917	0.968	0.920	−0.730	1.000	−0.772
qsec	0.751	−0.609	−0.754	0.230	−0.188	1.000

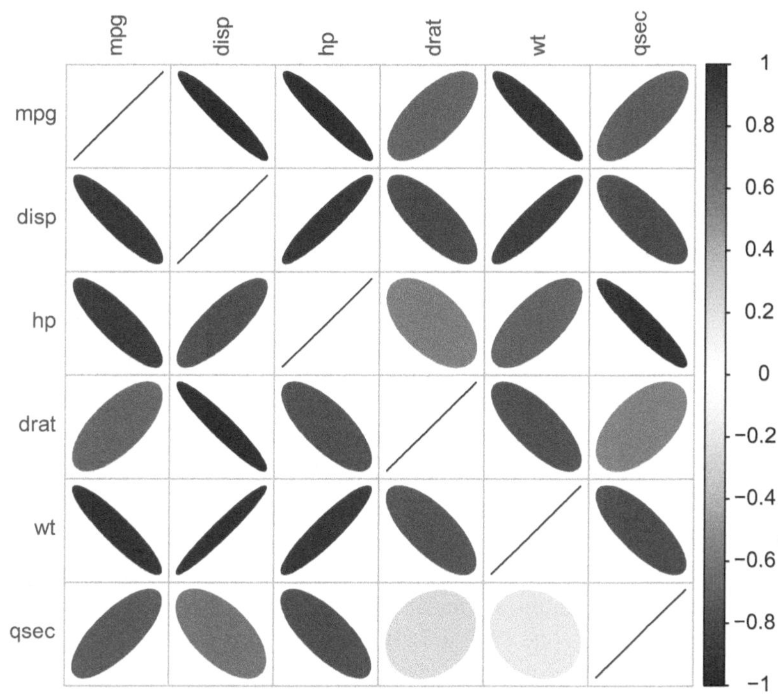

FIGURE 3 Color-coded generalized asymmetric correlations for cars data.

cause than the variable X_i. Hence, the new table and Fig. 3 represent useful supplements to the traditional table and Fig. 2. This has applications in all sciences including newer exploratory techniques using "Big Data."

8 MATRICES FOR POPULATION DYNAMICS

Births, deaths, and migrations affect the population dynamics of various creatures. Demographers have long known that age, size, or life history of individuals in any population influences the growth, survival, and reproduction of the population.

Let n_t denote a vector of populations at time t for various age, size, or stage categories. Its transition is given by

$$n_{t+1} = A\, n_t, \quad \text{where} \quad A = T + F, \tag{22}$$

where transitions matrix T represents growth and survival and where F represents fertilities or transitions due to reproduction. A powerful R package "popbio" by Stubben and Milligan (2007) is available for studying various transition matrices and models. It offers a basic function "pop.projection" for projection through Eq. (22). The demographers have developed sophisticated methods for reproductive value, damping ratio, sensitivity, and elasticity using eigenvalues of the matrix A, accomplished by the command "eigen.analysis(A)." The function "stoch.projection" can be used to simulate stochastic population growth.

This package is intended for R novices and includes a command "demo" which offers an overview.

```
require(popbio)
demo("fillmore")
years=unique(aq.trans$year)
sv<-table(aq.trans$stage, aq.trans$year)
addmargins(sv)
round(apply(sv, 1, mean),0)
stage.vector.plot(sv[-1,], prop=FALSE, col=rainbow(4),
     ylab = "Total number of plants",
     main = "Fillmore Canyon stage vectors")
```

Census data are in the form of transition data listing stages and fates from Aquilegia chrysantha in Fillmore Canyon, Organ Mountains, New Mexico, 1996–2003. One constructs and analyzes population projection matrices. Typical data are not available as matrices of the recurrence relation (22). The package 'popbio" converts such data into usable forms.

```
head2(aq.trans)
```

	Plot	year	plant	stage	leaf	rose	fruits	fate	rose2
1	903	1996	1	small	0	0	0	small	NA
2	903	1996	2	flower	NA	NA	1	large	NA
3	903	1996	3	small	0	0	0	large	NA
.	.	.	.	.	.	.	.	.	.
1637	930	2003	86	small	6	1	0	flower	1

```
addmargins(sv)
```

	1996	1997	1998	1999	2000	2001	2002	2003	Sum
seed	0	0	0	0	0	0	0	0	0
recruit	12	287	186	76	5	5	0	3	574
small	134	75	60	84	58	31	13	6	461
large	17	68	41	57	59	58	16	4	320
flower	62	6	80	74	52	8	0	0	282
Sum	225	436	367	291	174	102	29	13	1637

```
## mean stage vector
round(apply(sv, 1, mean),0)
```

seed	recruit	small	large	flower
0	72	58	40	35

Figure 4 plots the graphics produced by the code. The demo also provides code for sensitivity and elasticity matrices and many more figures.

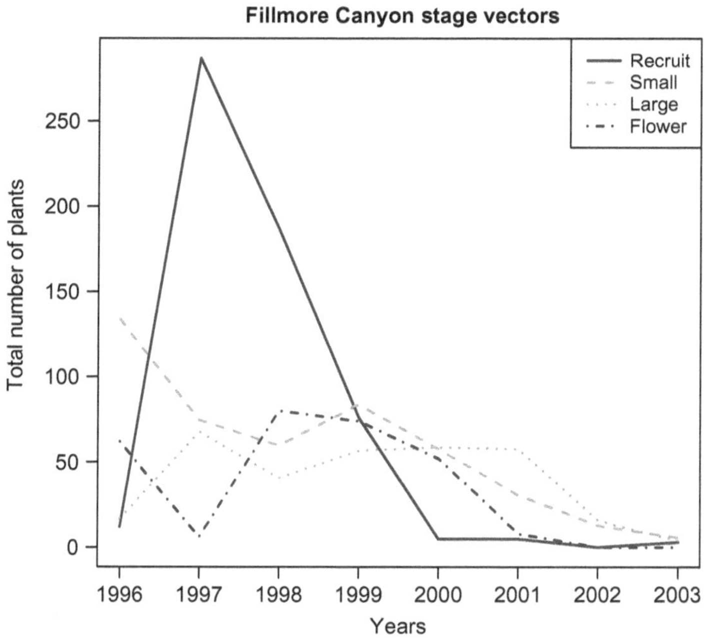

FIGURE 4 Population stages for Fillmore Canyon.

```
 $sensitivities

           seed  recruit   small   large      flower
seed    0.01112   0.0000  0.0000  0.0000   0.0002123
recruit 1.60604   0.0000  0.0000  0.0000   0.0306479
small   0.00000   0.2844  0.2122  0.2048   0.1466064
large   0.00000   0.0000  0.3793  0.3660   0.2619686
flower  0.00000   0.6814  0.5084  0.4906   0.3511985

 $ elasticities

            seed     recruit    small     large     flower
seed    0.002025   0.0000000  0.00000  0.00000   0.009098
recruit 0.009098   0.0000000  0.00000  0.00000   0.050362
small   0.000000   0.0587038  0.10262  0.03191   0.019009
large   0.000000   0.0000000  0.05277  0.17393   0.139268
flower  0.000000   0.0007563  0.05686  0.16013   0.133461

 $repro.value

seed  recruit  small   large  flower
 1.0    144.4  690.7  1234.2  1654.5
```

We have given an overview of recent R resources for matrices arising in demographics. The interested readers should consult further references by Stubben and Milligan (2007).

9 MULTIVARIATE COMPONENTS ANALYSIS

While correlation analysis studies joint dependence of two variables at a time, there is an obvious interest in extending it to several variables. Partial correlations study the relation between (*X*,*Y*), upon removing the effect of a third variable *Z*. Multivariate analysis can involve a study of joint dependence of one set of variables on another set of variables.

9.1 Projection Matrix: Generalized Canonical Correlations

R function "cancor" readily computes the canonical correlations between two data matrices. An application to estimation of joint production function (between wool and mutton on the output side and capital and labor on the input side) is discussed in (Vinod, 2008, section 5.2) . A generalized canonical correlation analysis is available under dependency modeling toolkit of the R package "dmt" by Lahti and Huovilainen (2013).

We illustrate it for the cars data with two sets of three variables in the following R code. An underlying latent variable model assumes that the two data sets, mtx1 and mtx2 can be decomposed in shared and dataset-specific components. In our artificial example from cars data, this assumption is likely to be invalid.

```
attach(mtcars); require(dmt)
mtx1=cbind(mpg,disp,hp)#first set of 3 variables
mtx2=cbind(drat,wt,qsec)
rc=regCCA(list(mtx1,mtx2))
print(head(rc$proj,3), digits=3)
matplot(rc$proj, typ="l", main="Projection matrix
of (mpg, disp, hp) against (drat, wt, qsec)")
drCCAcombine(list(mtx1,mtx2)) #dimension reduction
sharedVar(list(mtx1,mtx2),rc,3)#shared variation retained
fit.dependency.model(mtx1,
mtx2) #Bayesian with exponential priors
```

The detailed output is suppressed for brevity. We report only the first few lines and a plot of the six sets of components of the projection matrix.

```
print(head(rc$proj,3), digits=3)

        [,1]    [,2]    [,3]   [,4]   [,5]     [,6]
[1,]  -0.508  -0.498  -0.407  0.641  0.977  -0.0999
[2,]  -0.444  -0.259  -0.688  0.360  0.738  -0.1635
[3,]  -1.221   0.177  -0.417  0.915  0.120   0.0801
```

Figure 5 has six set of points numbered 1–6 for the six projected variables starting with two sets of three variables noted above. The cars data do not indicate important conclusions from these projection numbers. However, additional commands from the "dmt" package below are potentially useful in other applications.

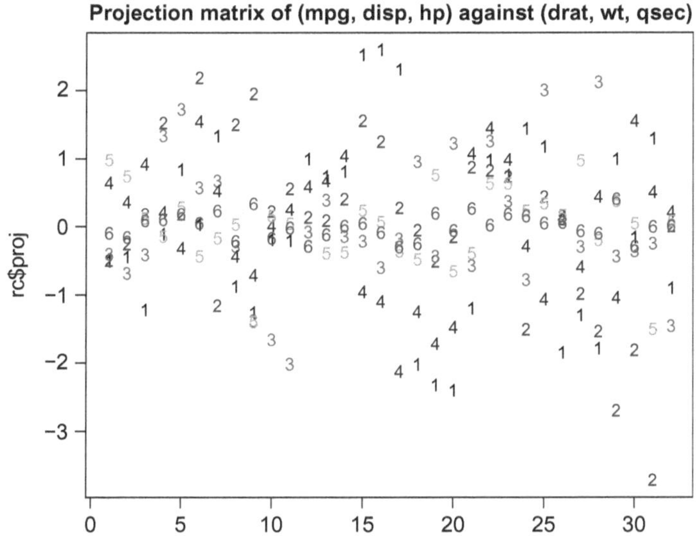

FIGURE 5 Projection matrix from cars data for two sets (mpg, disp, hp) and (drat, wt, qsec).

```
drCCAcombine(list(mtx1,mtx2)) #dimension reduction
sharedVar(list(mtx1,mtx2),rc,3)#shared variation retained
fit.dependency.model(mtx1,mtx2) #Bayesian, exponential priors
```

9.2 Invariant Coordinate Selection

Given the data matrix "A" of dimensions $n \times p$ the principal component analysis (PCA) is a well-known tool for rotating the data into orthogonal components. We have mentioned in Section 6.2.1 the matrix of eigenvectors. If G is a $p \times p$ matrix of orthogonal eigenvectors of the crossproduct matrix from A, then the matrix product AG has the principal components of A. The PCA being purely algebraic, it assumes no model with any subject-matter interpretation of the eigenvectors.

The R package "ICS," Nordhausen et al. (2008), allows deeper multivariate analyses for invariant coordinate selection (ICS) allowing interpretation of algebraic constructs. This is a two-step process. The first step standardizes the data wrt a covariance matrix, say S1. The second step performs a PCA transformation using a different covariance-type matrix S2. If B is some matrix which simultaneously diagonalizes both S1 and S2, then postmultiplying the data matrix by this B gives the ICS transform. The package authors cite some 12 choices of S2 available in various R packages.

The ICS transform uses two covariance-type scatter matrices. S1 is often the usual covariance matrix used in standardization of data. The second covariance S2 is defined differently from S1. The default for S2 is a robust estimate based on fourth order moments. It is used to find a rotation of the data obtained from a PCA of the standardized data. The new coordinate system is invariant (up to a sign change) under affine transformations mentioned above.

The "ics" function involves two covariance-type matrices, S1 and S2, and yields the invariant coordinates.

```
screeplot(princomp(A))#plot omitted for brevity
attach(mtcars);A=cbind(mpg,disp,hp)
require(ICS)
icsA=ics(A); icsA
```

The "ics" function also gives a generalized measure of the kurtosis computed by a ratio of quadratic forms and "Unmixing" matrix explained in Nordhausen et al. (2008). The output of the above code is next.

```
$gKurt
[1] 1.7133 0.8739 0.6679
$UnMix
           [,1]        [,2]        [,3]
[1,]    0.04161   -0.008423    0.024608
[2,]    0.26644    0.003913    0.005282
[3,]   -0.19138   -0.013723   -0.001489
```

The unmixing matrix allows one to get independent components by simply postmultiplying rows of data by this matrix. For our data unmixed data matrix is given by the commands:

```
Apost=A %*% icsA@UnMix
matplot(Apost)
```

The first line of the code shows postmultiplication of A by the unmixing matrix, where is accessed with the "at" (@) symbol–not the usual $ symbol. The plot is omitted for brevity, but it shows that unmixing is successful. The two covariance-type matrices S1 and S2 are accessed by the code below.

```
icsA@S1 #print basic covariance matrix
icsA@S2 #cov from fourth order moments.
icc=as.matrix(ics.components(icsA))
head(icc,3) #initial coordinate components
matplot(icc, main="Invariant Coordinates, cars data")
```

Now, we report the S1, S2 matrices, and initial independent components.

```
icsA@S1 #print basic covariance matrix

          mpg      disp       hp
mpg     36.32    -633.1   -320.7
disp  -633.10   15360.8   6721.2
hp    -320.73    6721.2   4700.9
```

```
icsA@S2 #cov from fourth order moments.
```

```
          mpg      disp       hp
mpg     30.49    -497.3   -292.1
disp  -497.35   11036.1   5266.4
hp    -292.10    5266.4   5362.6
```

```
icc=as.matrix(ics.components(icsA))
head(icc,3) #initial coordinates
```

```
       IC.1    IC.2     IC.3
[1,]  2.233   6.802   -6.378
[2,]  2.233   6.802   -6.378
[3,]  2.328   6.989   -5.984
```

The separation of coordinates obtained by the ICS package is seen in Fig. 6. A related package called "ICSNP" allows nonparametric (distribution-free) testing.

```
require(ICSNP)
rank.ctest(icc, scores="normal")
```

The output of above commands is as follows:

```
        Marginal One Sample Normal Scores Test
data: icc
T = 27.5234, df = 3, p-value = 4.573e-06
alternative hypothesis: true location is not equal to c(0,0,0)
```

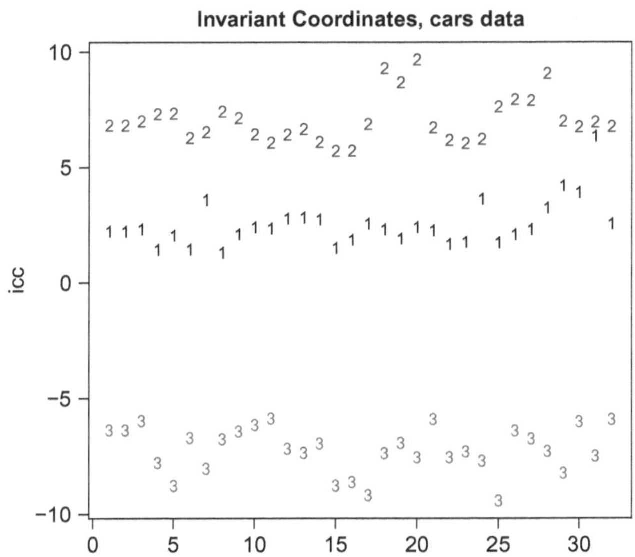

FIGURE 6 Invariant coordinates for *A* from cars data showing visible separation.

Not surprisingly, the null of zero means is rejected. These tools have also been applied to signal processing or image separation. The authors of the package have given simulation examples showing that their multivariate nonparametric distribution-free testing works well.

10 SPARSE MATRICES

There are many applications where the matrices involved have a large number of entries which are zero. This section reviews some R tools for efficiently handling them without burdening the R memory. Examples include indicator variable, design matrices of smoothing splines, fixed effects models, etc., as given in Koenker and Ng (2003).

We construct an artificial sparse matrix

```
set.seed(345)
a=sample(1:100)[1:(5*4)];a
a[a>25]=0;a
A=matrix(a,5,4);A
```

In this code, the line "a[a>25]=0" sets several elements to zero before constructing the matrix A of dimension 5×4 from the 20 numbers in "a," going columnwise.

```
a=sample(1:100)[1:(5*4)];a
[1] 22 28 39 64 42 77 37 78 44 9 24 85 16 55 83 8 80 21 84 71
a[a>25]=0;a
[1] 22 0 0 0 0 0 0 0 0 9 24 0 16 0 0 8 0 21 0 0
A=matrix(a,5,4);A

      [,1]  [,2]  [,3]  [,4]
[1,]    22     0    24     8
[2,]     0     0     0     0
[3,]     0     0    16    21
[4,]     0     0     0     0
[5,]     0     9     0     0
```

Now, let us use the function "as.matrix.csr" of the package "SparseM" to store it in a compressed sparse row (csr) format.

```
require(SparseM)
amc=as.matrix.csr(A)
myx=rbind(amc@ra,amc@ja)
image(myx, main="Visual location of nonzero entries")
rownames(myx)=c("ra","ja");myx
amc@ia
```

The csr format has four slots. The slot "ra" lists all nonzero values. The second slot "ja" lists the column indexes of the nonzero elements stored in "ra."

```
rownames(myx)=c("ra","ja");myx

     [,1]  [,2]  [,3]  [,4]  [,5]  [,6]
ra     22    24     8    16    21     9
ja      1     3     4     3     4     2

amc@ia
[1] 1 4 4 6 6 7
```

The third slot "ia" is the heart of compressing a vector of elements for economical storage. Unfortunately, its official description is hard to understand. I will give a new description below. We focus on nonzero locations only and write an *ad hoc* R function called "fc" to compute their count.

```
fc=function(x){length(x[x!=0])}
m=apply(A,1,fc);m
```

The output of the above code correctly counts the number of nonzero elements in each row as$m$=(3, 0, 2, 0, 1).

```
m=apply(A,1,fc);m
[1] 3 0 2 0 1
```

Now define m2 as one obtained by padding a one at the start, or $(1, m)$ vector. Then the compressed vector "ia" is the cumulative sum of m2 integers.

```
m2=c(1,m);m2
cumsum(m2)
```

The output is

```
m2=c(1,m);m2
[1] 1 3 0 2 0 1
cumsum(m2)
[1] 1 4 4 6 6 7
```

Verify that the output of the command "as.matrix.csr(A)" agrees with "cumsum(m2)." Since the compressed sparse row (csr) method applies to any pattern of nonzeros in A, it is commonly used for dealing with general sparse matrices. Matrix algebra for patterned matrices is discussed in Vinod, 2011, chapter 16).

The R package "SparseM" provides useful functions for various matrix operations including coercion and linear equation solving. Linear regression with sparse data is implemented by generalizing the "lm" function to achieve similar functionality with "slm, print.summary.slm" functions. Of course, the implementation uses more suitable Cholesky rather than QR methods in the sparse data context.

A regression example has a sparse matrix of regressors of dimension 1850×712.

```
require(SparseM);data(lsq)
X <- model.matrix(lsq) #extract th
y <- model.response(lsq) # extract the rhs
X1 <- as.matrix(X)
reg1=slm(y~X1-1)
su1=summary(reg1)
head(su1$coef)
su1$adj.r.squared
```

The command "`image(X,main="Visual location of nonzero entries")`" creates a graph displaying dots at all locations of nonzero entries in the large matrix seen in Fig. 7.

Note that the "slm" function is similar to the "lm." We report below a few lines from the 712 regression coefficients computed at great speed and the adjusted R^2.

```
head(su1$coef)
```

	Estimate	Std. Error	t value	Pr(>|t|)
[1,]	823.3613	0.1274477	6460.3857	0
[2,]	340.1156	0.1711477	1987.2631	0
[3,]	472.9760	0.1379109	3429.5758	0
[4,]	349.3175	0.1743084	2004.0201	0
[5,]	187.5595	0.2099702	893.2673	0
[6,]	159.0518	0.2201477	722.4776	0

FIGURE 7 The nonzero entries of a huge 1850 × 712 sparse regressor matrix shown as dots.

```
tail(sul$coef)

         Estimate    Std. Error    t value     Pr(>|t|)
[707,]  -2.0801136  0.13312708  -15.625022  0.00000000
[708,]  -6.4395314  0.14294089  -45.050310  0.00000000
[709,]  -0.1259875  0.05397642   -2.334121  0.01976273
[710,]  -0.1191570  0.10272585   -1.159952  0.24631169
[711,]  -2.0601158  0.05816518  -35.418367  0.00000000
[712,]  -7.8488311  0.18087842  -43.392856  0.00000000

sul$adj.r.squared #$
[1] 0.9999999
```

The R package "Matrix" also has several functions for sparse data. It has a great variety of sparse matrix operations and storage modes. For example, instead of "chol," it offers a sparse matrix version called "Cholesky." Considerable programming ingenuity is needed to work with sparse matrices. Our discussion can help an applied researcher who may not want to learn those intricacies. Another worthy R package "spam" uses available Fortran routines for sparse matrices and Cholesky factorization and extends them to sparse matrix algebra.

This chapter has reviewed several important topics from matrix algebra relevant for Statistics and Economics. We have provided explicit R software tools for all topics discussed along with illustrative examples.

REFERENCES

Gantmacher, F.R., 1959. The Theory of Matrices, vols. I and II. Chelsea Publishing, New York.

Hayfield, T., Racine, J.S., 2008. Nonparametric econometrics: the np package. J. Stat. Softw 27, 1–32. http://www.jstatsoft.org/v27/i05/.

Henderson, H.V., Velleman, P.F., 1981. Building multiple regression models interactively. Biometrics 37 (2), 391–411.

Koenker, R., Ng, P., 2003. SparseM: a sparse matrix package for R. J. Stat. Softw 8, 1–9. http://www.jstatsoft.org/v08/i06.

Lahti, L., Huovilainen, O.P., 2013. dmt: Dependency Modeling Toolkit. R Package Version 0.8.20. http://CRAN.R-project.org/package=dmt.

Nordhausen, K., Oja, H., Tyler, D.E., 2008. Tools for exploring multivariate data: the package ICS. J. Stat. Softw 28, 1–31. http://www.jstatsoft.org/v28/i06/.

Rao, C.R., 1973. Linear Statistical Inference and Its Applications, second ed. John Wiley and Sons, New York.

Serfling, R., 2009. Equivariance and invariance properties of multivariate quantile and related functions, and the role of standardization. University of Texas at Dallas. Technical report. http://www.utdallas.edu/serfling/papers/Equivariance:November2009.pdf.

Stubben, C.J., Milligan, B.G., 2007. Estimating and analyzing demographic models using the popbio package in R. J. Stat. Softw. 22, 1–23.

Venables, W.N., Ripley, B.D., 2002. Modern Applied Statistics with S, fourth ed. Springer, New York. ISBN 0-387-95457-0. http://www.stats.ox.ac.uk/pub/MASS4.

Vinod, H.D., 1978. Equivariance of ridge estimators through standardization: a note. Commun. Stat. A. 7 (12), 1159–1167.

Vinod, H.D., 2008. Hands-on Intermediate Econometrics Using R: Templates for Extending Dozens of Practical Examples. World Scientific, Hackensack, NJ. http://www.worldscibooks.com/economics/6895.html. ISBN 10-981-281-885-5.

Vinod, H.D., 2010. Superior estimation and inference avoiding heteroscedasticity and flawed pivots: R-example of inflation unemployment trade-off. In: Vinod, H.D. (Ed.), Advances in Social Science Research Using R. Springer, New York, pp. 39–63.

Vinod, H.D., 2011. Hands-on Intermediate Econometrics Using R: Templates for Extending Dozens of Practical Examples. World Scientific, Hackensack, NJ. http://www.worldscibooks.com/mathematics/7814.html, ISBN 978-981-4313-68-1.

Vinod, H.D., 2013. Generalized correlation and kernel causality with applications in development economics. SSRN eLibrary. http://ssrn.com/paper=2350592.

Wei, T., 2013. corrplot: Visualization of a Correlation Matrix. R Package Version 0.73. http://CRAN.R-project.org/package=corrplot.

Zeileis, A., 2004. Econometric computing with HC and HAC covariance matrix estimators. J. Stat. Softw 11, 1–17. http://www.jstatsoft.org/v11/i10/.

Zheng, S., Shi, N.Z., Zhang, Z., 2012. Generalized measures of correlation for asymmetry, nonlinearity, and beyond. J. Am. Stat. Assoc. 107, 1239–1252.

Chapter 5

Sample Size Calculations with R: Level 1

Marepalli B. Rao*[,1] and Subramanyam Kasala[†]
*Division of Biostatistics and Epidemiology, Department of Environmental Health, University of Cincinnati, Cincinnati, Ohio, USA
[†]Department of Mathematics and Statistics, University of North Carolina, Wilmington, North Carolina, USA
[1]Corresponding author: e-mail: Marepalli.rao@uc.edu

ABSTRACT
A basic question that crops up at the planning stage of an experiment is what should be the sample size. Inference based on too small a sample size may not have a stamp of authority of the conclusions drawn. A too large sample means wastage of valuable resources. It is imperative that the researcher involved takes into account all the relevant factors in order to come up with the right sample size. The purpose of this chapter is to explain what goes into a sample size calculation and lay out the relevant R code for getting the number. The goals in this chapter are modest in the sense we just focus on one-sample, two-sample, and multisample problems only, for quantitative variables.

Keywords: Level of significance, Power, One-sample t-test, Two-sample t-test, Paired t-test, Test statistic, Effect size, Error of estimation, Analysis of Variance

1 INTRODUCTION

Sample size determination is always at the forefront when planning an experiment and resources are limited. Most of the text books in Statistics and Biostatistics deal with sample size issues in various chapters meeting the needs of experiment planners. There are now specialized books devoted exclusively to sample size calculations. Cohen (1988) and Desu and Raghavarao (1990) are two of the earliest books on sample size calculations with substantial coverage but nothing on software usage. Cohen (1988) presented extensive tables of sample sizes to discount the need for software. Odeh and Fox (1975) also focused on sample size calculations but in a specific environment. A recent addition to the literature is Ryan (2013). In the context of clinical trials, Machin et al. (2009) and Chow et al. (2003) are notable

Handbook of Statistics, Vol. 32. http://dx.doi.org/10.1016/B978-0-444-63431-3.00005-X

contributions. In this chapter, sample size determination and the software R go hand in hand. This chapter is self-contained when working in the environment of scenarios outlined at the beginning. The goals are summarized succinctly below.

1.1 Goals

1. Understand what goes on behind the sample size calculations. Why do we need to spell out the specifications demanded?
2. Learn how to use R to get the required output.

1.2 Why Did We Choose R?

It is free and takes less than 5 min to download the software. A number of statistical software such as SAS, SPSS, BMDP, StatExact, Stata, PASS, nQuery, etc., do calculate sample sizes. All these software cost money. Commercial software such as SAS and SPSS produce a number of manuals to help you use their software. R does not do that. However, any one in the world can write a book on the usage of R, publish it commercially, and profit from it. Major commercial publishers are big in publishing books on R. You can start your own company helping small and big businesses, for a fee, by attending to their computational needs via R. The company "Revolution Analytics" is a prime example of such an endeavor.

If you develop a new methodology for analyzing data and your research is funded federally (NIH, NSF, etc.), the federal organizations now demand that you develop an R code for the implementation of the methodology.

2 GENERAL IDEAS ON SAMPLE SIZE CALCULATIONS

It will be good to bear in mind the following points:

1. The null hypothesis is a hypothesis of skepticism.
2. The alternative is a hypothesis of optimism.
3. Funding agencies put emphasis on the null hypothesis. Rejecting the null hypothesis when it is true is tantamount propagating a "false positive." Control false positives.
4. Researchers put emphasis on the alternative hypothesis. Not rejecting the null hypothesis when it is false is tantamount advocating a "false negative." Control false negatives.
5. We neither prove nor disprove the null hypothesis. We merely check whether the data collected are consistent with the null hypothesis.

There are exceptions. When testing a distributional assumption (in tests of goodness of fit), the null hypothesis is taken to be the distribution under contention. This is not really a hypothesis of skepticism.

2.1 Example

A targeted population has average systolic blood pressure $\mu_0 = 130$ and standard deviation $\sigma = 16$. The average is high. A researcher developed a drug that is specifically tailored to the target population. Her contention is that this drug will help bring the mean down. The problem as spelled out may not be natural, but it is good enough to illustrate salient features of sample size calculations.

Design: Take a random sample of n subjects from the target population and administer the drug for a prescribed period of time. Measure their systolic blood pressure. What should be the value of n? The sample that will be drawn is denoted by $X_1, X_2, \ldots, X_n$. We make an additional assumption that blood pressures in the population after the administration of the drug are normally distributed with mean μ and standard deviation $\sigma = 16$. This assumption will ease our computational problems. The hypothesis is a statement about the mean μ.

Null hypothesis: Skepticism: The drug has no effect.

$$H_0\colon \mu = \mu_0 = 130$$

Alternative: Optimism: The drug has indeed some effect.

$$H_1\colon \mu < \mu_0$$

We do not know which hypothesis is true? No amount of data collected will tell you that, unless you are a Bayesian. We can only check whether or not the data are consistent with the data collected. We use the test statistic $Z = \frac{\sqrt{n}(\overline{X} - \mu_0)}{\sigma}$, and a test is built on the test statistic Z. The entity $\overline{X}$ is the sample mean.

If we reject the null hypothesis when it is true, we are raising false hopes. We will be saying that the drug works in lowering blood pressure leading to a false positive. We need to control this type of error (Type I error). The Type I error probability is denoted by α. A typical value of α is 0.05, 0.01, or 0.10.

If we reject the null hypothesis when it is not true, this is a good thing. The researcher would love this scenario. She wants to control the chances of rejecting the null hypothesis when it is not true. This is power. Power is usually denoted by $1 - \beta$. The entity β has a special meaning. It is the probability of not rejecting the null hypothesis when the alternative is true (Type II error). The sample size calculation should take into account both types of error.

Controlling Type I error is easy. Meeting the power requirements is mathematically hard. In the sample size calculations, we need the distribution of the test statistic Z when the null hypothesis is true. This is usually easy. Using the distribution, we can obtain the critical value c of the test. In this example, the test would be of the form:

Reject the null hypothesis if $Z < c$.

The entity c comes from the equation

$$\Pr\ (\text{rejecting the null hypothesis} | H_0 \text{ is true}) = \Pr(Z < c | \mu = 130) = \alpha$$

The statistic Z has a standard normal distribution if the null hypothesis is true. We will be solving the integral equation to get c:

$$\int_{-\infty}^{c} \frac{1}{\sqrt{2\pi}} \exp\left(-\frac{x^2}{2}\right) dx = \alpha$$

We also need the distribution of Z when the alternative is true. This is where problems crop up. This is a technical issue. Our alternative is hazy. The mean is not clearly spelled out except that it should be $<\mu_0$. How can we find the distribution of Z when we do not know what the mean is? We have to be concrete. One reasonable choice, here, could be $\mu = \mu_1 = 120$ (normal blood pressure). We want people to have average normal blood pressure if they use the drug. When $\mu = 120$, we can find the distribution of Z. Incidentally, we define the effect size

$$\Delta = \left|\frac{\mu_1 - \mu_0}{\sigma}\right| = \left|\frac{120 - 130}{16}\right| = 0.625 = \left|\frac{\text{Alternative mean} - \text{status quo mean}}{\text{SD}}\right|$$

We might say that we want to see this effect size realized with specified power. Technically, the equation for sample size calculation would be

$$\text{Power} = \Pr\ (\text{reject } H_0 | \mu = 120) = \Pr(Z < c | \Delta = 0.625)$$

The entity Z has a normal distribution with mean $\frac{\sqrt{n}(120-130)}{16} = -\sqrt{n}{*}(0.625) = \sqrt{n}{*}\Delta$.

There will be another integral equation to solve

$$\int_{-\infty}^{c + \sqrt{n}*0.625} \frac{1}{\sqrt{2\pi}} \exp\left(-\frac{x^2}{2}\right) dx = \text{Power}$$

Solve this integral equation for n. Generally, a lot of computing effort is involved in solving integral equations of this type.

The solution n depends on α (via the critical value c), power, and alternative mean or, equivalently, on α, power, and effect size Δ.

Caution: The sample size depends on the choice of the test statistic too, besides the level of significance α, effect size Δ, and power.

2.2 FAQ and Pointers

1. How does one come up with the alternative mean? Equivalently, effect size? The researcher is responsible. Statistician has only a marginal role.

Conduct a pilot study. The pilot data might provide some clues. Survey the literature.

2. Most research proposals require sample size calculations. Typically, pilot studies do not ask for sample size calculations. Sample size is limited by the budget offered. Pilot studies, typically, are retrospective. How many records one extracts for a pilot study depends on time and budget.
3. Sample size calculations are required if one is conducting a prospective study. Usually, a pilot study is used to make a case for a prospective study.

2.3 Signal-to-Noise Ratio

Suppose we have a single population whose quantitative characteristic is normally distributed with mean μ and standard deviation (SD) σ. Engineers call the mean as "signal" and standard deviation "noise." For statisticians "noise" means "variation." Suppose mean = 1 and SD = 100. There is too much noise (too much variation) in the population values and the signal is weak. Look at the normal distribution curve with mean = 1 and standard deviation = 100. We use the "curve" command in R to lot the normal density function.

mean ± 3 * sigma limits: −299 to 301

```
> curve(dnorm(x, mean = 1, sd = 100), from = -299, to = 301,
+       xlab = "X", ylab = "Density", main = "Normal Density with
+       Mean = 1 SD = 100", col = "red")
```

2.4 Some Features of the Normal Distribution

1. 67% of the population values are in between mean − SD and mean + SD, i.e., 99 and 101.
2. 95% of the population values are in between mean—2*SD and mean+ *SD, i.e., 199 and 201.
3. 99% of the population values are in between mean—3*SD and mean+ *SD, i.e., −299 and 301.
4. The peak of the normal curve is at the "signal."

An illustration of a normal curve is given in Figure 1.

The signal-to-noise ratio is defined by (mean/SD), which, in our case, is 1/100 = 0.01. The signal-to-noise ratio measures how strong the signal is. A value close to zero indicates that the signal is buried deep in the noise.

Suppose we know the noise, SD, but not the signal. We want to extract the signal. We take a sample from the population. The sample mean is a good estimate of the signal. If the signal is weak, we need a very large sample to recognize the signal. Let us do a lab experiment.

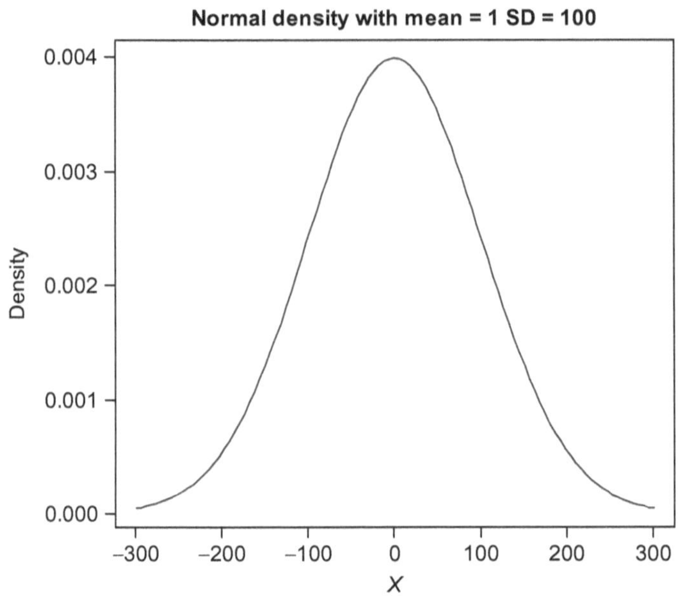

FIGURE 1 Normal density curve with mean = 1 and standard deviation = 100.

Draw a random sample of size 10 from $N(1, 100)$ with each observation rounded to three decimal places.

```
> x <- round(rnorm(10, 1, 100), 3)
> x
[1] −26.783 97.616  −36.035 −79.736 176.967 −110.166 32.604 −18.986
[9] −7.183  −58.410
> mean(x)
[1] −3.0112
> sd(x)
[1] 85.73381
```

The sample mean is off the mark (signal). This is just a single sample. The sample standard deviation, SD, 85.73381 gives an inkling that there is too much variation in the data and hence in the population.

Suppose we want to build a 95% confidence interval for the signal. Theory says that with 95% probability the random interval

$$[\text{Mean} - 2^*(\text{standard error of the mean}), \text{mean} + 2^*(\text{standard error of the mean})]$$

contains the population mean. Technically, the random interval

$$\left[\overline{X} - 2^*\left(\frac{\text{SD}}{\sqrt{n}}\right),\ \overline{X} + 2^*\left(\frac{\text{SD}}{\sqrt{n}}\right)\right]$$

contains the population mean with approximately 95% probability. (The number 2 that appears in the interval is not right within our premises. To get the right number and confidence interval with 95% coverage probability, we need to use Student's t-distribution with nine degrees of freedom. Further, we are using the sample SD in the construction of the interval.)

Let us calculate the confidence interval using our data on hand: [−57.23, 51.12].

Sure, this interval contains the population mean. It trapped the signal. We were lucky. Its utility is limited because the interval is very wide. Biomedical researchers use a measure to judge the value of a confidence interval in the context of odds ratio estimation and confidence interval. It is defined by

$$U = \frac{\text{Length of the interval}}{\text{Point estimate}}$$

The interval is deemed useful if $U < 1$. In our case, we calculate the utility of the interval by using the absolute value of the point estimate. The value $U = 36.01$.

Let us take a bigger sample. Now the sample size is 100.

```
> y <- round(rnorm(100, 1, 100), 3)
> y
 [1] 157.793   97.618   61.219  161.876   -3.316 -28.529   21.026   88.756
 [9]  29.196   60.545  -30.065  -42.276   17.944 -16.693  121.968   45.160
[17] 116.726   50.329   27.495 -200.384  -56.832 -47.813   65.768   39.445
[25]  68.762  -12.685  -98.909    2.028   21.909  65.423 -100.344   33.357
[33] -19.108  -34.252   80.272  -62.777   67.950  15.335    0.856 -140.248
[41] 159.324   52.562   40.661   50.240   83.867  43.863   98.458   92.784
[49]  -3.858  119.726   47.614  -44.094   24.885 200.171 -194.501  -41.512
[57] 111.389   69.714  118.845 -183.106  111.463   5.368 -138.438   21.821
[65] -69.850  243.863  -58.808   31.148  -70.995 -72.643 -159.951 -183.248
[73]  46.881 -132.611 -130.435  -49.650   42.142  61.817   30.484  -36.953
[81] -76.703   99.812   55.192    3.063 -149.835  -6.140   48.654    3.485
[89] -55.356  -32.341  -41.830  147.808   58.409 143.858   -8.832   -9.914
[97] -62.337    3.164   63.356  -89.980

> mean(y)
[1] 10.56495
> sd(y)
[1] 88.48868
The 95% confidence interval for the signal is:
[-7.13, 28.26]
U = 3.35
```

The interval is better but still wide. The interval trapped the signal. The utility of the interval has improved. This is just one sample. In repeated samples, the numbers are expected to be in the same ball park. There is a 5% chance that the interval fails to trap the signal.

Let us take much bigger samples with $n = 1000$ and $n = 10{,}000$.

```
> z <- round(rnorm(1000, 1, 100), 3)
> mean(z)
[1] 4.13059
> sd(z)
[1] 100.896
95% confidence interval: [-2.25, 10.51]
U = 3.089739
> u <- round(rnorm(10000, 1, 100), 3)
> mean(u)
[1] 3.237165
> sd(u)
[1] 99.8532
95% confidence interval: [1.24, 5.23]
U = 1.23
```

The important point is that quality of inference depends how much noise there is in the population. One needs really a very large sample to improve utility of inference.

The same argument works through sample size calculations. The effect size is akin to signal-to-noise ratio. The smaller the effect size is, the larger the sample required to detect the effect size.

3 SINGLE-SAMPLE PROBLEMS

3.1 Quantitative

The data variable is quantitative. Discrete variables are taken care of in the subsequent sections.

3.2 Testing of Hypotheses Environment

Sample size calculations are done in two different environments. Currently, the focus is on testing of hypotheses environment. The estimation environment will be taken up later in the section.

Setup: The random entity X has a normal distribution with mean μ and standard deviation σ. The null hypothesis is

H_0: $\mu = \mu_0$ (specified)

and the alternative

H_1: $\mu > \mu_0$ (one-sided alternative)

The population standard deviation σ is known.

Potential data: $X_1, X_2, \ldots, X_n$

Test statistic: $Z-$statistic $Z=\frac{\overline{X}-\mu_0}{(\sigma/\sqrt{n})}$

Prescribed level of significance: α

Test: Reject the null hypothesis H_0 in favor of the alternative if $Z>z_{1-\alpha}$

What is $z_{1-\alpha}$, the critical value of the test? The area under the standard normal curve to the left of $z_{1-\alpha}$ is $1-\alpha$.

Some examples of α:

α	Critical Value ($z_{1-\alpha}$)
0.05	1.645
0.01	2.326
0.10	1.282

What should be the sample size n?

3.3 Specifications

1. Level of significance: α (Type I error probability) (This provides the critical value c of the test.)
2. Power: $1-\beta$ ($1-$Type II error probability)
3. Effect size: $\Delta=\left|\frac{\mu_1-\mu_0}{\sigma}\right|$ (standardized difference) μ_1 is the mean under the alternative hypothesis.

What should be the sample size n? With false positive (Type I error) probability controlled at α, what should be the sample size n so that the test is able to detect the standardized difference Δ with power $1-\beta$?

How does a mathematical statistician come up with a formula for n?

He needs the distribution of the test statistics Z under the null hypothesis. $Z\sim N(0, 1)$

This is useful to calculate the critical value c of the test.

He needs the distribution of the test statistic Z under the alternative hypothesis. The alternative hypothesis is not specific. It stipulates the unknown mean $\mu>\mu_0$. One cannot find the distribution of Z. Stipulate the value of the mean μ. Set a value $\mu=\mu_1$ ($>\mu_0$). Under the alternative hypothesis,

$$Z\sim N\left(\sqrt{n}\left(\frac{\mu_1-\mu_0}{\sigma}\right), 1\right)$$

$$Z\sim N\left(\sqrt{n}\Delta, 1\right)$$

The mean of the normal distribution is a function of the standardized difference Δ and sample size.

An important lesson: One does not have to come up with a value of μ under the environment of the alternative hypothesis. It is enough to spell out the standardized difference Δ, i.e., the difference expressed in the units of the standard deviation.

3.4 Formula for Sample Size

The equation to solve is:

$$1-\beta=\Pr(Z>z_{1-\alpha}|\Delta)=\int_{z_{1-\alpha}}^{\infty} N\left(\sqrt{n}\Delta,1\right)\mathrm{d}x$$

For what value of n, this equation is satisfied. Fortunately, we have an explicit solution

$$n=\frac{\left(z_{1-\alpha}+z_{1-\beta}\right)^2}{\Delta^2}$$

Use R to calculate n.

Goal: For each of the levels (α) 0.01, 0.05, and 0.10, power ($1-\beta$) 0.80, 0.90, 0.95, and effect size (Δ) 0.1 (0.1) 0.9, calculate the required sample size.

Step 1: Calculate critical values: $z_{1-\alpha}$.

Use the loop command in R. The loop command is used when the same formula is used under different scenarios.

Create a vector of zeros. We will replace the zeros by critical values $z_{1-\alpha}$.

```
> Critical <- c(0, 0, 0)
```

Create a vector of levels of significance.

- `Level <- c(0.01, 0.05, 0.10)`

Start the loop command. Replace each entry in the folder "Critical" by an appropriate critical value. Replace the *i*th entry of "Critical," Critical[i], by calculating the critical value using the *i*th entry of "Level." The command "qnorm" does the trick.

```
> for (i in 1:3)
+ {
+ Critical[i] <- qnorm(Level[i], lower.tail = F)
+ }
> Critical
[1] 2.326348 1.644854 1.281552
```

The critical values are rounded to three decimal places.

```
> round(Critical, 3)
[1] 2.326 1.645 1.282
Let us calculate z1-β s for different choices of power.
> Power <- c(0.80, 0.90, 0.95)
> Powerz <- c(0, 0, 0)
> for (i in 1:3)
```

```
+ {
+ Powerz[i] <- qnorm(Power[i])
+ }
> Powerz
[1]  0.8416212  1.2815516  1.6448536
```

Let us experiment with different effect sizes. Create a folder consisting of numbers 0.1 through 0.9 in increments of 0.1.

```
Delta <- seq(0.1, 0.9, 0.1)
> Delta
[1]  0.1  0.2  0.3  0.4  0.5  0.6  0.7  0.8  0.9
```

We are ready to start another loop. Remember that the sample size is a function of significance level, power, and effect size. We have chosen three levels of significance, three choices of power, and nine choices of effect size. Let us create a three-dimensional array with nine slices and each slice is a matrix of three rows and three columns with all entries being zeros. Create a column vector of 81 zeros (rep = repeat).

```
> SampleSizes <- rep(0, 81)
Create the required array.
> dim(SampleSizes) <- c(3, 3, 9)
```

Start a loop. We are trying to fill the entries in the first slice of the loop using the sample size formula for $\Delta = 0.1$. The first line in the code achieves this objective for a general slice *k*.

```
> for (i in 1:3)
+ {
+ for (j in 1:3)
+ {
+ for (k in 1:9)
+ {
+ SampleSizes[i, j, k] <- ((Critical[i] + Powerz[j])^2)/Delta[k]^2
+ }
+ }
+ }
```

The sample sizes are rounded to integers.

```
> round(SampleSizes)

, , 1
      [1]   [2]   [3]
[1]  1004  1302  1577
[2]   618   856  1082
[3]   451   657   856
```

```
, , 2

      [1]  [2]  [3]
[1]   251  325  394
[2]   155  214  271
[3]   113  164  214

, , 3

      [1]  [2]  [3]
[1]   112  145  175
[2]    69   95  120
[3]    50   73   95

, , 4

      [1]  [2]  [3]
[1]    63   81   99
[2]    39   54   68
[3]    28   41   54

, , 5

      [1]  [2]  [3]
[1]    40   52   63
[2]    25   34   43
[3]    18   26   34

, , 6

      [1]  [2]  [3]
[1]    28   36   44
[2]    17   24   30
[3]    13   18   24

, , 7

      [1]  [2]  [3]
[1]    20   27   32
[2]    13   17   22
[3]     9   13   17

, , 8

      [1]  [2]  [3]
[1]    16   20   25
[2]    10   13   17
[3]     7   10   13

, , 9

      [1]  [2]  [3]
[1]    12   16   19
[2]     8   11   13
[3]     6    8   11
```

Let us name the rows, columns, and slices.

```
>dimnames(Sizes)[[1]]<-c("Level 0.01","Level 0.05","Level 0.10")
>dimnames(Sizes)[[2]]<-c("Power 0.80","Power 0.90","Power 0.95")
>dimnames(Sizes)[[3]]<-c("Effect 0.1","Effecr 0.2","Effect 0.3",
+                        "Effect 0.4","Effect 0.5","Effect 0.6",
+                        "Effect 0.7","Effect 0.8","Effect 0.9")
> round(SampleSizes)
, , Effect 0.1

           Power 0.80  Power 0.90  Power 0.95
Level 0.01       1004        1302        1577
Level 0.05        618         856        1082
Level 0.10        451         657         856

, , Effect 0.2
           Power 0.80  Power 0.90  Power 0.95
Level 0.01        251         325         394
Level 0.05        155         214         271
Level 0.10        113         164         214

, , Effect 0.3
           Power 0.80  Power 0.90  Power 0.95
Level 0.01        112         145         175
Level 0.05         69          95         120
Level 0.10         50          73          95

, , Effect 0.4
           Power 0.80  Power 0.90  Power 0.95
Level 0.01         63          81          99
Level 0.05         39          54          68
Level 0.10         28          41          54

, , Effect 0.5
           Power 0.80  Power 0.90  Power 0.95
Level 0.01         40          52          63
Level 0.05         25          34          43
Level 0.10         18          26          34

, , Effect 0.6
           Power 0.80  Power 0.90  Power 0.95
Level 0.01         28          36          44
Level 0.05         17          24          30
Level 0.10         13          18          24

, , Effect 0.7
           Power 0.80  Power 0.90  Power 0.95
Level 0.01         20          27          32
Level 0.05         13          17          22
Level 0.10          9          13          17
```

```
, , Effect 0.8
           Power 0.80  Power 0.90  Power 0.95
Level 0.01         16          20          25
Level 0.05         10          13          17
Level 0.10          7          10          13

, , Effect 0.9
           Power 0.80  Power 0.90  Power 0.95
Level 0.01         12          16          19
Level 0.05          8          11          13
Level 0.10          6           8          11
```

3.5 Comments

1. Fix the level and power. As the effect size increases, the sample size needed decreases. Fix $\alpha = 0.05$ and Power $= 0.80$.

Δ	0.1	0.2	0.3	0.4	0.5	0.6	0.7	0.8	0.9
n	618	155	69	39	25	17	13	10	8

2. Fix the level and effect size. As the power increases, the sample size increases. Set $\alpha = 0.05$ and $\Delta = 0.5$.

Power	80%	90%	95%
n	25	34	43

3. Fix the power and effect size. As the level increases, the sample size needed decreases. Set Power $= 80\%$ and $\Delta = 0.5$

Level	0.01	0.05	0.10
n	40	25	18

3.6 The Other Type of One-Sided Alternative

The null hypothesis is

H_0: $\mu = \mu_0$ (specified)

and the alternative

H_1: $\mu < \mu_0$ (one-sided alternative)

We use the same test statistic. The character of the test now changes. This scenario is discussed in detail in Section 1.

Test: Reject the null hypothesis in favor of the alternative if $Z < z_\alpha$. Sample size numbers will remain exactly the same.

3.7 The Case of Two-Sided Alternative

The null hypothesis is

H_0: $\mu = \mu_0$ (specified)

and the alternative

H_1: $\mu \neq \mu_0$ (two-sided alternative)

The test statistic Z remains the same.

Test: Reject the null hypothesis in favor of the alternative if $|Z| > z_{1-\alpha/2}$.

The critical value $z_{1-\alpha/2}$ is located on the following standard normal curve. The number c ($z_{1-\alpha/2}$) on the X-axis is the critical value.

The relevant R code is given by

```
> xv <- seq(-3, 3, 0.01)
> yv <- dnorm(xv)
> plot(xv, yv, type = "l", xlab = "X", ylab = "Density",
+       main = "Standard Normal Curve")
> axis(side = 1, at = c(-1.5, 1.5), labels = c("-c", "c"))
> polygon(c(xv[xv <= -1.5], -1.5), c(yv[xv <= - 1.5],
+           yv[xv == -3]), col = "lightgray")
> polygon(c(xv[xv >=1.5], 1.5), c(yv[xv >=1.5], yv[xv == 3]),
+          col ="lightgray")
> text(-2, 0.05, "Alpha/2")
> text(2, 0.05, "Alpha/2")
> text(0, 0.2, "1 - Alpha")
```

The final output of the code is presented in Figure 2.

All other features remain the same. The formula for the sample size changes slightly.

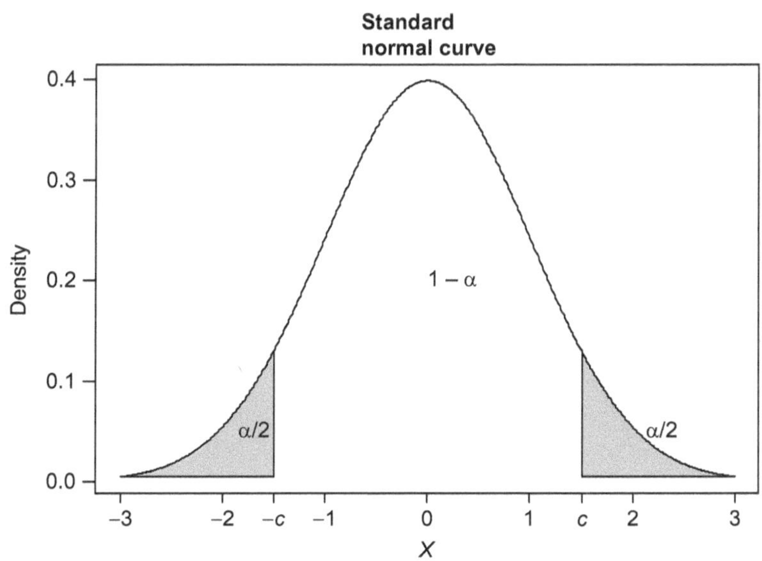

FIGURE 2 Standard normal curve with shaded area = α.

$$n = \frac{\left(z_{1-\alpha/2} + z_{1-\beta}\right)^2}{\Delta^2}$$

α	$z_{1-\alpha/2}$
0.05	1.96
0.01	2.576
0.10	1.645

Let us work out the sample size for a variety of configurations of level, power, and effect size. Use R.

Critical values are different now.

```
> Level <- c(0.01, 0.05, 0.10)
> Critical <- c(0, 0, 0)
> for (i in 1:3)
+ {
+ Critical[i] <- qnorm(Level[i]/2, lower.tail = F)
+ }
> Critical
[1] 2.575829 1.959964 1.644854
```

The power number $z_{1-\beta}$ remains the same. We will use the folder "Powerz" defined earlier. We will use the same "Delta" folder. Let us create an array of dimensions $3 \times 3 \times 9$ with every entry being zero.

```
> SampleSizes <- rep(0, 81)
> dim(SampleSizes) <- c(3, 3, 9)
> for (i in 1:3)
+ {
+ for (j in 1:3)
+ {
+ for (k in 1:9)
+ {
+ Sizes[i, j, k] <- ((Critical[i] + Powerz[j])^2)/0.1^2
+ }
+ }
+ }
> dimnames(SampleSizes)[[1]] <- c("Level 0.01", "Level 0.05",
+           "Level 0.10")
> dimnames(SampleSizes)[[2]] <- c("Power 0.80", "Power 0.90",
+           "Power 0.95")
> dimnames(SampleSizes)[[3]] <- c("Effect 0.1", "Effect 0.2",
+           "Effect 0.3", "Effect 0.4", "Effect 0.5", "Effect 0.6",
+           "Effect 0.7", "Effect 0.8", "Effect 0.9")
```

The sample sizes are given below:

```
> round(SampleSizes)
```

```
, , Effect 0.1
           Power 0.80 Power 0.90 Power 0.95
Level 0.01       1168       1488       1781
Level 0.05        785       1051       1299
Level 0.10        618        856       1082

, , Effect 0.2
           Power 0.80 Power 0.90 Power 0.95
Level 0.01        292        372        445
Level 0.05        196        263        325
Level 0.10        155        214        271

, , Effect 0.3
           Power 0.80 Power 0.90 Power 0.95
Level 0.01        130        165        198
Level 0.05         87        117        144
Level 0.10         69         95        120

, , Effect 0.4
           Power 0.80 Power 0.90 Power 0.95
Level 0.01         73         93        111
Level 0.05         49         66         81
Level 0.10         39         54         68

, , Effect 0.5
           Power 0.80 Power 0.90 Power 0.95
Level 0.01         47         60         71
Level 0.05         31         42         52
Level 0.10         25         34         43

, , Effect 0.6
           Power 0.80 Power 0.90 Power 0.95
Level 0.01         32         41         49
Level 0.05         22         29         36
Level 0.10         17         24         30

, , Effect 0.7
           Power 0.80 Power 0.90 Power 0.95
Level 0.01         24         30         36
Level 0.05         16         21         27
Level 0.10         13         17         22

, , Effect 0.8
           Power 0.80 Power 0.90 Power 0.95
Level 0.01         18         23         28
Level 0.05         12         16         20
Level 0.10         10         13         17
```

```
, , Effect 0.9
           Power 0.80  Power 0.90  Power 0.95
Level 0.01         14          18          22
Level 0.05         10          13          16
Level 0.10          8          11          13
```

3.8 Comments

Sample sizes are higher.

Set $\alpha=0.05$ and Power=80%.

3.9 One-Sided Alternative

Δ	0.1	0.2	0.3	0.4	0.5	0.6	0.7	0.8	0.9
n	618	155	69	39	25	17	13	10	8

3.10 Two-Sided Alternative

Δ	0.1	0.2	0.3	0.4	0.5	0.6	0.7	0.8	0.9
n	785	196	87	49	31	22	16	12	10

Maybe, we are not comparing apples with apples! The numbers for the two-sided alternative when $\alpha=0.10$ and Power=80% are exactly the same for the one-sided alternative when $\alpha=0.05$ and Power=80%.

3.11 The Case When the Population Standard Deviation σ Is Unknown

This scenario is more natural and common. The other features such as normality remain on hold. The test statistic used is now different. Let us check where the differences occur.

3.12 The Case of One-Sided Alternative

H_0: $\mu=\mu_0$ (specified)

and the alternative

H_1: $\mu>\mu_0$ (one-sided alternative)

The population standard deviation σ is unknown.

Potential data: $X_1, X_2, \ldots, X_n$

Test statistic: Student's t-statistic $t=\frac{\overline{X}-\mu_0}{(S/\sqrt{n})}$

S is the sample standard deviation. Here, the Z-statistic is modified by replacing the unknown population standard deviation σ by the sample standard deviation S.

Test: Reject the null hypothesis H_0 in favor of the alternative if $t>t_{n-1,1-\alpha}$.

The critical value $t_{n-1,1-\alpha}$ comes from the Student's t-distribution. The area under the Student's t-curve on $n-1$ degrees of freedom to the left of $t_{n-1,1-\alpha}$ is $1-\alpha$.

What should be the sample size n?

3.13 Specifications

1. Level of significance: α (Type I error probability)
2. Power: $1-\beta$ (1 − Type II error probability)
3. Effect size: $\Delta = \left|\frac{\mu_1-\mu_0}{\sigma}\right|$ (standardized difference)

The computation of Δ requires the population standard deviation σ!

With false positive (Type I error) probability controlled at α, what should be the sample size n so that the test is able to detect the standardized difference Δ with power $1-\beta$?

How does a mathematical statistician come up with a formula for n?

He needs the distribution of the test statistics t under the null hypothesis. This is simple.

$t \sim t$-distribution with $n-1$ degrees of freedom.

He needs the distribution of the test statistic t under the alternative hypothesis. The alternative stipulates that the unknown mean $\mu > \mu_0$. One cannot find the distribution of t. Stipulate the value of the mean μ. Set a value $\mu = \mu_1$ ($>\mu_0$). Under the alternative hypothesis, $t \sim$ noncentral t-distribution with $n-1$ degrees of freedom and noncentrality parameter Δ. We use the notation $t \sim t_{n-1,\Delta}$. The symbol $t_{n-1,\Delta}$ stands for the density function of the t-distribution.

The probability density function is very complicated and comes in the form of an infinite series. The calculation of the required sample size is much harder and one has to use quadrature formulas to evaluate integrals that show up in the calculations. Structurally, the power equation is the same but the normal distribution $N(\sqrt{n}\Delta, 1)$ in the integral is replaced by the t-distribution $t_{n-1,\Delta}$.

$$1-\beta = \Pr(t > t_{n-1,1-\alpha} \mid \Delta) = \int_{t_{n-1,1-\alpha}}^{\infty} t_{n-1,\Delta} \mathrm{d}x$$

The equation involves three entities: sample size n, power $1-\beta$, and effect size Δ. (The presence of the level α is felt in the critical value $t_{n-1,1-\alpha}$. We fix α.) If one specifies two of these three entities, one can solve the equation for the third one. In practice, we specify the power and effect size. The sample size n can be solved from the equation.

There is no explicit formula for the sample size. R has a package "pwr," which helps us in determining the sample size in this context. Download and activate this package.

Let us use this package. Look at the documentation of pwr.

```
> ?pwr
```

Look at the documentation of pwr.t.test.

```
> ?pwr.t.test
```

If we want sample size for one specific scenario, the following command would do:

```
Set Level = 0.05, effect size = 0.5, and power = 0.80.
> MB <- pwr.t.test(d = 0.5, sig.level = 0.05, power = 0.80,
+                      type = "one.sample", alternative = "greater")
> MB
One-sample t test power calculation
n = 26.13751
d = 0.5
sig.level = 0.05
power = 0.8
alternative = greater
```

What else is available in the folder "MB?"

```
> names(MB)
[1] "n" "d" "sig.level" "power" "alternative"
[6] "note" "method"
```

What is available under "*n*?" It is the sample size.

```
> MB$n
[1] 26.13751
```

We now prepare a table of sample sizes under a variety of scenarios.

Goal: Prepare tables of sample size under different combinations of level, power, and effect size.

```
Level = 0.01, 0.05, 0.10
Power = 0.80, 0.90, 0.95
Effect size = 0.1 (0.1) 0.9
Create the relevant folders.
> Level <- c(0.01, 0.05, 0.10)
> Power <- c(0.80, 0.90, 0.95)
> Delta <- seq(0.1, 0.9, 0.1)
> SampleSizes <- rep(0, 81)
> dim(SampleSizes) <- c(3, 3, 9)
```

Initiate the loop.

```
> for (i in 1:3)
+ {
+ for (j in 1:3)
```

```
+ {
+ for (k in 1:9)
+ {
+ SampleSizes[i, j, k] <- pwr.t.test(d = Delta[k],
+                                     sig.level = Level[i],
+                                     power = Power[j],
+                                     type = "one.sample",
+                                     alternative = "greater")$n
+ }
+ }
+ }
```

Let us name the rows, columns, and slices.

```
> dimnames(SampleSizes)[[1]] <- c("Level 0.01", "Level 0.05",
+                                 "Level 0.10")
> dimnames(SampleSizes)[[2]] <- c("Power 0.80", "Power 0.90",
+                                 "Power 0.95")
> dimnames(SampleSizes)[[3]] <- c("Effect 0.1", "Effect 0.2",
+                                 "Effect 0.3", "Effect 0.4",
+                                 "Effect 0.5", "Effect 0.6",
+                                 "Effect 0.7", "Effect 0.8",
+                                 "Effect 0.9")
The sample sizes are rounded to the nearest integers.
> round(SampleSizes)

, , Effect 0.1

           Power 0.80 Power 0.90 Power 0.95
Level 0.01       1006       1304       1580
Level 0.05        620        858       1084
Level 0.10        452        658        857

, , Effect 0.2

           Power 0.80 Power 0.90 Power 0.95
Level 0.01        254        328        397
Level 0.05        156        215        272
Level 0.10        114        165        215

, , Effect 0.3

           Power 0.80 Power 0.90 Power 0.95
Level 0.01        114        147        178
Level 0.05         70         97        122
Level 0.10         51         74         96

, , Effect 0.4

           Power 0.80 Power 0.90 Power 0.95
Level 0.01         65         84        101
Level 0.05         40         55         69
Level 0.10         29         42         54
```

```
, , Effect 0.5
            Power 0.80  Power 0.90  Power 0.95
Level 0.01          43          55          66
Level 0.05          26          36          45
Level 0.10          19          27          35

, , Effect 0.6
            Power 0.80  Power 0.90  Power 0.95
Level 0.01          31          39          47
Level 0.05          19          25          31
Level 0.10          13          19          25

, , Effect 0.7
            Power 0.80  Power 0.90  Power 0.95
Level 0.01          23          29          35
Level 0.05          14          19          24
Level 0.10          10          14          18

, , Effect 0.8
            Power 0.80  Power 0.90  Power 0.95
Level 0.01          18          23          27
Level 0.05          11          15          18
Level 0.10           8          11          14

, , Effect 0.9
            Power 0.80  Power 0.90  Power 0.95
Level 0.01          15          19          22
Level 0.05           9          12          15
Level 0.10           7           9          11
```

3.14 Comments

How do the sample sizes differ when knowing and not knowing the population standard deviation? At the outset, one notices that the test statistics are different: *Z*-statistic versus *t*-statistic. Sample sizes are a little higher when one uses the *t*-statistic. One is penalized for not knowing the population standard deviation and using its estimate.

3.15 One-Sample Problem: One-Sided Alternative: σ Is Known

Set $\alpha = 0.05$ and Power $= 0.80$

Δ	0.1	0.2	0.3	0.4	0.5	0.6	0.7	0.8	0.9
n	618	155	69	39	25	17	13	10	8

3.16 One-Sample Problem: One-Sided Alternative: σ Is Unknown

Set $\alpha = 0.05$ and Power $= 0.80$

Δ	0.1	0.2	0.3	0.4	0.5	0.6	0.7	0.8	0.9
n	620	156	70	40	26	19	14	11	9

The difference is at the most four observations under our list of scenarios.

The sample sizes for the other type of alternative $\mu < \mu_0$ are exactly the same.

Let us now focus on the two-sided alternative.

The null hypothesis is

H_0: $\mu = \mu_0$ (specified)

and the alternative

H_1: $\mu \neq \mu_0$ (two-sided alternative)

The population standard deviation σ is unknown.

The test statistic t remains the same. There is a modification to the test.

Test: Reject the null hypothesis if $|t| > t_{n-1,1-\alpha/2}$.

The area under the Student's t-curve with $n-1$ degrees of freedom to the left of the critical value $t_{n-1,1-\alpha/2}$ is $1-\alpha/2$.

What should be the sample size n? Let us focus on the same specifications.

Levels: 0.01, 0.05, 0.10
Power: 0.80, 0.90, 0.95
Effect size: 0.1 (0.1) 0.9

3.17 R Code

```
> Level <- c(0.01, 0.05, 0.10)
> Power <- c(0.80, 0.90, 0.95)
> Delta <- seq(0.1, 0.9, 0.1)
> SampleSizes <- rep(0, 81)
> dim(SampleSizes) <- c(3, 3, 9)
> for (i in 1:3)
+ {
+ for (j in 1:3)
+ {
+ for (k in 1:9)
+ {
+ SampleSizes[i, j, k] <- pwr.t.test(d = Delta[k],
+                                     sig.level = Level[i],
+                                     power = Power[j],
+                                     type = "one.sample",
+                                     alternative = "two.sided")$n
+ }
+ }
+ }
```

```
> dimnames(SampleSizes)[[1]] <- c("Level 0.01", "Level 0.05",
+                                 "Level 0.10")
> dimnames(SampleSizes)[[2]] <- c("Power 0.80", "Power 0.90",
+                                 "Power 0.95")
> dimnames(SampleSizes)[[3]] <- c("Effect 0.1", "Effect 0.2",
+                                 "Effect 0.3", "Effect 0.4",
+                                 "Effect 0.5", "Effect 0.6",
+                                 "Effect 0.7", "Effect 0.8",
+                                 "Effect 0.9")
> round(SampleSizes)
, , Effect 0.1

           Power 0.80 Power 0.90 Power 0.95
Level 0.01       1171       1491       1785
Level 0.05        787       1053       1301
Level 0.10        620        858       1084

, , Effect 0.2

           Power 0.80 Power 0.90 Power 0.95
Level 0.01        295        375        449
Level 0.05        198        265        327
Level 0.10        156        215        272

, , Effect 0.3

           Power 0.80 Power 0.90 Power 0.95
Level 0.01        133        169        201
Level 0.05         89        119        146
Level 0.10         70         97        122

, , Effect 0.4

           Power 0.80 Power 0.90 Power 0.95
Level 0.01         76         96        115
Level 0.05         51         68         83
Level 0.10         40         55         69

, , Effect 0.5

           Power 0.80 Power 0.90 Power 0.95
Level 0.01         50         63         75
Level 0.05         33         44         54
Level 0.10         26         36         45

, , Effect 0.6

           Power 0.80 Power 0.90 Power 0.95
Level 0.01         36         45         53
Level 0.05         24         31         38
Level 0.10         19         25         31
```

```
, , Effect 0.7

            Power 0.80  Power 0.90  Power 0.95
Level 0.01          27          34          40
Level 0.05          18          23          29
Level 0.10          14          19          24

, , Effect 0.8

            Power 0.80  Power 0.90  Power 0.95
Level 0.01          22          27          31
Level 0.05          14          18          22
Level 0.10          11          15          18

, , Effect 0.9

            Power 0.80  Power 0.90  Power 0.95
Level 0.01          18          22          25
Level 0.05          12          15          18
Level 0.10           9          12          15
```

It is instructive to compare sample sizes between the one-sided and two-sided alternatives.

So far, we have focused on sample size calculations in the environment of testing of hypotheses. Sample size can be computed in the environment of estimation.

3.18 One-Sample Problem

3.18.1 Estimation Environment

Setup: The observable entity X has a normal distribution with unknown mean μ and known standard deviation σ.

Potential data: $X_1, X_2, \ldots, X_n$

3.18.2 Error of Interval Estimation

Our focus is on the estimation of the unknown population mean μ. A 95% confidence interval is given by

$$\overline{X} - 1.96^{*}\frac{\sigma}{\sqrt{n}} \leq \mu \leq \overline{X} + 1.96^{*}\frac{\sigma}{\sqrt{n}}$$

The chances that the random interval covers the population mean are 95%.

$$\text{Margin of error (of interval estimation)} = \text{Length of the confidence intervals} = 2^{*}1.96^{*}\frac{\sigma}{\sqrt{n}}$$

We want to control the margin of error. Specify the margin of error d you are willing to tolerate. Set

$$d = 2*1.96*\frac{\sigma}{\sqrt{n}}$$

Solve for n

$$n = \left[\frac{2*1.96*\sigma}{d}\right]^2$$

3.19 Specifications

1. Confidence level (Here I have taken it to be 95%.)
2. Population standard deviation σ
3. Margin of error of interval estimation d

Some of the numerical examples are taken from Desu and Raghavarao (1990).

3.20 Example

The average miles per gallon (mpg) μ is under focus for a certain brand of cars. The standard deviation σ is 1 mpg. A 95% confidence interval is to be entertained. The margin of error of interval estimation is specified to be $d = 0.5$ mpg. What should be the sample size n?

$$n = \left[\frac{2*1.96*1}{0.5}\right]^2 = 61.4656 \sim 62$$

3.21 An Alternative Approach

3.21.1 Error of Estimation

A point estimate of the population mean μ is $\overline{X}$. Error of estimation is defined to be $|\overline{X} - \mu|$. One specifies how much d one is willing to tolerate the error with a specified probability 95%, say. Mathematically, this results in an equation.

$$\Pr\left(|\overline{X} - \mu| \leq d\right) = 0.95$$

The chances that sample mean is within d units of the population mean are 95%.

This gives a formula for n

$$n = \left[\frac{1.96*\sigma}{d}\right]^2$$

3.22 Example

The average miles per gallon (mpg) μ are under focus for a certain brand of cars. The standard deviation σ is 1 mpg. A 95% probability is to be entertained. The margin of error of estimation is specified to be $d=0.5$ mpg. What should be the sample size?

$$n=\left[\frac{1.96*1}{0.5}\right]^2=15.3664\sim 16$$

With this choice of n, we have $\Pr(\overline{X}-0.5\leq \mu \leq \overline{X}+0.5)=0.95$

The length of the interval is 1, as it should be.

With the choice of $n=62$,

$$\Pr\left(\overline{X}-1.96*\frac{\sigma}{\sqrt{n}}\leq \mu \leq \overline{X}+1.96*\frac{\sigma}{\sqrt{n}}\right)=0.95$$

$$=\Pr\left(\overline{X}-1.96*\frac{\sigma}{\sqrt{62}}\leq \mu \leq \overline{X}+1.96*\frac{\sigma}{\sqrt{62}}\right)$$

$$=\Pr(\overline{X}-0.25\leq \mu \leq \overline{X}+0.25)$$

The length of the interval is 0.5, as it should be.

No wonder we needed a bigger sample size. Note the difference between the margin of error and error of estimation.

What happens if the population standard deviation is unknown?

3.22.1 Error of Estimation

Stein's two-step procedure

3.22.2 Specificatioins

1. Error of estimation: d
2. Confidence level: $1-\alpha$

Goal: What should be the sample size n so that $\Pr(|\overline{X}-\mu|\leq d)=1-\alpha$.

Step 1: Take an initial sample of size m (>2). (Whatever your budget permits!) Let S^2 be the sample variance.

Step 2: Calculate $n=\max\left\{m,\left[\frac{S^2*t^2_{m-1,1-\alpha/2}}{d^2}\right]+1\right\}$. The symbol $[x]$ means the greatest integer $\leq x$.

Step 3: Take an additional sample of $n-m$ observations.

Step 4. Let $\overline{X}$ be the sample mean of all the n observations.

Stein (1945) has shown that the sample mean satisfies the equation

$$\Pr(|\overline{X}-\mu|\leq d)=1-\alpha$$

3.22.3 Illustration

Let us go back to the gas mileage problem. The population standard deviation σ is unknown.

The experimenter has taken a sample of $m=5$ observations. The sample data: 24.5, 26, 25, 24.9, and 24.6 mpg.

$S^2=0.355$

3.23 Specifications

$d=0.5$ mpg

$1-\alpha=0.95$

$t_{m-1,1-\alpha/2}=t_{4,0.975}=2.776$

Area under the Student's t-curve with four degrees of freedom to the left of 2.776 is 97.5%. The following R command gives the desired quantile:

```
> qt(0.975, 4)
[1] 2.776445
```

Calculate $\left[\frac{S^2 * t^2_{m-1,1-\alpha/2}}{d^2}\right] = \left[\frac{0.355*2.776^2}{0.5^2}\right] = [10.94] = 10$

$n=\max\{5, 11\}=11$

Conclusion: Take an additional sample of six observations.

How does one choose the initial sample size?

A discussion is presented in Moshman (1958) and Seelbinder (1953).

4 TWO-SAMPLE PROBLEMS: QUANTITATIVE RESPONSES

Setup:

Two independent populations are under focus. The same numerical characteristic of the populations is under scrutiny.

X = The value of the numerical characteristic of a random subject from Population 1. Assume $X \sim N(\mu_1, \sigma_1)$.

Y = The value of the numerical characteristic of a random subject from Population 2. Assume $Y \sim N(\mu_2, \sigma_2)$.

Null hypothesis: Hypothesis of skepticism: H_0: $\mu_1=\mu_2$

Alternative hypothesis: Hypothesis of optimism: H_1: $\mu_1>\mu_2$ (one-sided alternative)

Or

H_1: $\mu_1<\mu_2$ (one-sided alternative)

Or

H_1: $\mu_1 \neq \mu_2$ (two-sided alternative)

We want to draw a random sample of size m from Population 1 and of size n from Population 2. What should be the values of m and n?

4.1 Scenario 1

The standard deviations σ_1 and σ_2 are *known and equal.* Let the common value be σ. We will demand $m=n$. It can be shown that for fixed sum $m+n$, the power is maximum when $m=n$.

Look at the testing problem.

H_0: $\mu_1=\mu_2$ and H_1: $\mu_1>\mu_2$ (one-sided alternative)

4.2 Specifications

1. Level: α
2. Power: $1-\beta$
3. The absolute difference $|\mu_1-\mu_2|$ or the effect size $\Delta=\left|\frac{\mu_1-\mu_2}{\sigma}\right|$. Spell out the difference between the population means under the alternative hypothesis.

$$\text{Test statistic}: Z=\frac{\overline{X}-\overline{Y}}{\sqrt{\frac{2}{n}}*\sigma}$$

Test: Reject the null hypothesis if $Z>z_{1-\alpha}$.

Sample size per group

$$m=n=\frac{2*\left(z_{1-\alpha}+z_{1-\beta}\right)^2}{\Delta^2}$$

Compare the formula in the single-sample case for the one-sided alternative. The same formula works for the alternative H_1: $\mu_1<\mu_2$.

Look at the following testing problem.

H_0: $\mu_1=\mu_2$ versus H_1: $\mu_1\neq\mu_2$ (two-sided alternative)

The specifications are the same. The test statistic is the same. The test is slightly different and incorporates the two-sided nature of the alternative.

Test: $|Z|>z_{1-\alpha/2}$

The sample size formula is given by

$$m=n=\frac{2*\left(z_{1-\alpha/2}+z_{1-\beta}\right)^2}{\Delta^2}$$

Some reflections

1. When it comes to two populations, for the same specifications, sample size required per group is double of that of the single-sample case. Not only

one needs to take care of variation present in one population but also variation present in the other population.

2. More effort and more money are needed when comparing the means of two populations.

4.3 Scenario 2

The standard deviations σ_1 and σ_2 are *known but unequal.*

4.4 One-Sided Alternative

H_0: $\mu_1 = \mu_2$
H_1: $\mu_1 > \mu_2$ (one-sided alternative)

4.5 Specifications

1. Level: α
2. Power: $1-\beta$
3. The absolute difference $|\mu_1 - \mu_2|$. The difference between the population means under the alternative hypothesis.

$$\text{Test statistic: } Z = \frac{\overline{X} - \overline{Y}}{\sqrt{\frac{\sigma_1^2}{m} + \frac{\sigma_2^2}{n}}}$$

Test: Reject the null hypothesis if $Z > z_{1-\alpha}$.

Equal sample sizes are not recommended. Take a larger sample from the population which has more variation. One choice: let k be the ratio of the variances, i.e., $k = \frac{\sigma_1^2}{\sigma_2^2}$.

Formula for n, the sample size for Population 2

$$n = \frac{2^* \sigma_2^2}{\left(\frac{\mu_1 - \mu_2}{z_{1-\alpha} + z_{1-\beta}}\right)^2}$$

$$m = k^* n$$

One can choose the number k any way one wants. It does not have to be the ratio of the variances. For one's own choice of k, the formulas are:

$$n = \frac{\frac{\sigma_1^2}{k} + \sigma_2^2}{\left(\frac{\mu_1 - \mu_2}{z_{1-\alpha} + z_{1-\beta}}\right)^2}$$

$$m = k^* n$$

4.6 Scenario 3

The standard deviations σ_1 and σ_2 are un*known but equal*. Let the common value be σ, which is unknown.

This is more challenging and interesting.

4.7 One-Sided Alternative

Look at the testing problem.

H_0: $\mu_1 = \mu_2$ versus H_1: $\mu_1 > \mu_2$ (one-sided alternative)

4.8 Specifications

1. Level: α
2. Power: $1-\beta$
3. Effect size $\Delta = \left|\dfrac{\mu_1 - \mu_2}{\sigma}\right|$.

$$\text{Test statistic: } t = \frac{\overline{X} - \overline{Y}}{\sqrt{\left(\frac{1}{m} + \frac{1}{n}\right)} * S}$$

S is the pooled standard deviation of the data.

Test: Reject the null hypothesis if $t > t_{m+n-2,\ 1-\alpha}$.

Sample sizes $m = n$ are recommended.

Common sample size is hard to compute. The integral equation, which involves a noncentral t-distribution, is hard to solve by hand. We need software. We will use the "pwr" package.

4.9 Illustration

```
α = 0.05
Power = 0.90
Effect size = Δ = 0.5
> pwr.t.test(d = 0.4, sig.level = 0.05, power = 0.90, type = "two.
+             sample", alternative = "greater")
    Two-sample t test power calculation
        n = 107.7313
        d = 0.4
    sig.level = 0.05
      power = 0.9
    alternative = greater
```

Note: n is number in *each* group

One can take the required common sample size to be 108. The package "pwr" can be used to calculate the required common sample sizes under the one-sided alternative for various combination of specifications. It is not done here.

It will be instructive to understand the difference in knowing and not knowing the common standard deviation σ. If the common standard deviation σ is known, the common sample size is given explicitly by, for the same specifications,

$$m = n = \frac{2^*\left(z_{1-\alpha} + z_{1-\beta}\right)^2}{\Delta^2} = 107$$

4.10 Two-Sided Alternative

Look at the following testing problem.

H_0: $\mu_1 = \mu_2$ versus H_1: $\mu_1 \neq \mu_2$ (two-sided alternative)

4.11 Specifications

1. Level: α
2. Power: $1-\beta$
3. Effect size $\Delta = \left|\frac{\mu_1 - \mu_2}{\sigma}\right|$.

$$\text{Test statistic: } t = \frac{\overline{X} - \overline{Y}}{\sqrt{\left(\frac{1}{m} + \frac{1}{n}\right)}^{*}S}$$

S is the pooled standard deviation of the data.

Test: Reject the null hypothesis if $|t| > t_{m+n-2,1-\alpha/2}$.

Sample sizes $m = n$ are recommended. Common sample size is hard to compute. Again, the integral equation, which involves a noncentral t-distribution, is hard to solve by hand. We need software. We will use the "pwr" package.

4.12 An Illustration

```
Level = 0.05
Power = 0.90
Effect Size = 0.4
> pwr.t.test(d = 0.4, sig.level = 0.05, power = 0.90,
+             type = "two.sample", alternative = "two.sided")
   Two-sample t test power calculation
         n = 132.3105
         d = 0.4
   sig.level = 0.05
       power = 0.9
  alternative = two.sided
```

Note: n is number in *each* group

Sample size per group = 133

There is another command, "power.t.test," available in the "Base." Look at its documentation.

```
?power.t.test
> power.t.test(delta = 0.4, sig.level = 0.05, power = 0.90,
+                type = "two.sample", alternative = "two.sided")
    Two-sample t test power calculation
          n = 132.3106
      delta = 0.4
         sd = 1
  sig.level = 0.05
      power = 0.9
alternative = two.sided
```

Note: n is number in *each* group

The commands pwr.t.test from the package "pwr" and power.t.test from the "Base" produce the same output. Let us use R to compute sample sizes in the case of a two-sample problem under a variety of specifications.

Create three folders for each of the specifications "Level," "Power," and "Effect Size."

```
> Level <- c(0.01, 0.05, 0.10)
> Power <- c(0.80, 0.90, 0.95)
> Delta <- seq(0.1, 0.9, 0.1)
> SampleSizes <- rep(0, 81)
> dim(SampleSizes) <- c(3, 3, 9)
Initiate a loop.
> for (i in 1:3)
+ {
+ for (j in 1:3)
+ {
+ for (k in 1:9)
+ {
+ SampleSizes[i, j, k] <- pwr.t.test(d = Delta[k],
+                                    sig.level = Level[i],
+                                    power = Power[j],
+                                    type = "two.sample",
+                                    alternative = "two.sided")$n
+ }
+ }
+ }
> dimnames(SampleSizes)[[1]] <- c("Level 0.01", "Level 0.05",
+                                 "Level 0.10")
```

```
> dimnames(SampleSizes)[[2]] <- c("Power 0.80", "Power 0.90",
+                                 "Power 0.95")
> dimnames(SampleSizes)[[3]] <- c("Effect 0.1", "Effect 0.2",
+                                 "Effect 0.3", "Effect 0.4",
+                                 "Effect 0.5", "Effect 0.6",
+                                 "Effect 0.7", "Effect 0.8",
+                                 "Effect 0.9")
> round(SampleSizes)
, , Effect 0.1
            Power 0.80  Power 0.90  Power 0.95
Level 0.01        2337        2978        3564
Level 0.05        1571        2102        2600
Level 0.10        1237        1713        2165

, , Effect 0.2
            Power 0.80  Power 0.90  Power 0.95
Level 0.01         586         746         892
Level 0.05         393         526         651
Level 0.10         310         429         542

, , Effect 0.3
            Power 0.80  Power 0.90  Power 0.95
Level 0.01         261         332         398
Level 0.05         175         234         290
Level 0.10         138         191         241

, , Effect 0.4
            Power 0.80  Power 0.90  Power 0.95
Level 0.01         148         188         224
Level 0.05          99         132         163
Level 0.10          78         108         136

, , Effect 0.5
            Power 0.80  Power 0.90  Power 0.95
Level 0.01          95         121         144
Level 0.05          64          85         105
Level 0.10          50          69          87

, , Effect 0.6
            Power 0.80  Power 0.90  Power 0.95
Level 0.01          67          84         101
Level 0.05          45          59          73
Level 0.10          35          48          61

, , Effect 0.7
            Power 0.80  Power 0.90  Power 0.95
Level 0.01          49          62          74
Level 0.05          33          44          54
Level 0.10          26          36          45
```

```
, , Effect 0.8
           Power 0.80  Power 0.90  Power 0.95
Level 0.01         38          48          57
Level 0.05         26          34          42
Level 0.10         20          27          35

, , Effect 0.9
           Power 0.80  Power 0.90  Power 0.95
Level 0.01         31          38          46
Level 0.05         20          27          33
Level 0.10         16          22          27
```

Let us compare sample sizes vis-à-vis one-sample and two-sample cases:

One-Sample Case	Two-Sample Case
H_0: $\mu = \mu_0$	H_0: $\mu_1 = \mu_2$
H_1: $\mu \neq \mu_0$	H_1: $\mu_1 \neq \mu_2$

Common specifications:
Level = 0.05
Power = 0.90
Effect size = 0.4
$n = 68$, common sample size $= n = 132$

4.13 Scenario 4

The standard deviations σ_1 and σ_2 are un*known but possibly unequal.*

Welch test is used to test the equality of population means. No satisfactory treatment is available in the environment of the R software to calculate the required sample sizes.

So far we have focused on sample size calculations in the context of testing. We can calculate sample size from the estimation perspective.

4.14 Estimation Perspective

Absolute estimation error: $\left|\left(\overline{X} - \overline{Y}\right) - (\mu_1 - \mu_2)\right|$

4.15 Scenario 1

The population standard deviations σ_1 and σ_2 are known.

4.16 Specifications

1. Confidence level: $1 - \alpha$
2. Tolerable maximum error: d

Goal: Find the sample sizes m and n so that

$$\Pr\left(\left|\left(\overline{X}-\overline{Y}\right)-(\mu_1-\mu_2)\right|\le d\right)=1-\alpha$$

The chances that the difference in the sample means is within d units of the difference between the population means are $1-\alpha$.

The above equation cannot be solved. This is a single equation in two unknowns m and n. Choose a constant k and take $m=k*n$. Formulas

$$n=\left[\left(\frac{\sigma_1^2}{k}+\sigma_2^2\right)z_{1-\frac{\alpha}{2}}^2\right]/d^2$$

$$m=k*n$$

4.17 Example

Goal: Estimate the difference in mean pulse rates for men and women within one beat with probability at least 0.95. The common standard deviation σ is 2 beats.

Here, $d=1$ and $1-\alpha=0.95$.
Take $k=1$.
$m=n=31$

4.18 Scenario 2

The population standard deviations σ_1 and σ_2 are equal but unknown.

4.19 Specifications

1. Confidence level: $1-\alpha$
2. Tolerable maximum error: d

Goal: Find the sample sizes m and n so that

$$\Pr\left(\left|\left(\overline{X}-\overline{Y}\right)-(\mu_1-\mu_2)\right|\le d\right)=1-\alpha$$

The chances that the difference in the sample means is within d units of the difference between the populations' means are $1-\alpha$.

We take $m=n$.

No formula is available. A two-stage procedure has been worked out.

Step 1: Take initial samples of the same size s. Calculate the pooled variance S^2 of both the samples. Here, the pooled variance is the average of the individual variances.

Step 2: Calculate $c=\frac{1}{2}\left(\frac{d}{t_{2s-2,1-\alpha/2}}\right)^2$

Step 3: Calculate $n=\max\left\{s,\left[\frac{S^2}{c}\right]+1\right\}$

Step 4: Take additional samples of size $n - s$, if necessary, from each of the populations.

For mathematical details, see Koopmans and Qualls (1971).

4.20 Example

Goal: Estimate the difference in mean pulse rates for men and women within one beat with probability at least 0.95. The common standard deviation σ is unknown.

Here, $d = 1$ and $1 - \alpha = 0.95$.

A sample of five subjects from the men's and women's groups yielded the following data:

Men: 81, 79, 78, 82, 80
Women: 77, 76, 78, 79, 77

Pooled variance $= S^2 = 1.9$

```
> qt(0.975, 8)
[1] 2.306004
```

$$c = \frac{1}{2}\left(\frac{1}{2.306}\right)^2 = 0.094$$

$$n = \max\left\{5, \left[\frac{1.9}{0.094}\right] + 1\right\} = 21$$

An additional sample of 16 subjects from each group would do.

4.21 Scenario 3

The population standard deviations σ_1 and σ_2 are unequal but unknown.

Koopmans and Qualls (1971) discuss this scenario taking into account the cost function.

4.22 Paired *t*-Test

4.22.1 Setup

A subject in the population gives a pair (X, Y) of measurements on a numerical characteristic. Typically,

$X =$ Measurement before the treatment begins
$Y =$ Measurement after the treatment

Here X and Y are correlated. (In the traditional two-sample t-test, X and Y are independent.)

Assume X and Y have a bivariate normal distribution with means μ_1 and μ_2, standard deviations σ_1 and σ_2, respectively, and correlation coefficient

ρ. Let σ be the standard deviation of the difference $D = X - Y$. The hypothesis of skepticism is:

H_0: $\mu_1 = \mu_2$

The alternative is whatever you want: one-sided or two-sided.

4.22.2 Structure of the Data

$(X_1, Y_1), (X_2, Y_2), \ldots, (X_n, Y_n)$

The data come in pairs. What is n?

Let $D_i = X_i - Y_i$. Let $\overline{D}$ be the mean and S be the standard deviation of the differences.

Test statistic: $t = \dfrac{\overline{D}}{\sqrt{\dfrac{S^2}{n}}}$

If the alternative is

H_1: $\mu_1 \neq \mu_2$

The test is given by

Reject the null hypothesis if $|t| > t_{n-1,1-\alpha/2}$.

4.23 Specifications

1. Level: α
2. Power: $1 - \beta$
3. Effect size: $\Delta = \left|\frac{\mu_1 - \mu_2}{\sigma}\right|$

Since we are dealing with the difference D, it looks as though we are in the environment of a single-sample problem. Yes, it is indeed true. Let us look at an example.

```
Level = 0.05
Power = 0.80
Effect Size = 0.5
> pwr.t.test(d = 0.5, sig.level = 0.05, power = 0.80,
+                type = "paired", alternative = "two.sided")
   Paired t test power calculation
            n = 33.36713
            d = 0.5
   sig.level = 0.05
       power = 0.8
  alternative = two.sided
```

Note: n is number of *pairs*

```
> pwr.t.test(d = 0.5, sig.level = 0.05, power = 0.80,
+                type = "one.sample", alternative = "two.sided")
```

```
     One-sample t test power calculation
              n = 33.36713
              d = 0.5
      sig.level = 0.05
          power = 0.8
    alternative = two.sided
```

The important point is that in the context of paired data one could use either the two-sample paired t-test or one-sample test on the differences. The output is exactly the same.

5 MULTISAMPLE PROBLEM—QUANTITATIVE RESPONSES—ANALYSIS OF VARIANCE

Suppose we are testing the equality of k population means.

H_0: $\mu_1 = \mu_2 = \cdots = \mu_k$ versus H_1: H_0 not true.

Assumptions: All the k populations each is normally distributed with common standard deviation.

5.1 Specifications

For sample size calculations, specify level, power, alternative population means, and within population standard deviation σ. The between population variance is defined by

Between populations variance $\sigma_{\text{pop}}^2 = \frac{\sum_{i=1}^{k} (\mu_i - \overline{\mu})^2}{k}$, where $\overline{\mu} = \left(\frac{1}{k}\right) \sum_{i=1}^{k} \mu_i$ = average of the population means.

In the context of multiple populations, the effect size is defined by

$$\text{Effect size} = \Delta = \frac{\sigma_{\text{pop}}}{\sigma} = \text{bet.pop.SD/within.pop.SD}$$

If the number of populations is $k=2$, the effect size works out to be $\frac{|\mu_1 - \mu_2|}{2^* \sigma}$, which is different from $\frac{|\mu_1 - \mu_2|}{\sigma}$.

In engineering terminology, σ is the noise present in the populations and σ_{pop} is a measure of the signal. There are so many signals associated with different populations we need to measure how close they are. The standard deviation σ_{pop} of these signals is a measure how close they are. If they are very close, one needs a large sample to detect the differences. If the means are very close, the alternative and null hypotheses are very close. Let us find out how the effect size shapes in some examples.

5.2 Examples

No. of populations = 4

Within population standard deviation = σ = 1

Alternative Means				σ_{pop}	Effect Size
1	1	1	1.1	0.001875	0.001875
1	1	1	2	0.1875	0.1875
1	1	2	2	0.25	0.25
1	2	3	4	1.25	1.25

Look at Scenario 1 of the alternative means.

$$H_0: \mu_1 = \mu_2 = \mu_3 = \mu_4 \text{ versus } H_1: \mu_1 = 1,\ \mu_2 = 1,\ \mu_3 = 1,\ \mu_4 = 1.1.$$

H_0 and H_1 are almost similar. If we want to detect H_1 with a high level of probability, we will need a very, very large sample from each population.

5.3 Structure of the Data

We intend to draw samples from the populations with the same sample size n. We are investigating what the common sample size n is.

Potential samples with their means

Pop 1	Pop 2	...	Pop k
X_{11}	X_{21}	...	X_{k1}
X_{12}	X_{22}	...	X_{k2}
...	...	...	...
X_{1n}	X_{2n}	...	X_{kn}
$\overline{X_1}$	$\overline{X_2}$	...	$\overline{X_k}$

Let $\overline{X}$ be the grand mean.

$$\text{Test statistic}: F = \frac{\dfrac{\sum_{i=1}^{k}\left(\overline{X}_i - \overline{X}\right)^2}{k-1}}{\dfrac{\sum_{i=1}^{k}\sum_{j=1}^{n}\left(X_{ij} - \overline{X}_i\right)^2}{k(n-1)}}$$

Under the null hypothesis, the distribution of F statistic is F distribution with numerator degrees of freedom $k-1$ and denominator degrees of freedom $k(n-1)$.

Under the alternative hypothesis (with the specified means), the distribution of the F statistic is a noncentral F distribution with numerator degrees of freedom $k-1$ and denominator degrees of freedom $k(n-1)$ and noncentrality parameter Δ, the effect size. There is no need to specify the population means under the alternative hypothesis. It is enough to specify the effect size Δ. We can use the command "power.anova.test" from the "Base" or "pwr. anova.test" under the umbrella package of "pwr." The "power.anova.test" requires the specification of "Between Population Variance" and "Within Population Variance."

5.4 Specifications

```
Level = 0.05
Power = 0.80
No. of populations = 3
within.var = 1
between.var = 0.04
Effect size = sqrt(between.var)/sqrt(within.var) = 0.2
Use R. Let us try two scenarios.
> power.anova.test(groups = 3, between.var = 0.04, within.var = 1,
+                    sig.level = 0.05, power = 0.80)
   Balanced one-way analysis of variance power calculation
      groups = 3
           n = 121.4378
 between.var = 0.04
  within.var = 1
   sig.level = 0.05
       power = 0.8
```

Note: *n* is number in each group

5.5 Specifications

```
Level = 0.05
Power = 0.80
No. of populations = 3
within.var = 4
between.var = 0.16
Effect size = sqrt(between.var)/sqrt(within.var) = 0.2
> power.anova.test(groups = 3, between.var = 0.16, within.var = 4,
+                    sig.level= 0.05, power = 0.80)
  Balanced one-way analysis of variance power calculation
      groups = 3
           n = 121.4378
 between.var = 0.16
  within.var = 4
   sig.level = 0.05
       power = 0.8
```

Note: *n* is number in each group

Even though the specifications of "Between Population Variance" and "Within Population Variance" are different between the two scenarios, we get the same sample size per group. Why? The reason is that the effect size is the same.

Specification of alternative means is hard to come by for an experimenter. It is enough to specify the effect size. The following guidelines were set by Cohen (1988)

5.6 Some Guidelines from the Social Sciences and Psychology

Small effect size = 0.10
Medium effect size = 0.25
Large effect size = 0.40

The command "pwr.anova.test" requires the following specifications:

k = no. of populations; f = effect size; sig. level = α; and power

Let us get an array of required sample sizes under a number of configurations.

```
< Level <- c(0.01, 0.05, 0.10)
< Power <- c(0.80, 0.90, 0.95)
< Delta <- seq(0.1, 0.9, 0.1)
< Sizes <- rep(0, 81)
< dim(Sizes) <- c(3, 3, 9)
```

Populations: 3

```
Let us use R to get the sample sizes.
> for (i in 1:3)
+ {
+ for (j in 1:3)
+ {
+ for (k in 1:9)
+ {
+ Sizes[i, j, k] <- pwr.anova.test(k = 3, f = Delta[k], sig.level =
+                                   Level[i], power = Power[j])$n
+ }
+ }
+ }
>dimnames(Sizes)[[1]]<-c("Level 0.01", "Level 0.05", "Level 0.10")
>dimnames(Sizes)[[2]]<-c("Power 0.80", "Power 0.90", "Power 0.95")
> dimnames(Sizes)[[3]] <- c("Effect 0.1", "Effect 0.2",
+                           "Effect 0.3", "Effect 0.4",
+                           "Effect 0.5", "Effect 0.6",
+                           "Effect 0.7", "Effect 0.8",
+                           "Effect 0.9")
> round(Sizes)
```

```
, , Effect 0.1
           Power 0.80  Power 0.90  Power 0.95
Level 0.01        464         582         690
Level 0.05        322         423         516
Level 0.10        258         349         435

, , Effect 0.2
           Power 0.80  Power 0.90  Power 0.95
Level 0.01        117         147         174
Level 0.05         81         106         130
Level 0.10         65          88         109

, , Effect 0.3
           Power 0.80  Power 0.90  Power 0.95
Level 0.01         53          66          78
Level 0.05         37          48          58
Level 0.10         29          40          49

, , Effect 0.4
           Power 0.80  Power 0.90  Power 0.95
Level 0.01         30          38          45
Level 0.05         21          27          33
Level 0.10         17          23          28

, , Effect 0.5
           Power 0.80  Power 0.90  Power 0.95
Level 0.01         20          25          29
Level 0.05         14          18          22
Level 0.10         11          15          18

, , Effect 0.6
           Power 0.80  Power 0.90  Power 0.95
Level 0.01         14          18          21
Level 0.05         10          13          15
Level 0.10          8          11          13

, , Effect 0.7
           Power 0.80  Power 0.90  Power 0.95
Level 0.01         11          13          16
Level 0.05          8          10          12
Level 0.10          6           8          10

, , Effect 0.8
           Power 0.80  Power 0.90  Power 0.95
Level 0.01          9          11          12
Level 0.05          6           8           9
Level 0.10          5           6           8
```

```
, , Effect 0.9

           Power 0.80  Power 0.90  Power 0.95
Level 0.01          7           9          10
Level 0.05          5           6           7
Level 0.10          4           5           6
```

5.7 Comments

1. For fixed level and power, sample size decreases as the effect size increases.
2. For fixed level and effect size, sample size increases with increasing power.
3. For fixed power and effect size, sample size decreases with increasing level.

It is instructive to compare sample sizes for different numbers of populations for the same level, power, and effect size.

REFERENCES

Chow, S.-C., Shao, J., Wang, H., 2003. Sample Size Calculations in Clinical Research. Marcel-Dekker, New York.

Cohen, J., 1988. Statistical Power Analysis for the Behavioral Sciences, second ed. Lawrence Erlbaum Associates, Inc., Hillsdale, NJ.

Desu, M.M., Raghavarao, D., 1990. Sample Size Methodology. Academic Press, Inc., New York.

Koopmans, L., Qualls, C., 1971. Fixed length confidence intervals for parameters of the normal distribution based on two-stage sampling procedures. Rocky Mountain J. Math. 1, 587–602.

Machin, D., Campbell, M.J., Tan, S.B., Tan, S.H., 2009. Sample Size Tables for Clinical Studies, third ed. Wiley-Blackwell, New York.

Moshman, J., 1958. A method for selecting the size of the initial sample size in the Stein's two-sample procedure. Ann. Math. Stat. 29, 1271–1275.

Ryan, T.P., 2013. Sample Size Determination and Power. Wiley, New York.

Seelbinder, B.M., 1953. On Stein's two stage procedure. Ann. Math. Stat. 24, 640–649.

Stein, C., 1945. A two sample test for linear hypothesis whose power is independent of variance. Ann. Math. Stat. 16, 243–258.

Odeh, R.E., Fox, M., 1975. Sample Size Choice: Charts for Experiments with Linear Models. Marcel Dekker, Inc., New York.

Chapter 6

Sample Size Calculations with R: Level 2

Marepalli B. Rao[1] and Hansen Bannerman-Thompson[1]
Department of Environmental Health, University of Cincinnati, Cincinnati, Ohio, USA
[1]*Corresponding authors: e-mail: Marepalli.rao@uc.edu; Hansen.bthompson@gmail.com*

ABSTRACT

In Level 1, sample size calculations were arrayed in the environment of single-sample, two-sample, and multisample framework dealing with quantitative variables. In this chapter, we will focus on qualitative variables. In addition, correlation and survival analysis framework will be entertained.

Keywords: Single proportions, Two proportions, Multiple proportions, McNemar test, Hazard ratio, Survival analysis, R-squared in multiple regression, Effect size, Correlation, Arc Sine transformation, Fisher's z-transformation, Cox model, Chi-squared test

1 SINGLE PROPORTIONS

1.1 Problem

A drug is under scrutiny. If it is given to a patient, there are only two possible outcomes. Either the patient gets better (1 = Success) or fails to get better (0 = Failure). Let p stand as a generic symbol for the probability of success, which is unknown. We entertain the null hypothesis:

H_0: $p = p_0$ (specified).

The alternatives are of three kinds:

H_1: $p < p_0$ (one-sided alternative)
H_1: $p > p_0$ (one-sided alternative)
H_1: $p \neq p_0$ (two-sided alternative)

We need data to test H_0 versus H_1. A random sample of n patients is taken. Each patient is given the drug. Let X be the number of patients who got better on the drug. Note that X is our data. The possible values of X are 0, 1, 2, ..., n. The entity X is an obvious statistic to build a test on. An estimate of p is given by

Handbook of Statistics, Vol. 32. http://dx.doi.org/10.1016/B978-0-444-63431-3.00006-1

$$\hat{p} = \frac{X}{n}$$

which is the sample proportion of patients who get better.

For testing H_0 versus H_1: $p < p_0$, reject H_0 in favor of H_1 if

$$\frac{\frac{X}{n} - p_0}{\sqrt{\frac{p_0(1-p_0)}{n}}} < z_\alpha, \tag{1}$$

where z_α is such that area under the standard normal curve to the left of z_α is α and α is the chosen level of significance.

For testing H_0 versus H_1: $p > p_0$, reject H_0 in favor of H_1 if

$$\frac{\frac{X}{n} - p_0}{\sqrt{\frac{p_0(1-p_0)}{n}}} > z_{1-\alpha}, \tag{2}$$

where $z_{1-\alpha}$ is such that area under the standard normal curve to the left of $z_{1-\alpha}$ is $1-\alpha$ and α is the chosen level of significance.

For testing H_0 versus H_1: $p \neq p_0$, reject H_0 in favor of H_1 if

$$\left| \frac{\frac{X}{n} - p_0}{\sqrt{\frac{p_0(1-p_0)}{n}}} \right| > z_{1-\frac{\alpha}{2}} \tag{3}$$

where $z_{1-\alpha/2}$ is such that area under the standard normal curve to the left of $z_{1-\alpha/2}$ is $1-\alpha/2$ and α is the chosen level of significance.

All these tests are large-sample tests. Normal approximation is employed to calculate the critical values. The normal approximation works very well if the null proportion is in a close neighborhood of 0.5. If not, the arcsine square root transformation of the statistic X/n is recommended. Technically, calculate $\sin^{-1}(\text{sqrt}(X/n))$. Its large-sample standard deviation is given by $\text{sqrt}(1/4n)$. If p is the true proportion and n is large, the statistic

$$\sqrt{n}\left(2\sin^{-1}\left(\sqrt{\frac{X}{n}}\right) - 2\sin^{-1}(\sqrt{p})\right)$$

has a standard normal distribution approximately. One can build a test based on this statistic. For example, for testing H_0: $p = p_0$ versus H_1: $p < p_0$, reject the null hypothesis at the level of significance α if

$$\sqrt{n}\left(2\sin^{-1}\left(\sqrt{\frac{X}{n}}\right) - 2\sin^{-1}(\sqrt{p_0}) < z_\alpha\right) \tag{4}$$

For the other types of alternatives, the tests are appropriately modified.

For power calculations, we need to spell out the alternative value of p. Let $p_1 < p_0$ be the alternative value of interest. Let $h = 2\sin^{-1}(\sqrt{p_0}) - 2\sin^{-1}(\sqrt{p_1})$. If we are using the arc sine square root transformation for testing, we can call $|h|$ as the effect size. The following code and table gives the value of $|h|$ for a number of choices of p_0 and p_1. We create a loop for the calculation of the effect sizes h.

Create a column vector of proportions.

```
> Prop <- seq(0.1, 0.9, 0.1)
> Prop
[1] 0.1 0.2 0.3 0.4 0.5 0.6 0.7 0.8 0.9
```

Create a matrix of order 9×9 with each entry zero. We will replace the entries by the effect sizes $|h|$.

```
> Effect <- matrix(0, nrow = 9, ncol = 9)
```

Create an index set, which we will use in the loop.

```
> Index <- c(1:9)
> Index
[1] 1 2 3 4 5 6 7 8 9
```

Create a loop. We will round the effect sizes to three decimal places.

```
> for (i in Index)
+ {
+ for (j in Index)
+ {
+ Effect[i, j] <- round(abs(2*asin(sqrt(Prop[i]))
                          - 2*asin(sqrt(Prop[j]))), 3)
+ }
+ }
```

Label the rows and columns with the relevant proportions.

```
> rownames(Effect) <- c("0.1", "0.2", "0.3", "0.4", "0.5", "0.6",
+                       "0.7", "0.8", "0.9")
> colnames(Effect) <- c("0.1", "0.2", "0.3", "0.4", "0.5", "0.6",
+                       "0.7", "0.8", "0.9")
```

The following table calculates the effect size $h = 2\sin^{-1}(\sqrt{p_0}) - 2\sin^{-1}(\sqrt{p_1})$ in absolute terms. The columns designate various choices of p_0 and rows p_1.

```
> Effect
      0.1   0.2   0.3   0.4   0.5   0.6   0.7   0.8   0.9
0.1 0.000 0.284 0.516 0.726 0.927 1.129 1.339 1.571 1.855
0.2 0.284 0.000 0.232 0.442 0.644 0.845 1.055 1.287 1.571
0.3 0.516 0.232 0.000 0.210 0.412 0.613 0.823 1.055 1.339
0.4 0.726 0.442 0.210 0.000 0.201 0.403 0.613 0.845 1.129
0.5 0.927 0.644 0.412 0.201 0.000 0.201 0.412 0.644 0.927
0.6 1.129 0.845 0.613 0.403 0.201 0.000 0.210 0.442 0.726
0.7 1.339 1.055 0.823 0.613 0.412 0.210 0.000 0.232 0.516
0.8 1.571 1.287 1.055 0.845 0.644 0.442 0.232 0.000 0.284
0.9 1.855 1.571 1.339 1.129 0.927 0.726 0.516 0.284 0.000
```

Some comments are in order.

1. If the absolute difference $|p_1 - p_0| = 0.1$, the effect size varies from 0.201 to 0.284.
2. If the absolute difference $|p_1 - p_0| = 0.2$, the effect size varies from 0.403 to 0.516.
3. If the absolute difference $|p_1 - p_0| = 0.3$, the effect size varies from 0.613 to 0.726.
4. If the absolute difference $|p_1 - p_0| = 0.4$, the effect size varies from 0.823 to 0.927.
5. In sample size calculations using the arc sine transformation, as we shall see later, one could either spell out what p_1 and p_0 are or the effect size. It is instructive to see the connection between the proportions and effect size.

Formulas for sample size calculations with level set at α and power at $1-\beta$

Alternative	Alternative Value	Test	Sample Size
$H_1: p < p_0$	$p_1 < p_0$	(1)	$n = \left[\frac{z_\alpha\sqrt{p_0(1-p_0)} + z_\beta\sqrt{p_1(1-p_1)}}{p_1 - p_0}\right]^2$
$H_1: p < p_0$	$p_1 < p_0$	(4)	$n = \left[\frac{z_\alpha + z_\beta}{h}\right]^2$
$H_1: p > p_0$	$p_1 > p_0$	(1)	$n = \left[\frac{z_\alpha\sqrt{p_0(1-p_0)} + z_\beta\sqrt{p_1(1-p_1)}}{p_1 - p_0}\right]^2$
$H_1: p < p_0$	$p_1 > p_0$	(4)	$n = \left[\frac{z_\alpha + z_\beta}{h}\right]^2$
$H_1: p \neq p_0$	$p_1 \neq p_0$	(1)	$n = \left[\frac{z_{\alpha/2}\sqrt{p_0(1-p_0)} + z_\beta\sqrt{p_1(1-p_1)}}{p_1 - p_0}\right]^2$
$H_1: p \neq p_0$	$p_1 \neq p_0$	(4)	$n = \left[\frac{z_{\alpha/2} + z_\beta}{h}\right]^2$

Note that the sample size formula is the same for one-sided alternatives. For the two-sided alternative, z_α is replaced by $z_{\alpha/2}$. The sample size depends on the test statistic used. Normal approximation of the test statistic used in Eqs. (1), (2), or (3) is very good if p_1 and p_0 are close to 0.5. Otherwise, use of the test statistic in Eq. (4) is recommended. For further discussion, see Ryan (2013). We will calculate sample sizes using the loop command in R.

The levels of significance are chosen to be 1%, 5%, and 10%.

```
> Size
[1] 0.01 0.05 0.10
```

The critical values z_αs are computed.

```
> Critical <- qnorm(Size, lower.tail = T)
> Critical
[1] -2.326348 -1.644854 -1.281552
```

The powers are set at 80%, 90%, and 95%.

```
> Power <- c(0.80, 0.90, 0.95)
> Power
[1] 0.80 0.90 0.95
```

The quantiles z_βs are computed.

```
> Critical1 <- qnorm(1 - Power, lower.tail = T)
> Critical1
[1] -0.8416212 -1.2815516 -1.6448536
```

Create a 9×9 matrix with each entry zero.

```
> Sample <- matrix(0, nrow = 9, ncol = 9)
```

The proportions are chosen.

```
> Prop <- seq(0.1, 0.9, 0.1)
```

Set up a loop command to calculate sample sizes for H_0: $p=p_0$ versus H_1: $p=p_1$

```
> p0 and also for H0: p = p0 versus H1: p = p1 < p0 with level at 1% and
power 80%.
> for (i in 1:9)
+ {
+ for (j in 1:9)
+ {
+ Sample[i, j] <- round(((Critical[1]*sqrt(Prop[i]*(1-Prop[i])) +
Critical1[1]*
+ sqrt(Prop[j]*(1 - Prop[j])))^2)/(Prop[i] - Prop[j])^2)
+ }
+ }
> rownames(Sample) <- c("0.1", "0.2", "0.3", "0.4", "0.5", "0.6",
+                        "0.7", "0.8", "0.9")
> colnames(Sample) <- c("0.1", "0.2", "0.3", "0.4", "0.5", "0.6",
+                        "0.7", "0.8", "0.9")
```

In the following table, the rows are the null values p_0 and the columns alternative values p_1. The numbers in the lower triangle part of the matrix correspond to H_0: $p=p_0$ versus H_1: $p=p_1<p_0$. The numbers in the upper triangle part of the matrix correspond to H_0: $p=p_0$ versus H_1: $p=p_1>p_0$.

Level = 1% and Power = 80%

> Sample

	0.1	0.2	0.3	0.4	0.5	0.6	0.7	0.8	0.9
0.1	Inf	107	29	14	8	5	3	2	1
0.2	140	Inf	173	45	20	11	7	4	3
0.3	43	197	Inf	219	55	24	13	8	5
0.4	22	54	233	Inf	244	60	26	14	8
0.5	13	25	60	248	Inf	248	60	25	13
0.6	8	14	26	60	244	Inf	233	54	22
0.7	5	8	13	24	55	219	Inf	197	43
0.8	3	4	7	11	20	45	173	Inf	140
0.9	1	2	3	5	8	14	29	107	Inf

For other choices of level and power, the R code is modified at Critical[ ] and Critical1[ ]. The remaining tables are given below.

Level = 1% and Power = 90%

> Sample

	0.1	0.2	0.3	0.4	0.5	0.6	0.7	0.8	0.9
0.1	Inf	147	41	20	11	7	5	3	2
0.2	173	Inf	230	61	27	15	9	6	4
0.3	53	249	Inf	287	73	32	17	10	6
0.4	26	68	298	Inf	317	78	33	17	9
0.5	15	31	77	321	Inf	321	77	31	15
0.6	9	17	33	78	317	Inf	298	68	26
0.7	6	10	17	32	73	287	Inf	249	53
0.8	4	6	9	15	27	61	230	Inf	173
0.9	2	3	5	7	11	20	41	147	Inf

Level = 1% and Power = 95%

> Sample

	0.1	0.2	0.3	0.4	0.5	0.6	0.7	0.8	0.9
0.1	Inf	184	53	25	14	9	6	4	2
0.2	203	Inf	284	75	34	19	11	7	4
0.3	61	297	Inf	350	89	39	21	12	7
0.4	30	81	359	Inf	385	95	40	20	11
0.5	17	37	92	388	Inf	388	92	37	17
0.6	11	20	40	95	385	Inf	359	81	30
0.7	7	12	21	39	89	350	Inf	297	61
0.8	4	7	11	19	34	75	284	Inf	203
0.9	2	4	6	9	14	25	53	184	Inf

Level = 5% and Power = 80%

> Sample

	0.1	0.2	0.3	0.4	0.5	0.6	0.7	0.8	0.9
0.1	Inf	69	19	9	5	3	2	1	1
0.2	83	Inf	109	29	13	7	4	3	2

```
0.3    25   119   Inf   136    34    15     8     5     3
0.4    12    33   142   Inf   150    37    16     8     4
0.5     7    15    36   152   Inf   152    36    15     7
0.6     4     8    16    37   150   Inf   142    33    12
0.7     3     5     8    15    34   136   Inf   119    25
0.8     2     3     4     7    13    29   109   Inf    83
0.9     1     1     2     3     5     9    19    69   Inf
```

Level = 5% and Power = 90%

```
> Sample
      0.1   0.2   0.3   0.4   0.5   0.6   0.7   0.8   0.9
0.1   Inf   101    29    14     8     5     3     2     1
0.2   109   Inf   155    41    19    10     6     4     2
0.3    32   160   Inf   191    49    21    11     6     4
0.4    16    43   194   Inf   209    51    22    11     6
0.5     9    20    50   210   Inf   210    50    20     9
0.6     6    11    22    51   209   Inf   194    43    16
0.7     4     6    11    21    49   191   Inf   160    32
0.8     2     4     6    10    19    41   155   Inf   109
0.9     1     2     3     5     8    14    29   101   Inf
```

Level = 5% and Power = 95%

```
> Sample
      0.1   0.2   0.3   0.4   0.5   0.6   0.7   0.8   0.9
0.1   Inf   133    39    19    11     7     4     3     2
0.2   133   Inf   199    54    24    13     8     5     3
0.3    39   199   Inf   243    62    27    14     8     4
0.4    19    54   243   Inf   265    65    27    13     7
0.5    11    24    62   265   Inf   265    62    24    11
0.6     7    13    27    65   265   Inf   243    54    19
0.7     4     8    14    27    62   243   Inf   199    39
0.8     3     5     8    13    24    54   199   Inf   133
0.9     2     3     4     7    11    19    39   133   Inf
```

Level = 10% and Power = 80%

```
> Sample
      0.1   0.2   0.3   0.4   0.5   0.6   0.7   0.8   0.9
0.1   Inf    52    15     7     4     3     2     1     1
0.2    59   Inf    81    21    10     5     3     2     1
0.3    18    85   Inf   100    25    11     6     3     2
0.4     9    23   103   Inf   110    27    11     6     3
0.5     5    11    26   111   Inf   111    26    11     5
0.6     3     6    11    27   110   Inf   103    23     9
0.7     2     3     6    11    25   100   Inf    85    18
0.8     1     2     3     5    10    21    81   Inf    59
0.9     1     1     2     3     4     7    15    52   Inf
```

Level = 10% and Power = 90%

```
> Sample
      0.1   0.2   0.3   0.4   0.5   0.6   0.7   0.8   0.9
0.1   Inf    80    24    11     7     4     3     2     1
0.2    80   Inf   121    33    15     8     5     3     2
0.3    24   121   Inf   148    38    16     9     5     3
```

```
0.4    11    33   148   Inf   161    39    16     8     4
0.5     7    15    38   161   Inf   161    38    15     7
0.6     4     8    16    39   161   Inf   148    33    11
0.7     3     5     9    16    38   148   Inf   121    24
0.8     2     3     5     8    15    33   121   Inf    80
0.9     1     2     3     4     7    11    24    80   Inf
```

Level=10% and Power=95%

```
> Sample
      0.1   0.2   0.3   0.4   0.5   0.6   0.7   0.8   0.9
0.1   Inf   109    32    16     9     6     4     2     1
0.2   101   Inf   160    43    20    11     6     4     2
0.3    29   155   Inf   194    50    22    11     6     3
0.4    14    41   191   Inf   210    51    21    10     5
0.5     8    19    49   209   Inf   209    49    19     8
0.6     5    10    21    51   210   Inf   191    41    14
0.7     3     6    11    22    50   194   Inf   155    29
0.8     2     4     6    11    20    43   160   Inf   101
0.9     1     2     4     6     9    16    32   109   Inf
```

We will now use the test based on the arc sine square root transformation. We will work with one-sided alternatives. A simple modification of the statement inside the loop will give the needed results for each choice of the level and power. The code is presented again with necessary modifications.

Set the levels at 1%, 5%, and 10%.

```
> Size <- c(0.01, 0.05, 0.10)
> Size
[1] 0.01 0.05 0.10
```

Calculate the critical values z_αs.

```
> Critical <- qnorm(Size, lower.tail = T)
> Critical
[1] -2.326348 -1.644854 -1.281552
```

Set the power at 80%, 90%, and 95%.

```
> Power <- c(0.80, 0.90, 0.95)
> Power
[1] 0.80 0.90 0.95
```

Calculate the quantiles z_βs.

```
> Critical1 <- qnorm(1 - Power, lower.tail = T)
> Critical1
[1] -0.8416212 -1.2815516 -1.6448536
```

Create a 9×9 matrix with each entry zero.

```
> Sample <- matrix(0, nrow = 9, ncol = 9)
```

Choose the proportions.

```
> Prop <- seq(0.1, 0.9, 0.1)
```

Set up a loop command to calculate sample sizes for testing

```
> for (i in 1:9)
+ {
+ for (j in 1:9)
+ {
+ Sample[i, j] <- round(((Critical[1] + Critical1[1])^2)/(2*asin
                          (sqrt(Prop[i]))
+ - 2*asin(sqrt(Prop[j])))^2)
+ }
+ }
> rownames(Sample) <- c("0.1", "0.2", "0.3", "0.4", "0.5", "0.6",
+                        "0.7", "0.8", "0.9")
> colnames(Sample) <- c("0.1", "0.2", "0.3", "0.4", "0.5", "0.6",
+                        "0.7", "0.8", "0.9")
```

Level=1% and Power=80% (arc sine square root test)

```
> Sample
      0.1  0.2  0.3  0.4  0.5  0.6  0.7  0.8  0.9
0.1   Inf  125   38   19   12    8    6    4    3
0.2   125  Inf  186   51   24   14    9    6    4
0.3    38  186  Inf  227   59   27   15    9    6
0.4    19   51  227  Inf  248   62   27   14    8
0.5    12   24   59  248  Inf  248   59   24   12
0.6     8   14   27   62  248  Inf  227   51   19
0.7     6    9   15   27   59  227  Inf  186   38
0.8     4    6    9   14   24   51  186  Inf  125
0.9     3    4    6    8   12   19   38  125  Inf
```

One can calculate sample sizes using the "pwr" package in R. The package uses the test based on the arc sine square root transformation. One of the inputs of the command is the effect size *h*. Calculate the effect size first before using the pwr package and the command pwr.p.test. Suppose $p_0=0.1$ and $p_1=0.2$. We are testing H_0: $p=p_0=0.1$ versus H_1: $p>p_0$. The specific alternative value is 0.2. The effect size comes from:

```
> h <- ES.h(0.2, 0.1)
> h
[1] 0.2837941
> pwr.p.test(h = h, power = 0.80, sig.level = 0.01,
+               alternative = "greater")
  proportion power calculation for binomial distribution (arcsine
  transformation)
    h = 0.2837941
```

```
      n = 124.6106
  sig.level = 0.01
      power = 0.8
alternative = greater
```

The sample size here matches with what we have gotten earlier.
It is instructive to compare sample sizes under both tests.
Level: 1% and Power: 80%

$p_0=0.1$

p_1	0.2	0.3	0.4	0.5	0.6	0.7	0.8	0.9
Traditional	107	29	14	8	5	3	2	1
Arc sine	125	38	19	12	8	6	4	3

Level = 1% and Power = 90% (arc sine square root transformation)

```
> Sample
     0.1  0.2  0.3  0.4  0.5  0.6  0.7  0.8  0.9
0.1  Inf  162   49   25   15   10    7    5    4
0.2  162  Inf  242   67   31   18   12    8    5
0.3   49  242  Inf  295   77   35   19   12    7
0.4   25   67  295  Inf  321   80   35   18   10
0.5   15   31   77  321  Inf  321   77   31   15
0.6   10   18   35   80  321  Inf  295   67   25
0.7    7   12   19   35   77  295  Inf  242   49
0.8    5    8   12   18   31   67  242  Inf  162
0.9    4    5    7   10   15   25   49  162  Inf
```

Level = 1% and Power = 95% (arc sine square root transformation)

```
> Sample
     0.1  0.2  0.3  0.4  0.5  0.6  0.7  0.8  0.9
0.1  Inf  196   59   30   18   12    9    6    5
0.2  196  Inf  293   81   38   22   14   10    6
0.3   59  293  Inf  357   93   42   23   14    9
0.4   30   81  357  Inf  389   97   42   22   12
0.5   18   38   93  389  Inf  389   93   38   18
0.6   12   22   42   97  389  Inf  357   81   30
0.7    9   14   23   42   93  357  Inf  293   59
0.8    6   10   14   22   38   81  293  Inf  196
0.9    5    6    9   12   18   30   59  196  Inf
```

Level = 5% and Power = 80% (arc sine square root transformation)

```
> Sample
     0.1  0.2  0.3  0.4  0.5  0.6  0.7  0.8  0.9
0.1  Inf   77   23   12    7    5    3    3    2
0.2   77  Inf  115   32   15    9    6    4    3
0.3   23  115  Inf  140   37   16    9    6    3
0.4   12   32  140  Inf  152   38   16    9    5
0.5    7   15   37  152  Inf  152   37   15    7
0.6    5    9   16   38  152  Inf  140   32   12
0.7    3    6    9   16   37  140  Inf  115   23
0.8    3    4    6    9   15   32  115  Inf   77
0.9    2    3    3    5    7   12   23   77  Inf
```

Level = 5% and Power = 90% (arc sine square root transformation)

> Sample

	0.1	0.2	0.3	0.4	0.5	0.6	0.7	0.8	0.9
0.1	Inf	106	32	16	10	7	5	3	2
0.2	106	Inf	159	44	21	12	8	5	3
0.3	32	159	Inf	194	51	23	13	8	5
0.4	16	44	194	Inf	211	53	23	12	7
0.5	10	21	51	211	Inf	211	51	21	10
0.6	7	12	23	53	211	Inf	194	44	16
0.7	5	8	13	23	51	194	Inf	159	32
0.8	3	5	8	12	21	44	159	Inf	106
0.9	2	3	5	7	10	16	32	106	Inf

Level = 5% and Power = 95% (arc sine square root transformation)

> Sample

	0.1	0.2	0.3	0.4	0.5	0.6	0.7	0.8	0.9
0.1	Inf	134	41	21	13	8	6	4	3
0.2	134	Inf	201	55	26	15	10	7	4
0.3	41	201	Inf	245	64	29	16	10	6
0.4	21	55	245	Inf	267	67	29	15	8
0.5	13	26	64	267	Inf	267	64	26	13
0.6	8	15	29	67	267	Inf	245	55	21
0.7	6	10	16	29	64	245	Inf	201	41
0.8	4	7	10	15	26	55	201	Inf	134
0.9	3	4	6	8	13	21	41	134	Inf

Level = 10% and Power = 80% (arc sine square root transformation)

> Sample

	0.1	0.2	0.3	0.4	0.5	0.6	0.7	0.8	0.9
0.1	Inf	56	17	9	5	4	3	2	1
0.2	56	Inf	84	23	11	6	4	3	2
0.3	17	84	Inf	102	27	12	7	4	3
0.4	9	23	102	Inf	111	28	12	6	4
0.5	5	11	27	111	Inf	111	27	11	5
0.6	4	6	12	28	111	Inf	102	23	9
0.7	3	4	7	12	27	102	Inf	84	17
0.8	2	3	4	6	11	23	84	Inf	56
0.9	1	2	3	4	5	9	17	56	Inf

Level = 10% and Power = 90% (arc sine square root transformation)

> Sample

	0.1	0.2	0.3	0.4	0.5	0.6	0.7	0.8	0.9
0.1	Inf	82	25	12	8	5	4	3	2
0.2	82	Inf	122	34	16	9	6	4	3
0.3	25	122	Inf	149	39	17	10	6	4
0.4	12	34	149	Inf	162	41	17	9	5
0.5	8	16	39	162	Inf	162	39	16	8
0.6	5	9	17	41	162	Inf	149	34	12
0.7	4	6	10	17	39	149	Inf	122	25
0.8	3	4	6	9	16	34	122	Inf	82
0.9	2	3	4	5	8	12	25	82	Inf

Level = 10% and Power = 95% (arc sine square root transformation)

```
> Sample
     0.1  0.2  0.3  0.4  0.5  0.6  0.7  0.8  0.9
0.1  Inf  106   32   16   10    7    5    3    2
0.2  106  Inf  159   44   21   12    8    5    3
0.3   32  159  Inf  194   51   23   13    8    5
0.4   16   44  194  Inf  211   53   23   12    7
0.5   10   21   51  211  Inf  211   51   21   10
0.6    7   12   23   53  211  Inf  194   44   16
0.7    5    8   13   23   51  194  Inf  159   32
0.8    3    5    8   12   21   44  159  Inf  106
0.9    2    3    5    7   10   16   32  106  Inf
```

Comments: Sample sizes are heavily reliant on how good the normal approximation is. Further, they do depend on the choice of the test statistic. Sample sizes are provided for one-sided alternatives only. The R code can be modified to accommodate two-sided alternatives.

A more reliable method is via simulations. The idea runs as follows. Suppose we want to test H_0: $p = 0.15$ against H_1: $p \neq 0.15$. Suppose the alternative value is 0.20.

1. Choose a level of significance 5%, say.
2. We use the test: Reject the null hypothesis if

$$\left|\sqrt{n}\left(2\sin^{-1}\left(\sqrt{\frac{X}{n}}\right) - 2\sin^{-1}\left(\sqrt{0.15}\right)\right)\right| > 1.96$$

3. Choose and fix $n = 100$, say.
4. Draw a random sample of size 100 from the Bernoulli population with $p = 0.20$. Samples are drawn under the alternative hypothesis. The null hypothesis is not true.
5. Apply the test. Check whether or not the null hypothesis is rejected.
6. Repeat the procedure 10,000 times.
7. Calculate what percentage of times the null hypothesis is rejected. This percentage is the power of the test.
8. If the power is below what was sought, increase the sample size.
9. If the power is above what was sought, decrease the sample size.
10. This is a computer-intensive procedure. However, computing is cheap. One can write an R code to achieve economies in computing.

2 TWO-SAMPLE PROPORTIONS

In the section, we focus on two independent populations and an attribute A of interest. Let p_1 be the proportion of individuals in Population 1 having the attribute A and p_2 be that of Population 2. We test

H_0: $p_1 = p_2$ versus H_1: $p_1 \neq p_2$ (two-sided alternative).

Let n be the common sample size. Let $\hat{p}_1$ and $\hat{p}_2$ be the respective sample proportions. The estimated standard error of the difference under the null hypothesis is $\sqrt{\frac{2\overline{p}(1-\overline{p})}{n}}$, where $\overline{p}=\frac{\hat{p}_1+\hat{p}_2}{2}$.

2.1 Traditional Test

A test of H_0 versus H_1 is given by: reject H_0 if

$$\left|\frac{\hat{p}_1-\hat{p}_2}{\sqrt{\frac{2\overline{p}(1-\overline{p})}{n}}}\right|>z_{1-\alpha/2}$$

where the critical value $z_{1-\alpha/2}$ is such that the area under standard normal curve to the left of $z_{1-\alpha/2}$ is $1-\alpha/2$. The determination of the critical value hinges on the result that the test statistic $\frac{\hat{p}_1-\hat{p}_2}{\sqrt{\frac{2\overline{p}(1-\overline{p})}{n}}}$ has a standard normal distribution under the null hypothesis if the common sample size is large. The approximation is good if both p_1 and p_2 are close to 0.5. Under these circumstances, the sample size is determined for a specified power $1-\beta$ and alternative values of p_1 and p_2. There is an explicit formula for the common sample size n.

$$n=\left(\frac{z_{\alpha/2}{}^{*}\sqrt{\frac{(p_1+p_2)(2-p_1-p_2)}{2}}+z_{\beta}{}^{*}\sqrt{(p_1(1-p_1)+p_2(1-p_2)})}{p_1-p_2}\right)^2$$

For one-sided alternatives, in the formula $z_{\alpha/2}$ is replaced by z_α.

We have calculated the common sample size n under two scenarios of p_1 and p_2 ($p_1=0.1$, $p_2=0.2$ and $p_1=0.45$, $p_2=0.55$) for several combinations of level α and power $1-\beta$ using R. The R code and annotation are provided.

Three levels of significance are entertained.

```
> Size <- c(0.01, 0.05, 0.10)
> Size
[1] 0.01 0.05 0.10
```

The corresponding $z_{\alpha/2}$s are computed.

```
> Critical <- qnorm(Size/2, lower.tail = T)
> Critical
[1] -2.575829 -1.959964 -1.644854
```

Three choices of power are made.

```
> Power <- c(0.80, 0.90, 0.95)
```

```
> Power
[1] 0.80 0.90 0.95
```

The corresponding z_βs are computed.

```
> Critical1 <- qnorm(1 - Power, lower.tail = T)
> Critical1
[1] -0.8416212 -1.2815516 -1.6448536
```

A 3 × 3 matrix with zeros is created.

```
> Sample <- matrix(0, nrow = 3, ncol = 3)
```

The alternative values of the population proportions are chosen and their average is computed.

```
> AveProp <- (0.1 + 0.2)/2
```

A loop is created for the calculation of sample sizes.

```
> for (i in 1:3)
+ {
+ for (j in 1:3)
+ {
+ Sample[i, j] <- round(((Critical[i]*sqrt(2*AveProp*(1-AveProp))
+ Critical1[j]* sqrt(0.1*0.9 + 0.2*0.8))^2)/(0.1 - 0.2)^2)
+ }
+ }
> rownames(Sample) <- c("Size 0.01", "Size 0.05", "Size 0.10")
> colnames(Sample) <- c("Power 80%", "Power 90%", "Power 95%")
> Sample # Scenario: p1 = 0.1 and p2 = 0.2

          Power 80%   Power 90%   Power 95%
Size 0.01       296         377         451
Size 0.05       199         266         328
Size 0.10       157         216         273
```

A simple modification of the R code produces the following table of sample sizes:

```
> Sample # Scenario: p1 = 0.45 and p2 = 0.55
          Power 80%   Power 90%   Power 95%
Size 0.01       583         741         887
Size 0.05       391         523         647
Size 0.10       308         426         538
```

2.2 Arcsine Square Root Transformation

In many applications, the population proportions are far from 0.5. For a better normal approximation, the arc sine square root transformation is recommended. The test statistic is given by

$$\frac{\sin^{-1}\left(\sqrt{\hat{p}_1}\right) - \sin^{-1}\left(\sqrt{\hat{p}_2}\right)}{\sqrt{\frac{1}{4n} + \frac{1}{4n}}}$$

For a two-sided alternative, with size at α, power $1-\beta$, and specified alternatives p_1 and p_2, the common sample size is given by

$$n = \frac{1}{2}\left(\frac{z_{\alpha/2} + z_\beta}{\arcsin\left(\sqrt{p_1}\right) - \arcsin\left(\sqrt{p_2}\right)}\right)^2$$

The difference between the single proportion case and two-proportion case is that the sample size in the later case is the double of that of the first case. The command "pwr.2p.test" in the "pwr" package accomplishes calculation of sample size in this case. An example follows:

Specifications $p_1 = 0.3$ and $p_2 = 0.4$.
Test: Two-sided proportion test based on the arcsine transformation
Level = 0.05
Power = 0.80

Calculate the effect size.
Effect size = $h = 2 * \sin^{-1}(\text{sqrt}(p_2)) - 2 * \sin^{-1}(\text{sqrt}(p_1))$

```
> h <- ES.h(0.4, 0.3)
> h
[1] 0.2101589
```

We do a two-sided test. The default is the two-sided test. It does not matter whether we take h or $-h$.

```
> pwr.2p.test(h = h, sig.level = 0.05, power = 0.80)
```

Difference of proportion power calculation for binomial distribution (arcsine transformation)

```
h = 0.2101589
n = 355.4192
sig.level = 0.05
power = 0.8
alternative = two.sided
```

Note: Same sample sizes.

Comment. We need a sample size of 356 subjects from each group in order to detect the alternative proportions 0.3 and 0.4 with probability 0.80.

We now compute sample sizes under different combinations of size and power for two different scenarios of the proportions used earlier. The goal is to examine how the choice of test influences sample sizes. The code and tables follow:

```
> Size <- c(0.01, 0.05, 0.10)
> Size
[1] 0.01 0.05 0.10
> Critical <- qnorm(Size/2, lower.tail = T)
> Critical
[1] -2.575829 -1.959964 -1.644854
> Power <- c(0.80, 0.90, 0.95)
> Power
[1] 0.80 0.90 0.95
> Critical1 <- qnorm(1 - Power, lower.tail = T)
> Critical1
[1] -0.8416212 -1.2815516 -1.6448536
> Sample <- matrix(0, nrow = 3, ncol = 3)
> for (i in 1:3)
+ {
+ for (j in 1:3)
+ {
+ Sample[i, j] <- round(((Critical[i] + Critical1[j])^2)/
+ (2*asin(sqrt(0.1))- 2*asin(sqrt(0.2)))^2)
+ }
+ }
> rownames(Sample) <- c("Size 0.01", "Size 0.05", "Size 0.10")
> colnames(Sample) <- c("Power 80%", "Power 90%", "Power 95%")
> Sample # Scenario: p1 = 0.1 and p2 = 0.2

          Power 80%  Power 90%  Power 95%
Size 0.01       290        369        442
Size 0.05       195        261        323
Size 0.10       154        213        269

> Sample # Scenario: p1 = 0.45 and p2 = 0.55

          Power 80%  Power 90%  Power 95%
Size 0.01       582        741        888
Size 0.05       391        524        648
Size 0.10       308        427        539
```

Comments

Let us compare the sample sizes under both the tests.

1. When the alternative proportions are close to 0.5, the sample sizes produced by the traditional test and arcsine test are more or less equal.
2. When the alternative proportions are far away from 0.5, the arcsine test produces lower sample sizes but not a lot lower.

3 EFFECT SIZES

Social scientists and psychologists provide ballpark numbers as to what constitute small, medium, and large effect sizes. The Guru of these ideas is Jacob Cohen (1988). The package "pwr," maintained by Champely (2009), provides the relevant numbers.

The command "cohen.ES" within the "pwr" package provides all the answers as illustrated below.

3.1 The Case of Proportions

```
> cohen.ES(test = "p", size = c("small"))
   Conventional effect size from Cohen (1982)
      test = p
      size = small
  effect.size = 0.2
> cohen.ES(test = "p", size = c("medium"))
   Conventional effect size from Cohen (1982)
      test = p
      size = medium
  effect.size = 0.5
> cohen.ES(test = "p", size = c("large"))
   Conventional effect size from Cohen (1982)
      test = p
      size = large
  effect.size = 0.8
```

Note that the test statistic used for testing hypotheses on proportions is built on the arcsine square root transformation. Some wisdom about what constitutes a small effect size in the context of *t* tests is also provided under the umbrella of "cohen.ES" command.

3.2 The Case of *t*-Test

What are small, medium, and large effect sizes?

```
> cohen.ES(test = "t", size = "small")
   Conventional effect size from Cohen (1982)
      test = t
      size = small
  effect.size = 0.2
> cohen.ES(test = "t", size = "medium")
   Conventional effect size from Cohen (1982)
      test = t
      size = medium
  effect.size = 0.5
```

```
> cohen.ES(test = "t", size = "large")
   Conventional effect size from Cohen (1982)
      test = t
      size = large
 effect.size = 0.8
```

3.3 The Case of Correlation

Let ρ be the population correlation coefficient between two quantitative variables.
H_0: $\rho = 0$
H_1: $\rho \neq 0$; $\rho = \rho_1 = 0.3$, say.
What are the small, medium, and large effect sizes?
$\mathrm{ES} = 0.5^* \ln\frac{1+\rho_1}{1-\rho_1}$

```
> cohen.ES(test = "r", size = "small")
   Conventional effect size from Cohen (1982)
      test = r
      size = small
 effect.size = 0.1
> cohen.ES(test = "r", size = "medium")
   Conventional effect size from Cohen (1982)
      test = r
      size = medium
 effect.size = 0.3
> cohen.ES(test = "r", size = "large")
   Conventional effect size from Cohen (1982)
      test = r
      size = large
 effect.size = 0.5
```

3.4 Analysis of Variance

Small, medium, and large effect sizes

```
> cohen.ES(test = "anov", size = "small")
   Conventional effect size from Cohen (1982)
      test = anov
      size = small
 effect.size = 0.1
> cohen.ES(test = "anov", size = "medium")
   Conventional effect size from Cohen (1982)
      test = anov
      size = medium
 effect.size = 0.25
```

```
> cohen.ES(test = "anov", size = "large")
   Conventional effect size from Cohen (1982)
      test = anov
      size = large
 effect.size = 0.4
```

4 MULTISAMPLE PROPORTIONS

4.1 Testing Equality of Several Population Proportions

For testing the equality of the means of several independent populations, we use the method of Analysis of Variance. In this section, we deal with a problem similar in spirit. The focus is on several independent populations and an attribute A. Let p_i be the proportion of the ith population having the attribute A. We test the null hypothesis H_0: $p_1 = p_2 = \cdots = p_k$. Let n be the common sample size. The count data that will be collected can be arranged in the form of a $2 \times k$ contingency table.

	Populations			
Attribute A	**1**	**2**	**...**	***k***
Present	O_{11}	O_{12}	...	O_{1k}
Absent	O_{21}	O_{22}	...	O_{2k}
Total	n	n	...	n

The entity O_{1j} is the number of members with the attribute A in the jth sample and O_{2j} the number with the attribute A absent. The hypothesis is tested using the chi-squared test statistic χ^2. The large-sample distribution of the statistic is chi-squared with $k-1$ degrees of freedom, from which the critical value calculated. For power calculations, we need the large-sample distribution of the chi-squared statistic under the alternative hypothesis. For this, we need the alternative values of the proportions. Let the specified values of the proportions be

$p_1 = p_{11}, p_2 = p_{12}, \ldots, p_k = p_{1k}$.

The effect size w is defined by

$$w = \sqrt{\frac{(p_{11} - p_o)^2}{p_0} + \frac{(p_{12} - p_0)^2}{p_0} + \cdots + \frac{(p_{1k} - p_0)^2}{p_0}}$$

The entity p_0 is the common value of the proportions under the null hypothesis, which is the average of the specified proportions. The large-sample distribution of the chi-squared statistic under the alternative hypothesis depends on the effect size w through the alternatives specified. The distribution is a noncentral chi-squared with $k-1$ degrees of freedom and noncentrality parameter w. For power calculations, it is sufficient to specify the effect size w.

Let us find out what small, medium, and large effect sizes are in this environment?

```
> cohen.ES(test = "chisq", size = "small")
    Conventional effect size from Cohen (1982)
       test = chisq
       size = small
  effect.size = 0.1
> cohen.ES(test = "chisq", size = "medium")
    Conventional effect size from Cohen (1982)
       test = chisq
       size = medium
    effect.size = 0.3
> cohen.ES(test = "chisq", size = "large")
    Conventional effect size from Cohen (1982)
       test = chisq
       size = large
    effect.size = 0.5
```

Let us find the sample size when $k=4$ and $w=0.289$.

```
> pwr.chisq.test(w = 0.289, df = 3, sig.level = 0.05, power = 0.80)
  Chi squared power calculation
       w = 0.289
       N = 130.5368
       df = 3
  sig.level = 0.05
       power = 0.8
```

Note: N is the number of observations.

We need to survey 131 consumers.

Using R, we prepare a table of sample sizes for number of populations $k=3$, 4, and 5 for several combinations of level and power.

Number of populations entertained: 3, 4, and 5

```
> k <- c(3, 4, 5)
```

Effect sizes entertained: small, medium, and large

```
> w <- c(0.1, 0.3, 0.5)
```

Levels of significance entertained: 0.01, 0.05, and 0.10

```
> Level <- c(0.01, 0.05, 0.10)
```

Powers entertained: 80%, 90%, and 95%

```
> Power <- c(0.80, 0.90, 0.95)
```

Create an array of zeros of dimension $3 \times 3 \times 3 \times 3$. Label the dimensions.

```
> MB <- rep(0, 81)
```

```
> dim(MB) <- c(3, 3, 3, 3)
> dimnames(MB) <- list(c("Level 0.01", "Level 0.05", "Level 0.10"),
+                      c("Power 0.80", "Power 0.90", "Power 0.95"),
+                      c("3 Pop", "4 Pop", "5 Pop"),
+                      c("Effect 0.1", "Effect 0.3", "Effect 0.5"))
```

Create a 4-level loop.

```
> for (i in 1:3)
+ {
+ for (j in 1:3)
+ {
+ for (u in 1:3)
+ {
+ for (v in 1:3)
+ {
+ MB[i, j, u, v] <- round(pwr.chisq.test(sig.level = Level[i],
power = Power[j], df = k[u]-1, w = w[v])$N)
+ }
+ }
+ }
+ }
> MB
, , 3 Pop, Effect 0.1

           Power 0.80  Power 0.90  Power 0.95
Level 0.01       1388        1743        2065
Level 0.05        963        1265        1544
Level 0.10        771        1046        1302

, , 4 Pop, Effect 0.1
           Power 0.80  Power 0.90  Power 0.95
Level 0.01       1546        1925        2267
Level 0.05       1090        1417        1717
Level 0.10        880        1180        1457

, , 5 Pop, Effect 0.1

           Power 0.80  Power 0.90  Power 0.95
Level 0.01       1675        2074        2433
Level 0.05       1194        1541        1857
Level 0.10        968        1288        1583

, , 3 Pop, Effect 0.3

           Power 0.80  Power 0.90  Power 0.95
Level 0.01        154         194         229
Level 0.05        107         141         172
Level 0.10         86         116         145
```

```
, , 4 Pop, Effect 0.3

           Power 0.80  Power 0.90  Power 0.95
Level 0.01        172         214         252
Level 0.05        121         157         191
Level 0.10         98         131         162

, , 5 Pop, Effect 0.3

           Power 0.80  Power 0.90  Power 0.95
Level 0.01        186         230         270
Level 0.05        133         171         206
Level 0.10        108         143         176

, , 3 Pop, Effect 0.5

           Power 0.80  Power 0.90  Power 0.95
Level 0.01         56          70          83
Level 0.05         39          51          62
Level 0.10         31          42          52

, , 4 Pop, Effect 0.5

           Power 0.80  Power 0.90  Power 0.95
Level 0.01         62          77          91
Level 0.05         44          57          69
Level 0.10         35          47          58

, , 5 Pop, Effect 0.5

           Power 0.80  Power 0.90  Power 0.95
Level 0.01         67          83          97
Level 0.05         48          62          74
Level 0.10         39          52          63
```

5 McNemar TEST

This is similar to the paired t-test for a pair of binary variables. For every subject in the target population, the response is the outcome of a pair (X, Y) of binary variables.

An example from toxicology

Two medical devices are being compared for the presence of a particular type of bacteria in human beings.

Device 1	Dosimeter
Device 2	Nasal swab

Design

Take a sample of n subjects. On each subject apply both the devices.

X = Verdict from the dosimeter: Presence of bacteria: yes or no

Y = Verdict from the nasal swab: Presence of bacteria: yes or no

X and Y are correlated because the responses are coming from the same subject.

Joint distribution of responses from a random subject

	Nasal Swab		
Dosimeter	**Yes**	**No**	**Marginal**
Yes	p_{11}	p_{12}	p_1
No	p_{21}	p_{22}	p_2
Marginal	q_1	q_2	1

p_1 = Population probability that Dosimeter gives the verdict that bacteria is present.

q_1 = Population probability that Nasal swab gives the verdict that bacteria is present.

Hypothesis of skepticism: No difference

H_0: $p_1 = q_1$

H_1: $p_1 \neq q_1$

Mathematical aside: $p_1 = q_1$ if and only if $p_{12} = p_{21}$. This observation has an important bearing in the development of a test for this testing problem.

A sample of n subjects gives data: (X_1, Y_1), (X_2, Y_2), …, (X_n, Y_n)

Each X_i is yes or no. Each Y_i is yes or no.

Arrange the data in the form of a 2×2 contingency table.

	Nasal Swab		
Dosimeter	**Yes**	**No**	**Marginal**
Yes	n_{11}	n_{12}	n_1
No	n_{21}	n_{22}	n_2
Marginal	m_1	m_2	n

Test statistic: $\chi^2 = \frac{(n_{12} - n_{21})^2}{n_{12} + n_{21}}$

If the null hypothesis is true, the chi-squared statistic has a chi-squared distribution with one degree of freedom. This is important for the computation of the critical value or p-value.

McNemar test: Reject the null hypothesis if $\chi^2 > \chi^2_{1,1-\alpha}$.

The area under the chi-squared distribution with one degree of freedom to the left of $\chi^2_{1,1-\alpha}$ is $1 - \alpha$.

We need the distribution of the chi-squared statistic under the alternative hypothesis for power computation. What is the effect size here?

Specifications

1. Level α
2. Power $1 - \beta$
3. Values of p_{12} and p_{21} (from the joint distribution table)

The closer the values of p_{12} and p_{21} are, the closer the values of p_1 and q_1. The closer the values of p_{12} and p_{21} are, the closer the alternative and null hypotheses are. The closer the values of p_{12} and p_{21} are, the larger the sample size is.

An R package "TrialSize" does sample size calculations for the McNemar test.

Download and then activate the package.

Specifications

1. Level: 0.05
2. Power $=0.80$ or $\beta=0.20$
3. $p_{12}=0.2$ and $p_{21}=0.5$. The command needs the ratio "psai" of these two numbers and the sum "paid" of these numbers.

```
> mb <- McNemar.Test(alpha = 0.05, beta = 0.20, psai = 0.2/0.5,
+                    paid = 0.7)
> mb
[1] 58.63224
```

The required sample size is 59.

Using R one can prepare tables of sample sizes for combinations of specifications.

6 CORRELATIONS

A subject in the targeted population gives a pair of measurement X and Y on possibly different numerical characteristics. For example, let $X=$weight and $Y=$height, or $X=$SBP and $Y=$DBP. Assume bivariate normality of the joint distribution of X and Y. Let ρ be the correlation between X and Y. The entity ρ is a measure of association between X and Y. The hypothesis of skepticism is:

H_0: $\rho=0$ (X and Y are not associated. X and Y are statistically independent.)

Structure of the data: $(X_1, Y_1), (X_2, Y_2), \ldots, (X_n, Y_n)$

Let r be the sample correlation coefficient.

Test statistic: $t=\sqrt{n-2}*\frac{r}{\sqrt{1-r^2}}$

Under the null hypothesis, the statistic t has a t-distribution with $n-2$ degrees of freedom.

The alternative is: H_1: $\rho\neq 0$

Test: Reject the null hypothesis if $|t|>t_{n-2,1-\alpha/2}$.

Specifications for sample size calculation:

1. Level
2. Power
3. The value of ρ under the alternative

The distribution of the t-statistic under the alternative hypothesis is Student's noncentral t with $n-2$ degrees of freedom and noncentrality parameter is the specified value of ρ. The "pwr" package lets us compute sample size.

Specifications: $\alpha=0.05$; Power$=0.80$; $\rho=0.3$

```
> pwr.r.test(r = 0.3, sig.level = 0.05, power = 0.80,
+            alternative = "two.sided")
```

```
approximate correlation power calculation (arctangh
 transformation)
   n = 84.74891
   r = 0.3
sig.level = 0.05
   power = 0.8
alternative = two.sided
```

Under the two-sided alternative, what happens if $\rho = -0.3$? We get exactly the same number.

```
> pwr.r.test(r = -0.3, sig.level = 0.05, power = 0.80,
              alternative = "two.sided")
approximate correlation power calculation (arctangh
 transformation)
    n = 84.74891
    r = 0.3
sig.level = 0.05
   power = 0.8
alternative = two.sided
```

We can have the whole array of sample sizes to cover a variety of configurations of the specifications.

```
Specifications
Level: 0.01, 0.05, 0.10
Power: 0.80, 0.90, 0.95
Rho: 0.1(0.1)0.9
The following is the R code.
> Level <- c(0.01, 0.05, 0.10)
> Power <- c(0.80, 0.90, 0.95)
> Rho <- seq(0.1, 0.9, 0.1)
> Sizes <- rep(0, 3*3*9)
> dim(Sizes) <- c(3, 3, 9)
> for (i in 1:3)
+ {
+ for (j in 1:3)
+ {
+ for (k in 1:9)
+ {
+ Sizes[i, j, k] <- pwr.r.test(r = Rho[k], sig.level = Level[i],
+                                   power = Power[j],
+ alternative = "two.sided")$n
+ }
+ }
+ }
```

```
> dimnames(Sizes)[[1]] <- c("Level 0.01", "Level 0.05",
+                           "Level 0.10")
> dimnames(Sizes)[[2]] <- c("Power 80%", "Power 90%", "Power 95%")
> dimnames(Sizes)[[3]] <- c("Rho 0.1", "Rho 0.2", "Rho 0.3",
+                           "Rho 0.4", "Rho 0.5", "Rho 0.6",
+                           "Rho 0.7", "Rho 0.8", "Rho 0.9")
> round(Sizes)
, , Rho 0.1

           Power 80% Power 90% Power 95%
Level 0.01      1163      1481      1773
Level 0.05       782      1046      1293
Level 0.10       617       853      1077

, , Rho 0.2

           Power 80% Power 90% Power 95%
Level 0.01       287       365       436
Level 0.05       194       258       319
Level 0.10       153       211       266

, , Rho 0.3

           Power 80% Power 90% Power 95%
Level 0.01       125       158       189
Level 0.05        85       112       138
Level 0.10        67        92       115

, , Rho 0.4

           Power 80% Power 90% Power 95%
Level 0.01        68        86       102
Level 0.05        47        61        75
Level 0.10        37        50        63

, , Rho 0.5

           Power 80% Power 90% Power 95%
Level 0.01        42        52        62
Level 0.05        29        38        46
Level 0.10        23        31        38

, , Rho 0.6

           Power 80% Power 90% Power 95%
Level 0.01        28        34        40
Level 0.05        19        25        30
Level 0.10        16        20        25

, , Rho 0.7

           Power 80% Power 90% Power 95%
Level 0.01        19        23        27
Level 0.05        13        17        20
Level 0.10        11        14        17
```

```
, , Rho 0.8
           Power 80%  Power 90%  Power 95%
Level 0.01        13         16         18
Level 0.05         9         12         14
Level 0.10         8         10         12

, , Rho 0.9

           Power 80%  Power 90%  Power 95%
Level 0.01         9         10         11
Level 0.05         7          8          9
Level 0.10         6          7          8
```

7 HAZARD RATIO IN SURVIVAL ANALYSIS

What is hazard ratio?

A motivating example: the following example is used to lay out the key ideas in survival analysis.

The target population consists of those patients with retinitis pigmentosa. The goal is the prevention of visual loss. The treatment proposed is heavy intake of vitamin A supplement. Look at the clinical trial reported in Berson et al. (1993).

Design

Take a sample of *m* patients. Each one takes 15,000 IU of vitamin A per day. Follow each one of them for six years. This is the experimental group.

Take a sample of n patients. Each one takes 75 IU of vitamin A per day. Follow each one of them for 6 years. This is the control group.

The response variable is the time *T* at which vision loss occurs. In the survival analysis jargon, this is called survival time. The time *T* is measured in years.

In studies like this, censoring occurs. Censoring (dropouts) within the period of study could occur for several reasons.

1. Death
2. Stops taking medication because of other diseases.
3. Stops taking medication because of side effects.
4. Unwilling to continue take medications.
5. No vision loss occurs during the entire period of time.

Data structure

Subject	Times	Status	Group	
1	2	1	E	
2	3	0	C	etc.

Interpretation

The first subject was in the experimental group (group = E). Vision loss occurred in Year 2. This is a genuine observation (status = 1) of the variable T.

The second subject was in the Control group (group = C). He dropped out in the Year 3 (status = 0). This is a censored observation of T. He took vitamin A for 3 years and no vision loss occurred until he was in the study. Then he is lost.

The goal is whether taking a mega dose of vitamin A postpones vision loss significantly. This question can be answered via a model building endeavor. We use Cox Proportional Hazards model. The covariate is "group."

$h(t;\ \text{group}) = h_0(t) * \exp(\gamma * \text{group}),\ t > 0$

$h(t)$ is the hazard function of the variable T at time t.

$h_0(t)$ is the baseline hazard function.

"group" is codified as 1 (Experimental) or 0 (Control).

HR = hazard ratio = $\exp(\gamma)$.

Interpretation

HR = 1 if and only if $\gamma = 0$, which means the mega treatment is not effective.

HR < 1 means $\gamma < 0$, which means the mega treatment is more effective than the control treatment.

There is a statistical interpretation. Let T_1 be the time at which vision loss occurs in the experimental group and T_2 in the control group. If the distributions of T_1 and T_2 are exponential, the hazard ratio is precisely the ratio of the median vision loss times. The statement that HR = 1 is tantamount to the statement that the median vision loss times are the same. See Bedrick (2014). Thus, the statement HR = 1 is the key one for testing.

Test

H_0: HR = 1

H_1: HR $\neq$ 1

How many patients to recruit under the experimental treatment and how many under the control treatment?

The answer is complicated by the threat of censoring. We need to know rates of censoring under both scenarios.

At the fundamental level, we need to specify the level, power, and the HR we are shooting for under the alternative. Suppose the specifications are:

1. Level: 0.05
2. Power: 0.80 or $\beta = 0.2$
3. HR = 0.8

We need other information such as censoring rates before we can calculate sample sizes. We need data from a pilot study if we wish to use the R package "powerSurvEpi." The R command "ssizeCT" uses the pilot study data thrown at it to calculate sample sizes. Download this package and activate it. In order to illustrate the use of the package, we will use the example provided in the package. An outline of the study was already presented above. Other information is provided below.

7.1 A Pilot Study

Berson et al. (1993) initiated a study to examine the effectiveness of the mega vitamin A treatment. Recruitment was carried out over a 2-year period 1984–1987. The study was terminated in September 1991. The data are available in the package under the name "Oph." Let us download the data and understand it.

```
> data(Oph)
> dim(Oph)
[1] 354 3
> head(Oph)
    times  status  group
1       1       1      E
2       1       1      E
3       1       1      E
4       1       0      E
5       1       0      E
6       1       0      E

> summary(Oph)
           Times            status    group
Min.    :1.000   Min.    :0.000   C:182
1st Qu. :4.000   1st Qu.:0.000    E:172
Median  :5.000   Median :0.000
Mean    :4.401   Mean    :0.435
3rd Qu. :5.000   3rd Qu.:1.000
Max.    :6.000   Max.    :1.000

> table(Oph$status)
  0    1
200  154
```

There were 172 patients in the experimental group and 182 in the control group. One hundred fifty-four of the subjects provided genuine observations on vision loss. The rest censored observations.

Overall censoring rate

```
> 200/354
[1] 0.5649718
```

Isolate the experimental group

```
> OphE <- subset(Oph, Oph$group == "E")
```

```
> head(OphE)
   times  status  group
1      1       1      E
2      1       1      E
3      1       1      E
4      1       0      E
5      1       0      E
6      1       0      E
> table(OphE$status)
  0   1
107  65
```

Censoring rate under the experimental group:

```
> 107/172
[1] 0.622093
```

Isolate the control group.

```
> OphC <- subset(Oph, Oph$group == "C")
> head(OphC)
      times  status  group
173       1       1      C
174       1       1      C
175       1       1      C
176       1       1      C
177       1       1      C
178       1       1      C
> table(OphC$status)
 0   1
93  89
```

Censoring rate under the control group

```
> 93/182
[1] 0.510989
```

Fit the Cox model.

```
> MB <- coxph(Surv(times, status) ~ group, data = Oph)
> summary(MB)
Call:
coxph(formula = Surv(times, status) ~ group, data = Oph)
n= 354, number of events= 154
            coef exp(coef ) se(coef )       z Pr(>|z|)
groupE -0.3735     0.6883   0.1633 -2.288  0.0222 *
---
Signif. codes: 0 '***' 0.001 '**' 0.01 '*' 0.05 '.' 0.1 ' ' 1
       exp(coef ) exp(-coef ) lower .95 upper .95
```

```
groupE     0.6883        1.453    0.4998    0.9479
Concordance= 0.548 (se = 0.024 )
Rsquare= 0.015 (max possible= 0.991 )
Likelihood ratio test= 5.31 on 1 df, p=0.02126
Wald test = 5.23 on 1 df, p=0.02216
Score (logrank) test = 5.29 on 1 df, p=0.0214
```

Use this pilot data to calculate sample sizes. The stipulation $k=1$ means we want equal sample sizes.

```
> MB1 <- ssizeCT(Surv(times, status) ~ group, dat = Oph,
+                  power = 0.8, k = 1, RR = 0.8, alpha = 0.05)
> names(MB1)
[1] "mat.lambda" "mat.event" "pC" "pE" "ssize"
> MB1$ssize
nE nC
706 706
```

We need to recruit 706 patients under each treatment regimen so that we can be sure with 80% probability that a hazard ratio of 0.8 is detectable if it is true.

8 MULTIPLE REGRESSION

Basic formulation:

Y: Quantitative response variable

Predictors: $X_1, X_2, \ldots, X_p$

Multiple regression model

$Y=\beta_0+\beta_1 * X_1+\beta_2 * X_2+\cdots+\beta_p * X_p+\varepsilon$

$\varepsilon \sim N(0, \sigma)$

Assumptions

The vector $(Y, X_1, X_2, \ldots, X_p)$ has a multivariate normal distribution.

This assumption implies that the conditional distribution

$Y \mid X_1, X_2, \ldots, X_p \sim N(\beta_0+\beta_1 * \mathrm{X}_1+\beta_2 * X_2+\cdots+\beta_p * X_p, \sigma)$

This also implies that

$E(Y \mid X_1, X_2, \ldots, X_p)=\beta_0+\beta_1 * X_1+\beta_2 * X_2+\cdots+\beta_p * X_p$

$\mathrm{Var}(Y \mid X_1, X_2, \ldots, X_p)=\sigma^2$

The standard deviation σ does not depend on the covariates $X_1, X_2, \ldots, X_p$ (homoscedasticity)

The population multiple correlation coefficient between Y and $X=(X_1, X_2, \ldots, X_p)$ is usually denoted by $P^2_{Y.X}$ or simply, P^2.

A standard interpretation is that the percentage of variation in Y accounted by the predictors is given by $100 * P^2$. A mathematical formula is available for P^2.

We want to test the null hypothesis
H_0: $P^2 = 0$ against the alternative H_1: $P^2 > 0$.
The null hypothesis and $\beta_1 = \beta_2 = \cdots \beta_p = 0$ are equivalent. The alternative can be interpreted as at least one $\beta_i \neq 0$.
Let R^2 be the sample multiple correlation coefficient.

$$\text{Test statistic: } F = \frac{\frac{R^2}{p}}{\frac{1-R^2}{n-p-1}}$$

If the null hypothesis is true $F \sim F$ distribution with numerator degrees of freedom p and denominator degrees of freedom $n-p-1$. The critical value for a given level can be found from the F distribution in R.

What should be the sample size n? Spell out the alternative value of P^2.

Under this given alternative value of P^2, F has a noncentral F distribution with numerator degrees of freedom p and denominator degrees of freedom $n-p-1$ and noncentrality parameter $n * \frac{P^2}{1-P^2}$.

We can develop an integral equation to solve for the desired sample size. The package "MBESS" has a command "ss.power.R2," which gives the sample size.

Specifications

1. Level
2. Power
3. Number of predictors
4. Alternative value of the population P^2.

Let us calculate the sample size for the following specifications:

```
Level = 0.05; Power = 0.80; P2 = 0.25; Number of predictors = 5
> MB <- ss.power.R2(Population.R2 = 0.25, alpha.level = 0.05,
+                   desired.power =0.80, p = 5)
> MB
$Necessary.Sample.Size
[1] 45
$Actual.Power
[1] 0.8075842
$Noncentral.F.Parm
[1] 15
$Effect.Size
[1] 0.3333333
```

Let us calculate the sample size for each of the configurations coming from:
Level: 0.01, 0.05, 0.10

Power: 0.80, 0.90, 0.95
Alternative P^2: 0.2, 0.5, 0.8
Number of variables: 3, 4, 5
There are $3*3*3*3=81$ configurations.
We need a four-dimensional array.

```
> Level <- c(0.01, 0.05, 0.10)
> Power <- c(0.80, 0.90, 0.95)
> RSquared <- c(0.2, 0.5, 0.8)
> Variables <- c(3, 4, 5)
Create a four-dimensional array of zeros.
> Sizes <- rep(0, 81)
> dim(Sizes) <- c(3, 3, 3, 3)
```

Label the rows, columns, and doubly indexed slices.

```
> dimnames(Sizes)[[1]] <- c("Level 0.01", "Level 0.05",
+                             "Level 0.10")
> dimnames(Sizes)[[2]] <- c("Power 80%", "Power 90%", "Power 95%")
> dimnames(Sizes)[[3]] <- c("R2 = 0.2", "R2 = 0.5", "R2 = 0.8")
> dimnames(Sizes)[[4]] <- c("p = 3", "p = 4", "p = 5")
```

Start four loops.

```
> for (i in 1:3)
+ {
+ for (j in 1:3)
+ {
+ for (k in 1:3)
+ {
+ for (l in 1:3)
+ {
+ Sizes[i, j, k, l] <- ss.power.R2(Population.R2 = RSquared[k],
+ alpha.level =Level[i], desired.power = Power[j],
p = Variables[l])$Necessary.Sample.Size
+ }
+ }
+ }
+ }
> Sizes
, , R2 = 0.2, p = 3
           Power 80%  Power 90%  Power 95%
Level 0.01        68         83         97
Level 0.05        48         61         73
Level 0.10        39         51         62
```

```
, , R2 = 0.5, p = 3

           Power 80% Power 90% Power 95%
Level 0.01        22        26        29
Level 0.05        16        19        22
Level 0.10        13        16        19

, , R2 = 0.8, p = 3

           Power 80% Power 90% Power 95%
Level 0.01        11        12        13
Level 0.05         8         9        10
Level 0.10         7         8         9

, , R2 = 0.2, p = 4

           Power 80% Power 90% Power 95%
Level 0.01        74        90       105
Level 0.05        53        67        80
Level 0.10        43        56        68

, , R2 = 0.5, p = 4

           Power 80% Power 90% Power 95%
Level 0.01        24        28        32
Level 0.05        18        21        24
Level 0.10        15        18        21

, , R2 = 0.8, p = 4

           Power 80% Power 90% Power 95%
Level 0.01        12        13        14
Level 0.05        10        11        11
Level 0.10         9         9        10

, , R2 = 0.2, p = 5

           Power 80% Power 90% Power 95%
Level 0.01        80        96       111
Level 0.05        58        72        85
Level 0.10        47        61        73

, , R2 = 0.5, p = 5

           Power 80% Power 90% Power 95%
Level 0.01        26        31        34
Level 0.05        20        23        27
Level 0.10        17        20        23

, , R2 = 0.8, p = 5

           Power 80% Power 90% Power 95%
Level 0.01        13        15        16
Level 0.05        11        12        13
Level 0.10        10        10        11
```

REFERENCES

Bedrick, Edward, 2014. Two useful reformulations of the hazard ratio. Am. Stat. 68, 36–41.

Berson, E.L., Rosner, B., Sandberg, M.A., Hayes, K.C., Nicholson, B.W., Weigel-DiFranco, C., Willett, W., 1993. A randomized trial of vitamin A and vitamin E supplementation for retinitis pigmentosa. Arch. Ophthalmol. 111, 761–772.

Champely, Stephane, 2009. 'pwr' R package.

Cohen, Jacob, 1988. Statistical Power Analysis for the Behavioral Sciences, second ed. Lawrence Erlbaum Associates, London.

Ryan, Thomas, 2013. Sample Size Determination and Power. Wiley, New York.

Chapter 7

Binomial Regression in R

John Muschelli*[,2], Joshua Betz*[,2] and Ravi Varadhan[†,1,2]
*Department of Biostatistics, Johns Hopkins Bloomberg School of Public Health, Baltimore, Maryland, USA
[†]Division of Biostatistics and Bioinformatics, Department of Oncology, Johns Hopkins University School of Medicine, Baltimore, Maryland, USA
[1]Corresponding author: e-mail: Ravi.varadhan@jhu.edu

ABSTRACT

Binomial regression is used to assess the relationship between a binary response variable and other explanatory variables. Popular instances of binomial regression include examination of the etiology of adverse health states using a case–control study and development of prediction algorithms for assessing the risk of adverse health outcomes (e.g., risk of a heart attack). In R, a binomial regression model can be fit using the `glm()` function. In this chapter, we demonstrate the following aspects of binomial regression, with R code, using real data examples:

- To highlight the main components of a binomial model fitting using the `glm()` function
- How to evaluate the modeling assumptions in binomial regression?
- How to relax the assumptions when they are violated?
- How to fit binomial models for non-independent data?
- How to develop and evaluate prediction models for binary response?

The chapter is meant to be a quick, practical guide to binomial regression using R. We particularly envision the accompanying *task view* to be a useful resource on all topics closely related to binomial regression.

Keywords: Binary response, Logistic regression, Model assumptions, Diagnostics, GLMM, GEE, Prediction

2. All authors contributed equally

Handbook of Statistics, Vol. 32. http://dx.doi.org/10.1016/B978-0-444-63431-3.00007-3

1 BINOMIAL REGRESSION IN THE GENERALIZED LINEAR MODEL

Binomial regression is used to assess the relationship between a binary response variable and other explanatory variables. Popular instances of binomial regression include examination of the etiology of adverse health states using a case–control study and development of prediction algorithms for assessing the risk of adverse health outcomes (e.g., risk of a heart attack). Nelder and Wedderburn (1972) provided a unified framework for the estimation of regression models for different types of response variables including normal, binary, and counts. They called this unified regression modeling framework, the *generalized linear model* (GLM). In R, a GLM can be fit using the `glm()` function, which has the essential syntax:

```
glm(formula, family, data)
```

where `formula` specifies the linear regression part of the model, `family` specifies the type of probability distribution for the response, and `data` specifies the `data.frame` containing the data. Commonly used families are `gaussian`, `binomial`, and `poisson`. For binary regression models using `glm()` and many other methods, specifying `family=binomial` is used.

The standard problem for a binary GLM is as follows: suppose we have data on N independent units: $(\mathbf{x}_i, y_i), i = 1, \ldots, N$, where $\mathbf{x}_i \in \mathcal{R}^p$ is a p-dimensional vector of covariate values and $y_i \in \{0,1\}$ is the binary response for unit i. The binomial model fit by `glm()` can be written as

$$\pi_i := \text{Prob}(y_i = 1) = h(\mathbf{x}_i'\beta) \tag{1}$$

where $h(\cdot)$ is the inverse-link function. The default model in `glm`, which is also the most widely used for binary data as described, is the logistic regression, for which the inverse link is $h(\mathbf{x}_i'\beta) = \exp(\mathbf{x}_i'\beta)/(1 + \exp(\mathbf{x}_i'\beta))$.

In this chapter, we wish to demonstrate the following, with R code, using real data examples:

1. To highlight the main components of a binomial model fitting using the `glm()` function
2. How to evaluate the modeling assumptions in binomial regression?
3. How to relax the assumptions when they are violated?
4. How to fit binomial models for nonindependent data?
5. How to develop and evaluate prediction models for binary response?
6. How to incorporate study design (e.g., case–control, matched case–control, complex survey) in binomial regression modeling?

Our chapter is meant to be a quick, practical guide to binomial regression using R. We particularly envision the accompanying *task view* to be a useful resource on all topics closely related to binomial regression. For more detailed

TABLE 1 Datasets Used in This Chapter

Dataset	Description	Rows	Columns	Source
`aSAH`	Subarachnoid hemorrhage data	113	7	`pROC` package
`beetle`	Beetles exposed to fumigant	10	3	`faraway` package
`ROCR.simple`	Simple artificial prediction data for use with ROCR	200	2	`ROCR` package
`phoneme`	Data from a acoustic–phonetic continuous speech corpus	4509	258	`phoneme` package
`aom`	Acute otitis media in children	8	5	Rovers et al. (2006)
`schiz`	National Institute of Mental Health schizophrenia study	1603	9	Hedeker (2005)[a]

[a]*http://tigger.uic.edu/~hedeker/SCHIZX1.DAT.txt.*

discussions of binomial regression and its theory, we refer the reader to texts such as Hosmer et al. (2000), Collett (2003), and McCullagh and Nelder (1989). The book by Faraway (2005) provides an excellent, practical guide to GLM and its extensions using R.

The datasets used in this chapter are listed in Table 1.

2 STANDARD LOGISTIC REGRESSION

Response variables in binomial regression can be in two formats: (i) as a binomial variable where each row contains the number of successes, and number of failures from a group of Bernoulli experiments, and (ii) as a Bernoulli variable, where each row represents the outcome of a single Bernoulli experiment taking one of two values, usually 0 or 1. We start with a dataset in the format (i) using the `beetle` dataset from the `faraway` package, which contains 3 variables and 10 rows. Each row represents an independent binomial experiment, where a certain number of beetles were exposed to a particular concentration of an insecticide to see how many were affected. These data can be fitted using `glm()` in two ways, both of which are equivalent, as shown below.

```
require(faraway)
data(beetle)
# two ways of fitting
ans1 <- glm(cbind(affected, exposed-affected) ~ conc,
            data=beetle, family=binomial)
```

```
ans2 <- glm(affected/exposed ~ conc, weights=exposed,
            data=beetle, family=binomial)
all.equal(coef(ans1), coef(ans2))

[1] TRUE

summary(ans1)

Call:
glm(formula = cbind(affected, exposed - affected) ~ conc,
    family = binomial, data = beetle)

Deviance Residuals:
    Min       1Q   Median       3Q      Max
-3.067   -1.712    0.149    1.734    2.415

Coefficients:
                Estimate  Std. Error  z value    Pr(>|z|)
(Intercept)     -6.0105      0.7794     -7.71    1.20e-14   ***
conc             0.3413      0.0415      8.22     < 2e-16   ***
---
Signif. codes: 0 '***' 0.001 '**' 0.01 '*' 0.05 '.' 0.1 ' ' 1

(Dispersion parameter for binomial family taken to be 1)

    Null deviance: 138.001 on 9 degrees of freedom
Residual deviance: 37.697 on 8 degrees of freedom
AIC: 69.27

Number of Fisher Scoring iterations: 5
```

In ans1, the left hand of the formula (left of the ~) is a two-column matrix with affected in the first column and not affected in the second. In ans2, we model the proportion affected, weighted by the number of exposed. The summary command produces the table of coefficients, standard errors, z-statistics, and p-values. The coefficients are either log-odds ratios (conc) or, in the case of the (Intercept), log-odds. Exponentiating these log-odds ratios will give the odds ratio, which is the form commonly reported.

We can then get confidence intervals (CIs) for the coefficients and use exp to get a CI for the odds ratio. Note that we calculated the CI *before* exponentiating and not vice versa.

```
confint.default(ans2)
                  2.5%     97.5%
(Intercept)    -7.5380   -4.4829
conc            0.2599    0.4227

exp(confint.default(ans2))

                    2.5%    97.5%
(Intercept)    0.0005325   0.0113
conc           1.2967612   1.5260
```

3 ASSUMPTIONS INVOLVED IN THE STANDARD LOGISTIC REGRESSION MODEL

GLMs provide a framework for modeling many different types of outcomes, but assumptions underlying the GLM are often overlooked and the impact of violating these assumptions is underappreciated. According to Breslow (1996), conclusions from the analyses may be seriously compromised when one or more of the following conditions are present:

- incorrect form of the explanatory variables
- undue influence of individual observations on the model fit
- incorrect specification of the variance function, of the dispersion factor or of the link function
- statistical independence of the n observations

These conditions represent the possible ways in which the main assumptions of GLM can be violated. We will discuss how these assumption violations are presented in practice, how to test for these (along with some rules-of-thumb judgments), and strategies to account for these violations.

4 RESIDUALS

After fitting the model, we can see how well our model predicts the outcome. Residuals are the differences between the observed data and the expected value from the fitted model. By examining how the model predictions depart from the data, we can determine whether the assumptions of the model are being met, identify influential observations, and find ways to improve various aspects of the model.

We will examine a logistic regression model with `age`, `s100b`, and `NDKA` in a logistic regression model from the `aSAH` dataset (run `?aSAH` for definitions):

```
library(pROC)
data(aSAH)
sah.all <- glm(outcome=="Good"~s100b+age+ndka, data=aSAH,
               family=binomial)
```

The working residuals (r_{W_i}) are residuals scaled by the variance of the observation, in order to account for the fact that the variance of the response is not independent of the mean:

$$r_{W_i} = \frac{y_i - E[y_i]}{Var(y_i)} = \frac{y_i - n_i\hat{\pi}_i}{n_i\hat{\pi}_i(1-\hat{\pi}_i)} \quad (2)$$

The Pearson residuals (r_{P_i}) are standardized by the square root of the variance:

$$r_{P_i} = \frac{y_i - E[y_i]}{\sqrt{Var(y_i)}} = \frac{y_i - n_i\hat{\pi}_i}{\sqrt{n_i\hat{\pi}_i(1-\hat{\pi}_i)}} \quad (3)$$

As such, these can be considered similar to t-statistics, allowing us to identify observations that fall outside the range predicted by the t-distribution.

Just as the residual sum of squares can be used to indicate goodness of fit in a linear model (LM), the sum of squared Pearson residuals gives an indication of goodness of fit in a GLM, the Pearson's χ^2 statistic (χ_P^2).

$$\chi_P^2 = \sum_{i=1}^{n} \frac{(y_i - n_i\hat{\pi}_i)^2}{\sqrt{n_i\hat{\pi}_i(1-\hat{\pi}_i)}} \tag{4}$$

The deviance (D) is another model fit measure based on residuals, based on the log-likelihood of the fitted model. It can be decomposed into the sum of squared deviance residuals, separating the how well the data fit for the $y_i = 0$ versus $y_i = 1$:

$$D = 2\sum_{i=1}^{n} y_i \log\left(\frac{y_i}{n_i\hat{\pi}_i}\right) + (n_i - y_i)\log\left(\frac{n_i - y_i}{n_i(1-\hat{\pi}_i)}\right) \tag{5}$$

The sign of the residuals can be obtained by the difference between the observed and fitted values, which indicates if the model over- or underpredicts y_i:

$$r_{D_i} = \text{sign}(y_i - n_i\hat{\pi}_i)\sqrt{2\left(\log\left(\frac{y_i}{n_i\hat{\pi}_i}\right) + (n_i - y_i)\log\left(\frac{n_i - y_i}{n_i(1-\hat{\pi}_i)}\right)\right)} \tag{6}$$

Under the null hypothesis that the model is correctly specified, the Pearson χ^2 and deviance both asymptotically follow a chi-square distribution with $N - p$ degrees of freedom, where N is the number of unique covariate patterns (e.g., if there are 100 people and 2 people have the exact same covariate, then N is 99 since there is a duplicate) and p is the number of parameters estimated. Large values of these statistics relative to the degrees of freedom can indicate poor fit, but failing to reject the null hypothesis does not necessarily indicate adequate model fit. We can then use these to test if the model fit is inadequate:

```
# Calculate Pearson and Deviance Chi-Square:
1-pchisq(sum(resid(sah.all, type="pearson")^2),
         sah.all$df.residual)

[1] 0.4931

1-pchisq(sum(resid(sah.all, type="deviance")^2),
         sah.all$df.residual)

[1] 0.2609
```

Leverage is a measure of how different an observation is in their covariate pattern irrespective of the outcome. The leverage for the ith observation is given by $1 - h_{ii}$, where h_{ii} is the ith element of the diagonal of the hat matrix $\mathbf{H} = \mathbf{X}(\mathbf{X}^{\mathrm{T}}\mathbf{X})^{-1}\mathbf{X}^{\mathrm{T}}$. Pearson and deviance residuals can be scaled using the leverage; these scaled residuals have variance 1 and asymptotically have a standard normal distribution if the number of observations corresponding to each pattern of covariates is large enough:

$$r_{S_i} = r_i / \sqrt{1 - h_{ii}} \tag{7}$$

In R, both scaled and unscaled residuals, using `resid()` and `rstandard()`, respectively, default to deviance residuals, yet the residuals inside the `glm` object are actually the working residuals. Externally studentized residuals, residuals obtained from the model leaving out observation i can be obtained using `rstudent()`.

4.1 Interpreting Residuals

In the `aSAH` data, since the three covariates are continuous, the conditions for asymptotic normality are unlikely to be met in this dataset. Even if asymptotic normality has not been achieved, comparing the residual distribution to normal or half-normal quantiles can still be used to help detect potential lack of fit. Observations that do not follow a general trend may indicate of lack of fit (Faraway, 2005). We present a quantile–quantile (QQ) plot of the half-normal quantiles compared to the residuals (Fig. 1).

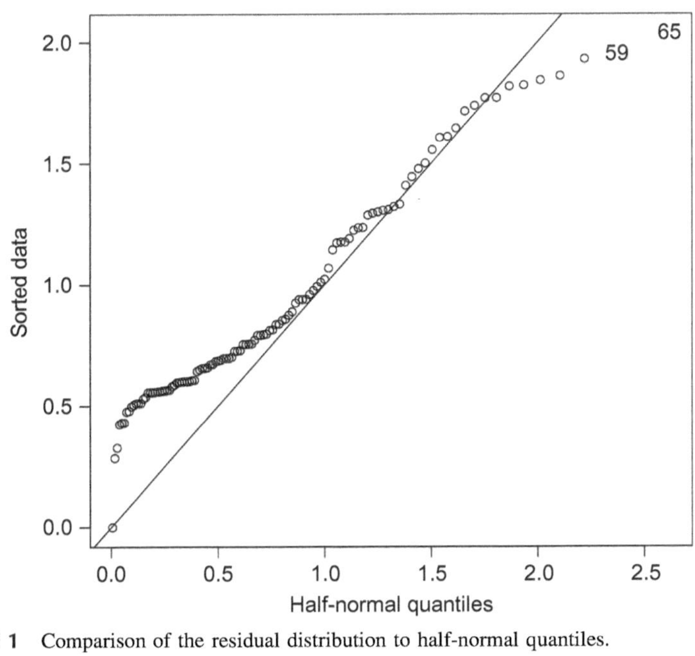

FIGURE 1 Comparison of the residual distribution to half-normal quantiles.

```
# Half Normal Plot
library(faraway)
halfnorm(resid(sah.all))
abline(a=0, b=1)
```

While the residuals do not exactly follow the expected quantiles based on the half-normal distribution, indicated by the diagonal line in Fig. 1, all of the residuals follow a generally contiguous trend. In this dataset, the half-normal residual plot does not indicate any unusual patterns in the residuals.

Just as in ordinary least squares regression, we should also plot the residuals against the fitted values and covariates (Fig. 2).

```
par(mfrow=c(2,1), mar=c(4, 4, 1, 1))
plot(rstandard(sah.all)~fitted(sah.all),
     ylab="Standardized Deviance Residual", xlab="Fitted Value")
# Predict LOESS fit over range of fitted values
fit.range <- seq(-.99, .99, 0.01)
```

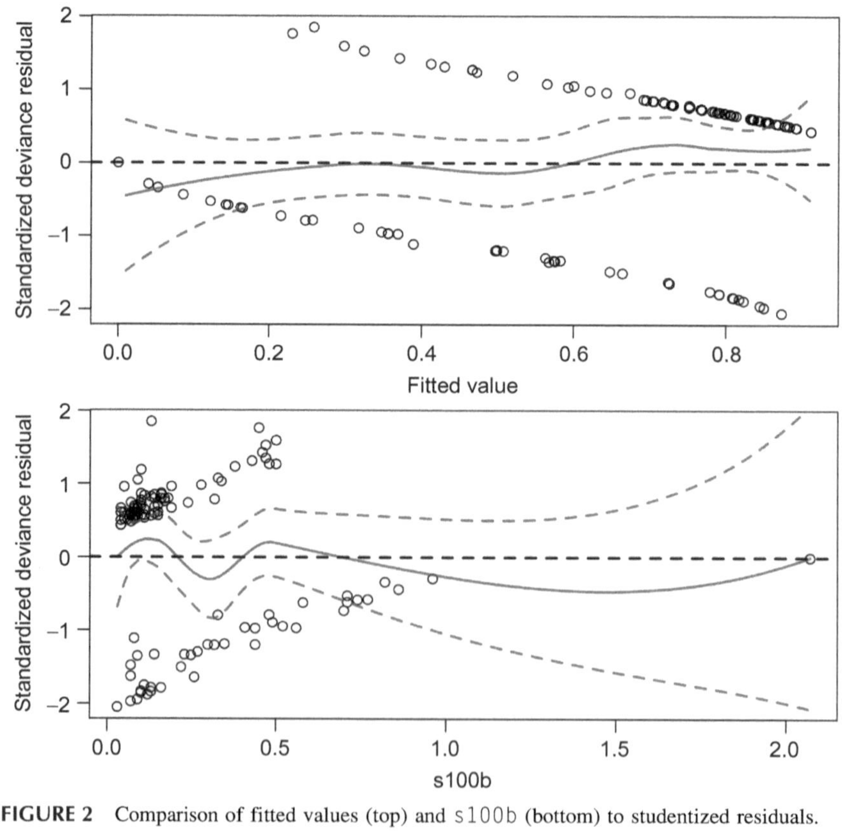

FIGURE 2 Comparison of fitted values (top) and `s100b` (bottom) to studentized residuals.

```
rloess <-predict(loess(rstandard(sah.all)~fitted(sah.all)),
                     newdata=fit.range, se=TRUE)
lines(fit.range, rloess$fit, lwd=2, col=rgb(1,0,0,.75))
lines(fit.range, rloess$fit - qt(0.975, rloess$df )*rloess$se,
       lty=2, lwd=2, col=rgb(1,0,0,.75))
lines(fit.range, rloess$fit + qt(0.975, rloess$df )*rloess$se,
       lty=2, lwd=2, col=rgb(1,0,0,.75))
abline(h=0, lty=2, lwd=2)

plot(rstandard(sah.all)~sah.all$model$s100b,
     ylab="Standardized Deviance Residual", xlab="s100b")
fit.range <- seq(min(sah.all$model$s100b),
                    max(sah.all$model$s100b), length.out=100)
rloess <-predict(loess(rstandard(sah.all)~aSAH$s100b),
                     newdata=fit.range, se=TRUE)
lines(fit.range, rloess$fit, lwd=2, col=rgb(1,0,0,.75))
lines(fit.range, rloess$fit - qt(0.975, rloess$df )*rloess$se,
       lty=2, lwd=2, col=rgb(1,0,0,.75))
lines(fit.range, rloess$fit + qt(0.975, rloess$df )*rloess$se,
       lty=2, lwd=2, col=rgb(1,0,0,.75))
abline(h=0, lty=2, lwd=2)
```

The LOESS (local regression) (Cleveland et al., 1992) of the residuals on the fitted values is roughly flat, around zero, and smooth with no discernible pattern. If a pattern is observed, the model may be misspecified. Using transformations of the explanatory variables may be needed in that case. We see the LOESS of the residuals on `s100b` has a discernable pattern within `s100b` from 0 to 1. This suggests that there may be a nonlinear relationship between `s100b` and the log-odds of a good functional outcome ($Y = 1$). We also see one point far away (`s100b` > 2) from the majority of the `s100b` data, which may be an influential or outlying point, which we will investigate later using influential point measures. Instead of looking for lack of linearity with the standardized residuals, the correct functional form may be easier to see with partial residuals, which are the residuals from a model without a given covariate in the model (Fig. 3).

```
plot(data.frame(resid(sah.all, type="partial"))$age~sah.all
     $model$age, ylab="Partial Residual: Age", xlab="Age")
fit.range<-seq(min(sah.all$model$age),max(sah.all$model$age),
                   length.out=100)
rloess <-predict(loess(data.frame(resid(sah.all, type
                     ="partial"))$age~aSAH$age),     newdata=fit.
                     range, se=TRUE)
lines(fit.range, rloess$fit, lwd=2, col=rgb(1,0,0,.75))
lines(fit.range, rloess$fit - qt(0.975, rloess$df )*rloess$se,
       lty=2, lwd=2, col=rgb(1,0,0,.75))
lines(fit.range, rloess$fit + qt(0.975, rloess$df )*rloess$se,
       lty=2, lwd=2, col=rgb(1,0,0,.75))
abline(h=0, lty=2, lwd=2)
```

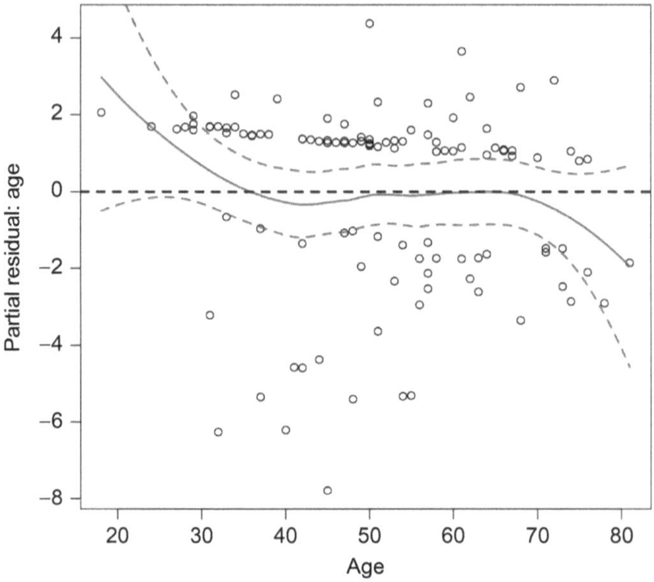

FIGURE 3 Partial residual plot for age in `sah.all` logistic regression model.

4.2 Influential Points

While lack of fit can be caused by an improperly specified model, it is also possible that highly influential observations are exerting undue influence over parameter estimates. Influential observations have a large effect on model-based estimation and inference. Outlying observations are ones that are not consistent with general trends observed in the data. An outlier is not always influential, as its removal may not alter the estimates or inference, and an influential point is not always an outlier.

Again, leverage is a measure of how different an observation is in their covariate pattern irrespective of the outcome. The standardized residuals indicate a discrepancy between the model prediction and the observed response. The DFFITS (Belsey et al., 1980) and Cook's distance (Cook, 1977) combine information from the leverage $(1 - h_{ii})$ and the standardized residuals (r_{D_i}) to obtain information about which observations may be highly influential:

$$\text{DFFITS}_i = r_{D_i}\sqrt{\frac{h_{ii}}{1-h_{ii}}} \qquad D_{\text{Cook}_i} = \frac{1}{p}\left(\frac{h_{ii}}{1-h_{ii}}\right) r_{D_i}^2 \tag{8}$$

We can also determine highly influential data points by removing observations one at a time, reobtaining the maximum likelihood estimates, and see how and how much they change. This gives the DFBETA (Belsey et al., 1980) statistics for each parameter. Since outliers can also influence the standard errors of parameters, the covariance ratio looks at the covariance matrix of the estimated parameters with and without an observation in the dataset.

We can obtain all of these along with indications of whether an observation is influential based on each measure using guidelines proposed by Belsey et al. 1980) and Cook and Weisberg 1980). These guidelines are a function of the number covariates and overall sample size and are implemented on `influence.measures`.

```
sah.influence <- influence.measures(sah.all)
# sah.influence\$is.inf gives a matrix of logical values for
whether a point was
# influential by each measure - can sum by row to find influential
points
inf.points <- apply(1*sah.influence$is.inf, 1, sum)
inf.points <- inf.points[which(inf.points > 0)]

# Find which influence measures were notable, and print their values
sah.influence$is.inf[which(rownames(sah.influence$infmat) %in%
                                        names(inf.points)),]

    dfb.1_ dfb.s100 dfb.age dfb.ndka dffit cov.r cook.d   hat
48   FALSE    FALSE   FALSE    FALSE FALSE  TRUE  FALSE FALSE
57   FALSE    FALSE   FALSE    FALSE  TRUE  TRUE  FALSE  TRUE
90   FALSE    FALSE   FALSE    FALSE FALSE  TRUE  FALSE  TRUE
97   FALSE    FALSE   FALSE    FALSE FALSE  TRUE  FALSE FALSE
101  FALSE    FALSE   FALSE    FALSE FALSE  TRUE  FALSE  TRUE

sah.influence$infmat[which(rownames(sah.influence$infmat) %in%
                                        names(inf.points)),]

       dfb.1_  dfb.s100   dfb.age dfb.ndka   dffit cov.r   cook.d     hat
48   0.07577 -0.08465 -0.007937 -0.22042 -0.2371 1.115 0.008833 0.08915
57  -0.36534  0.06070  0.161358  0.89564  0.9355 1.171 0.246251 0.21264
90   0.09693 -0.01882 -0.165304  0.29004  0.3684 1.167 0.022556 0.13839
97  -0.04776 -0.18076  0.100803 -0.01528 -0.2097 1.111 0.006768 0.08293
101  0.16150  0.00633 -0.059282 -0.50549 -0.5360 1.245 0.050198 0.20108
```

In the presence of influential points, it may be wise to conduct a sensitivity analysis, using robust estimates of the regression coefficients, obtained using robust/resistant estimation techniques (Cantoni, 2004; Cantoni and Ronchetti, 2001, 2006; Heritier et al., 2009; Valdora and Yohai, 2014; Wang et al., 2014). These estimates are less prone to bias from influential data.

```
library(robust)
sah.all.rob <- glmrob(outcome=="Good"~s100b+age+ndka,
                      data=aSAH, family=binomial)
```

```
summary(sah.all)$coefficients

               Estimate   Std. Error   z value     Pr(>|z|)
(Intercept)     3.45858      0.99544     3.474   5.120e-04
s100b          -5.04662      1.25878    -4.009   6.094e-05
age            -0.02202      0.01693    -1.301   1.934e-01
ndka           -0.03131      0.01631    -1.920   5.484e-02

summary(sah.all.rob)$coefficients

               Estimate   Std. Error   z value     Pr(>|z|)
(Intercept)     3.53962      1.02418     3.456   0.0005482
s100b           4.74478      1.26221    -3.759   0.0001705
age            -0.02457      0.01729    -1.421   0.1552039
ndka           -0.03157      0.01650    -1.913   0.0557637
```

Since the differences in parameters are small relative to their standard errors, and the standard errors are consistent between models, it looks like the influential points are not unduly influential.

5 OVERDISPERSION

Another important aspect of model fit is evidence of overdispersion, which indicates how well the variance in a binomial GLM is specified. Overdispersion is present when the deviance is larger than the residual degrees of freedom. Revisiting our `beetle` example:

```
ans1 <- glm(cbind(affected, exposed-affected) ~ conc, data=beetle,
            family=binomial)
summary(ans1)

Call:
glm(formula = cbind(affected, exposed - affected) ~ conc,
    family = binomial, data = beetle)

Deviance Residuals:
   Min       1Q   Median      3Q     Max
-3.067   -1.712    0.149   1.734   2.415

Coefficients:
             Estimate   Std. Error   z value   Pr(>|z|)
(Intercept)   -6.0105      0.7794     -7.71    1.2e-14    ***
conc           0.3413      0.0415      8.22    < 2e-16    ***
---
Signif. codes: 0 '***' 0.001 '**' 0.01 '*' 0.05 '.' 0.1 ' ' 1

(Dispersion parameter for binomial family taken to be 1)

    Null deviance: 138.001 on 9 degrees of freedom
Residual deviance: 37.697 on 8 degrees of freedom
AIC: 69.27

Number of Fisher Scoring iterations: 5
```

The residual deviance is 37.7 on 8 degrees of freedom, clearly suggesting that overdispersion is present. A rough estimate of overdispersion is given by 37.7/8 = 4.7. When overdispersion is present, the standard errors of regression coefficients are underestimated, roughly by a factor of $\sqrt{\hat{\phi}}$, where $\hat{\phi}$ is the estimate of overdispersion. Thus, we would expect the standard error to be deflated roughly by a factor of 2.2.

Overdispersion can arise from many ways such as model misspecification and correlation among responses. To properly address overdispersion it is best to try to identify which sources of overdispersion exist.

We can account for overdispersion in several different ways:

1. changing model specification: adding more explanatory variables, if available, to the model or transforming explanatory variables (e.g., `log`, `sqrt`, etc.),
2. estimate using quasilikelihood or a generalized estimating equation (GEE) approach, with robust standard errors,
3. fit a generalized linear mixed model (GLMM).

We will first illustrate the quasilikelihood approach using the `beetle` dataset and then illustrate adding explanatory variables using another dataset. We will illustrate GEE and GLMM approaches in Section 8 on modeling longitudinal (repeated measures) binary data.

5.1 Estimation Using Quasilikelihood

To estimate the model using quasilikelihood, we can use `glm` as before. Instead of using `family=binomial`, we fit a quasibinomial family (`family=quasibinomial`), which estimates the dispersion parameter (ϕ).

```
qbinmod = glm(cbind(affected, exposed-affected) ~ conc,
              data=beetle, family=quasibinomial)
summary(qbinmod)
```

```
Call:
glm(formula = cbind(affected, exposed - affected) ~ conc,
    family = quasibinomial, data = beetle)

Deviance Residuals:
   Min       1Q   Median      3Q     Max
-3.067   -1.712    0.149   1.734   2.415

Coefficients:
              Estimate  Std. Error  t value  Pr(>|t|)
(Intercept)   -6.0105      1.6407    -3.66    0.0064   **
conc           0.3413      0.0874     3.90    0.0045   **
---
Signif. codes: 0 '***' 0.001 '**' 0.01 '*' 0.05 '.' 0.1 ' ' 1
```

```
(Dispersion parameter for quasibinomial family taken to be 4.432)

    Null deviance: 138.001 on 9 degrees of freedom
Residual deviance: 37.697 on 8 degrees of freedom
AIC: NA

Number of Fisher Scoring iterations: 5
```

When we compare this result from the quasibinomial model to that previously obtained from the binomial fit, `ans1`, we observe some important differences: (i) the parameter estimates are identical; (ii) an estimate of dispersion parameter is given; (iii) standard errors are increased by a factor equal to the square root of the dispersion parameter; and (iv) the Akaike information criterion (AIC) is not shown. The estimate of overdispersion, $\hat{\phi} = 4.43$, is not too far from our approximation of 4.7. The ratio of standard errors of the estimate of coefficient of `conc` is 0.0874/0.0415 = 2.10, which is equal to $\sqrt{\hat{\phi}} = (4.43)^{1/2}$. The AIC is not displayed for quasibinomial models because there is no explicit log-likelihood to be maximized as in the case of binomial models. Thus, this limits approaches for model selection that rely on the log-likelihood, discussed in Section 9.

Let us look at the accuracy of standard error from the quasilikelihood approach by comparing it to bootstrap (with 1000 replications), using the `boot` package (Canty and Ripley, 2014; Davison and Hinkley, 1997).

```
# Bootstrap using 'boot' library
require(boot)
binfun <- function(dat, inds)
{
coefs <- coef(glm(cbind(affected, exposed-affected) ~ conc,
                    data=dat[inds,], family=binomial))
coefs
}
system.time(bin.boot <- boot(beetle, binfun, R = 1000))

   user  system  elapsed
  1.944   0.005    1.953

bin.boot

ORDINARY NONPARAMETRIC BOOTSTRAP

Call:
boot(data = beetle, statistic = binfun, R = 1000)

Bootstrap Statistics :
      original       bias   std. error
t1*    -6.0105   -0.42349      1.68094
t2*     0.3413    0.02349      0.08941
```

Standard errors given by the bootstrap are larger than those from the quasilikelihood approach (Bootstrap: 1.681 vs. Quasibinomial: 1.641 for the intercept, Bootstrap: 0.089 vs. Quasibinomial: 0.087 for the slope). This behavior is quite typical. See Breslow (1996) for a detailed discussion of this issue. Using the `boot::boot.ci` function, we can compute several different types of CIs for the parameter estimates. We show the CI estimated using the bias-corrected bootstrap.

```
boot.ci(bin.boot, index=2)

BOOTSTRAP CONFIDENCE INTERVAL CALCULATIONS
Based on 1000 bootstrap replicates

CALL :
boot.ci(boot.out = bin.boot, index = 2)

Intervals :
Level      Normal              Basic
 95 %   (0.1425, 0.4930)   (0.1212, 0.4718)

Level     Percentile            BCa
 95 %   (0.2107, 0.5613)   (0.1714, 0.5076)

Calculations and Intervals on Original Scale
Some BCa intervals may be unstable

# confidence interval: bias-corrected bootstrap
round(boot.ci(bin.boot, index=2)$bca[c(4,5)],3)

[1] 0.171 0.508
```

The bootstrap CI is slightly larger (on the left side) compared to the CI using the quasibinomial model:

```
round(confint.default(qbinmod, parm = "conc"), 3)

       2.5 %   97.5 %
conc   0.17    0.513
```

5.2 Adding Explanatory Terms to the Model

We use a different dataset (`aom`) to demonstrate how overdispersion may be eliminated by using a richer (i.e., larger) regression model. Let us review the data structure of `aom`:

```
# see Table 3 on p. 1432 of Rovers et al.
aom <- read.csv("aom.csv")
print(aom)

     no   yes   trt   ageG2   bilateral
1    87    42     0       0           0
2    91    41     1       0           0
3   249    59     0       1           0
4   259    44     1       1           0
```

```
5    72   62   0    0    1
6   107   32   1    0    1
7    61   26   0    1    1
8    80   16   1    1    1
```

Let us fit a binomial model with only the main effects.

```
summary(ans.aom <- glm(cbind(yes, no) ~ trt + ageG2 + bilateral,
                       data=aom, family=binomial))
Call:
glm(formula = cbind(yes, no) ~ trt + ageG2 + bilateral,
    family = binomial, data = aom)

Deviance Residuals:
     1       2       3       4       5       6       7       8
-0.785   1.535  -0.820   0.365   1.073  -1.848   0.941  -0.083

Coefficients:
             Estimate  Std. Error  z value  Pr(>|z|)
(Intercept)    -0.582       0.133    -4.38   1.2e-05  ***
trt            -0.510       0.132    -3.87   0.00011  ***
ageG2          -0.741       0.136    -5.43   5.7e-08  ***
bilateral       0.246       0.140     1.76   0.07833  .
---
Signif. codes: 0 '***' 0.001 '**' 0.01 '*' 0.05 '.' 0.1 ' ' 1

(Dispersion parameter for binomial family taken to be 1)

    Null deviance: 65.2030 on 7 degrees of freedom
Residual deviance: 9.2405 on 4 degrees of freedom
AIC: 58.37

Number of Fisher Scoring iterations: 4
```

Since the residual deviance is more than twice as large as the degrees of freedom, the reported standard errors would be too small by a factor of $\sqrt{2} = 1.41$; thus, the data may exhibit overdispersion. To rule out model misspecification as a source of the overdispersion, let us try adding an interaction term, the interaction of the treatment with `bilateral` status variable.

```
summary(ans.aom2 <- glm(cbind(yes, no) ~ ageG2 + trt*bilateral,
                        data=aom, family=binomial))

Call:
glm(formula = cbind(yes, no) ~ ageG2 + trt * bilateral,
    family = binomial, data = aom)

Deviance Residuals:
      1       2       3       4       5       6       7       8
-0.0978 0.7235 0.0757 -0.6020 -0.0768 -0.5556 0.1042 0.8109
```

```
Coefficients:
                Estimate  Std. Error  z value   Pr(>|z|)
  (Intercept)    -0.710      0.143     -4.96      7e-07   ***
        ageG2    -0.741      0.137     -5.42      6e-08   ***
          trt    -0.225      0.168     -1.34     0.1802
    bilateral     0.574      0.184      3.11     0.0019    **
trt:bilateral    -0.735      0.273     -2.70     0.0070    **
---
Signif. codes: 0 '***' 0.001 '**' 0.01 '*' 0.05 '.' 0.1 ' ' 1

(Dispersion parameter for binomial family taken to be 1)

    Null deviance: 65.203 on 7 degrees of freedom
Residual deviance: 1.884 on 3 degrees of freedom
AIC: 53.01

Number of Fisher Scoring iterations: 4
```

The residual deviance is small relative to the residual degrees of freedom. Thus, changing the model specification has eliminated the overdispersion.

6 HYPOTHESIS TESTING AND INFERENCE

Now that we have performed diagnostics to check model assumptions, we can test hypotheses about the model. Let us start with the summary of the multivariate model from the SAH data:

```
summary(sah.all)

Call:
glm(formula = outcome == "Good" ~ s100b + age + ndka,
    family = binomial, data = aSAH)

Deviance Residuals:
   Min       1Q   Median      3Q     Max
-2.034   -0.874    0.564   0.755   1.716

Coefficients:
             Estimate  Std. Error  z value   Pr(>|z|)
(Intercept)    3.4586     0.9954     3.47    0.00051   ***
s100b         -5.0466     1.2588    -4.01    6.1e-05   ***
age           -0.0220     0.0169    -1.30    0.19340
ndka          -0.0313     0.0163    -1.92    0.05484    .

---
Signif. codes: 0 '***' 0.001 '**' 0.01 '*' 0.05 '.' 0.1 ' ' 1

(Dispersion parameter for binomial family taken to be 1)

    Null deviance: 148.04 on 112 degrees of freedom
Residual deviance: 118.03 on 109 degrees of freedom
AIC: 126

Number of Fisher Scoring iterations: 5
```

The output from `summary.glm()` gives the Wald hypothesis tests, based on the asymptotic normality of maximum likelihood estimates. The *p*-values from these tests are located in the column `Pr(>|z|)`. To obtain a comparable test using the deviance chi-square, we can use the `drop1` function:

```
drop1(sah.all, test="Chisq")

Single term deletions
Model:
outcome == "Good" ~ s100b + age + ndka
         Df   Deviance   AIC     LRT   Pr(>Chi)
<none>             118   126
s100b     1        140   146   21.63    3.3e-06   ***
age       1        120   126    1.73      0.189
ndka      1        122   128    3.88      0.049   *
---
Signif. codes: 0 '***' 0.001 '**' 0.01 '*' 0.05 '.' 0.1 ' ' 1
```

The `anova()` function gives sequential tests, which look at the incremental change in the likelihood with the addition of each successive term. In these tests, the order of the covariates can play an important difference. The `Anova()` function in the `car` (Fox and Weisberg, 2011) package gives results similar to `drop1`, but adds additional features, such as the ability to use empirical standard errors using the `white.adjust` option, which uses heteroscedasticity-corrected or "Huber-White" covariance matrices (Huber, 1967; White, 1980).

```
library(car)
Anova(sah.all, white.adjust=TRUE)

Analysis of Deviance Table (Type II tests)

Response: outcome == "Good"
        LR Chisq   Df   Pr(>Chisq)
s100b      21.63    1      3.3e-06   ***
age         1.73    1        0.189
ndka        3.88    1        0.049   *
---
Signif. codes: 0 '***' 0.001 '**' 0.01 '*' 0.05 '.' 0.1 ' ' 1
```

The general linear hypothesis test, `glht()` from `multcomp` package (Hothorn et al., 2008), allows us to test hypotheses about linear combinations of parameters. This is especially relevant when comparing levels of categorical variables, as it allows us to test if parameters are different, and obtain CIs for these differences:

```
library(multcomp)
# Make WFNS an unordered factor
aSAH$wfns <- factor(aSAH$wfns, ordered=FALSE)
sah.wfns <- glm(outcome=="Good"~wfns, data=aSAH,
                family=binomial)
```

```
summary(glht(sah.wfns, linfct = c("wfns2 - wfns3 = 0")))

Simultaneous Tests for General Linear Hypotheses

Fit: glm(formula = outcome == "Good" ~ wfns, family = binomial,
      data = aSAH)

Linear Hypotheses:
                     Estimate Std. Error  z value Pr(>|z|)
wfns2 - wfns3 == 0    -0.588     1.211     -0.49    0.63
(Adjusted p values reported - single-step method)

confint(glht(sah.wfns, linfct = c("wfns2 - wfns3 = 0")))

Simultaneous Confidence Intervals

Fit: glm(formula = outcome == "Good" ~ wfns, family = binomial,
      data = aSAH)

Quantile = 1.96
95% family-wise confidence level

Linear Hypotheses:
                     Estimate lwr     upr
wfns2 - wfns3 == 0 -0.588   -2.961 1.786
```

The estimate (– 0.588) represents the log-odds of a good outcome when comparing the WFNS (World Federation of Neurological Surgeons) of a level 2 to a level 3 (level 3 is better). Thus, the estimated log-odds ratio of a good outcome comparing level 2 to level 3 is $e^{-0.588}$, or 0.555, indicating level 3 is better. This may seem surprising, but only 4 patients had a level 3; this demonstrates the fact that inferential statements need to be made considering many factors.

7 MODEL PERFORMANCE

In order to assess the performance of a method employed with binary outcomes, we must define which criteria we wish to maximize. For example, if we have a strong assumption that the data are truly logistic, using the model AIC or deviance may be criteria to use to choose one model over the other. In other situations, our end goal may be different, such as prediction accuracy, and likelihood-based measures may not be the most appropriate. Receiver operating characteristic (ROC) curves are commonly used to assess the predictive performance of a model/classifier at a series of cutoffs of the predictor. Area under the curve (AUC) is also used as a one-dimensional summary of the entire ROC curve.

7.1 ROC Curves/Sensitivity/Specificity/Accuracy

Here, we introduce basic notation to define the terminology used in ROC/AUC analysis. Let x_i be some measure used to classify Y_i, where i refers to observation. For a given cutoff c, $\hat{Y}_i$, the predicted value of Y_i, is given by the following rule:

$$\hat{Y}_i = \begin{cases} 1 & \text{if } x_i \geq c \\ 0 & \text{if } x_i < c \end{cases}$$

From these definitions, we can estimate the sensitivity of the predictor (also called true positive rate), which is the probability of accurately predicting a positive outcome ($\hat{Y}_i = 1$), conditional that the observation is truly positive ($Y_i = 1$):

$$\hat{P}(\hat{Y}_i = 1 | Y_i = 1) = \frac{1}{n} \sum_i I(\hat{Y}_i = 1 | Y_i = 1)$$

the specificity (or the true negative rate) of the predictor, defined as the probability of predicting a negative outcome when Y_i is truly negative:

$$\hat{P}(\hat{Y}_i = 0 | Y_i = 0) = \frac{1}{n} \sum_i I(\hat{Y}_i = 0 | Y_i = 0)$$

or the overall accuracy of the predictor:

$$\hat{P}(\hat{Y}_i = Y_i) = \frac{1}{n} \sum_i I(\hat{Y}_i = Y_i)$$

where n is the number of observations. The ROC curve plots 1 − specificity versus the sensitivity for all possible cut points (Fig. 4).

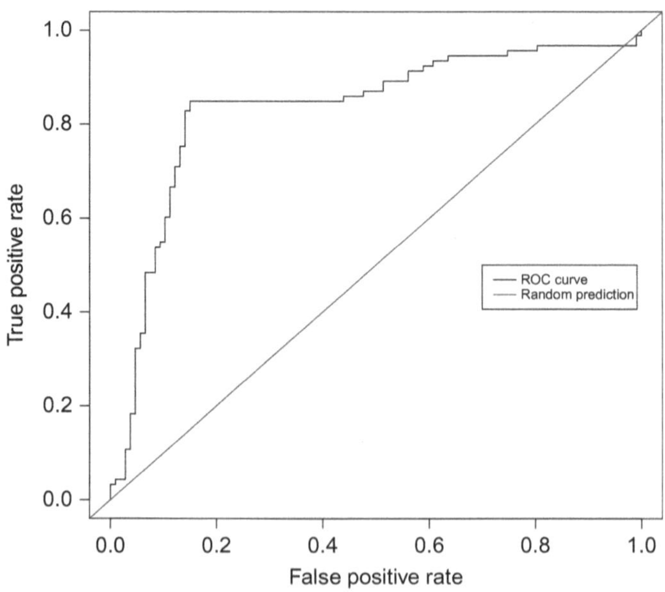

FIGURE 4 Example of an ROC curve.

ROCR (Sing et al., 2005) and pROC (Robin et al., 2011) are two commonly used packages for ROC analysis. In order to create an ROC curve in ROCR, you must create a prediction object, which takes in a prediction from a model/algorithm and the true Y values (referred to as "labels" in ROCR). The prediction object is then passed into a performance object, which can create many metrics of performance. In the case below, we specify that we would like to assess the true positive rate (tpr) versus the false positive rate (fpr = 1 − specificity) and plot the result (Fig. 4).

```
library(ROCR)
data(ROCR.simple)
pred <- prediction( ROCR.simple$predictions, ROCR.simple$labels)
perf <- performance(pred,"tpr","fpr")
plot(perf)
abline(a=0, b=1, col="red")
legend(x=.7, y=.5, legend=c("ROC Curve", "Random prediction"),
lty=c(1,1), col=c("black", "red"), cex=.75)
```

7.2 Area Under the Curve

Strength of ROC curves is that they show the estimates of criteria over all cut points, do not rely on the outcome prevalence, and can weigh sensitivity and specificity separately. One drawback is we usually want a one-dimensional summary measure of performance. Finding the highest accuracy, the value that has the highest proportion of correct classification, can be determined using the performance object:

```
accur = performance(pred, "acc")
max(accur@y.values[[1]])
```

```
[1] 0.85
```

Note that the objects ROCR are S4 classes in R and are referenced with an @ symbol. We may want higher sensitivity or specificity (they cannot be jointly maximized) but not overall accuracy, but this depends on the problem.

The area under the ROC curve (AUC) provides a one-dimensional metric between 0.5 and 1 where 1 implies perfect prediction, where the curve shoots up a sensitivity of 1 with specificity of 1. Any AUC below 0.5 implies that a larger measure in the prediction implies a **lower** probability of $Y = 1$, so if these predictions are multiplied by -1, then the new AUC is 1 minus the original AUC and is thus above 0.5. The AUC of 0.5 is shown by a 45-degree line (Fig. 4) and represents where the true positive rate equals the false positive rate (FPR). In essence, this classifier is doing as well, on average, as random prediction. The AUC can be calculated as follows:

```
auc = performance(pred,"auc")
auc@y.values[[1]]
```

```
[1] 0.8342
```

While the AUC provides a one-dimensional measure, it computes an area over a range of FPRs that are unacceptable in practice. For example, if a classifier had high sensitivity, but 50% specificity, it would not likely be used due to the potential for a large number of false positives. Therefore, the partial AUC (pAUC) is usually a better measure of model fit. The pAUC can be calculated in `ROCR` by using the `fpr.stop`, which calculates the AUC up to a 10% FPR:

```
pauc = performance(pred,"auc", fpr.stop = .1)
print(pauc@y.values[[1]])
```

```
[1] 0.02781
```

Note that the pAUC given is not scaled, so the maximum is 1, which can be done dividing by the `fpr.stop`:

```
print(pauc@y.values[[1]] / .1)
```

```
[1] 0.2781
```

We can plot this ROC for the constrained FPR (Fig. 5):

```
plot(perf, xlim=c(0, .1))
abline(a=0, b=1, col="red")
```

The `pROC` package has functions for testing if ROC or partial ROC curves are statistically different. The standard test uses the method from DeLong et al. (1988):

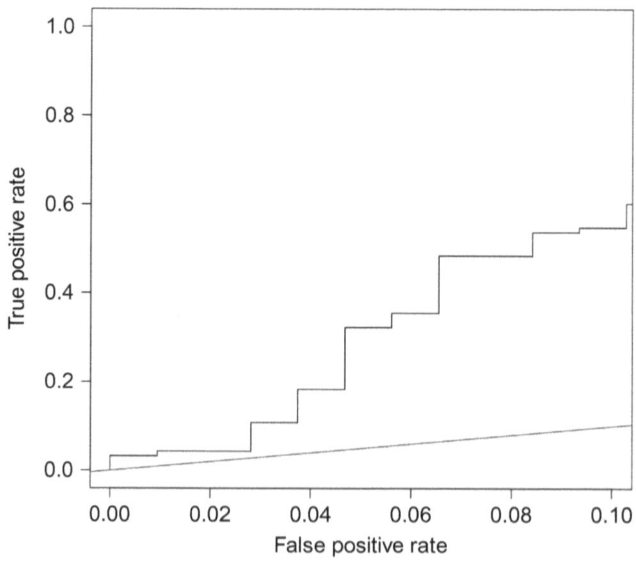

FIGURE 5 Example of partial ROC curve, limiting to a false positive rate of 0.1.

```
library(pROC)
data(aSAH)
roc1 <- roc(aSAH$outcome, aSAH$s100b)
roc2 <- roc(aSAH$outcome, aSAH$wfns)
roc.test(roc1, roc2, paired=TRUE)

DeLong's test for two correlated ROC curves

data: roc1 and roc2
Z = -2.209, p-value = 0.02718
alternative hypothesis: true difference in AUC is not equal to 0
sample estimates:
AUC of roc1   AUC of roc2
     0.7314        0.8237
```

Other methods for testing are implemented and we can also test for differences in pAUC (for a given sensitivity range of 0.8 to 1):

```
roc3 <- roc(aSAH$outcome, aSAH$s100b, partial.auc=c(0.8, 1),
            partial.auc.correct=TRUE)
roc4 <- roc(aSAH$outcome, aSAH$wfns, partial.auc=c(0.8 ,1),
            partial.auc.correct=TRUE)
roc.test(roc3, roc4, progress="none", paired=TRUE)

Bootstrap test for two correlated ROC curves

data: roc3 and roc4
D = -0.8116, boot.n = 2000, boot.stratified = 1, p-value = 0.417
alternative hypothesis: true difference in AUC is not equal to 0
sample estimates:
Corrected pAUC (1-0.8 specificity) of roc1
                                     0.6683
Corrected pAUC (1-0.8 specificity) of roc2
                                     0.7036
```

Above we showed how to use any measure as the cut points for an ROC. More commonly, one uses the predictions from model fits with different variables. Let us show how this can be applied to the previous model predictions.

```
library(pROC)
sah.step <- glm(outcome=="Good"~wfns + ndka, data=aSAH,
                family=binomial)
roc.all <- roc(sah.all$y, fitted(sah.all))
roc.step <- roc(sah.step$y, fitted(sah.step))

# Test for Equal AUC
roc.test(roc.all, roc.step, paired=TRUE)

DeLong's test for two correlated ROC curves

data: roc.all and roc.step
```

```
Z = -1.856, p-value = 0.0634
alternative hypothesis: true difference in AUC is not equal to 0
sample estimates:
AUC of roc1   AUC of roc2
     0.7751        0.8621
```

In the output from roc.test, we see DeLong's test for two correlated ROC curves returns a p-value of 0.06, which is higher than the threshold of $\alpha = 0.05$ required to reject the equivalence of the ROC curves. We do see the AUC for the regression model with the variables wfns and ndka in the model has a higher AUC (0.8621) compared to the model will the s100b, age, and ndka included in the model (AUC 0.7751). So in practice, we would likely chose the model with higher AUC.

7.3 Selecting a Cut Point

Since the ROC curve depicts the performance of a classifier across a range of possible cutoffs of a predictor, finding a cutoff that is "optimal" in some sense is often of interest. It is important to note that the optimal decision rule depends on the "costs" (monetary value or utility) associated with misclassifications, which vary substantially across different applications. When the costs of false positives and false negatives are equal, one optimal cutoff is associated with the point on the ROC curve that jointly maximizes sensitivity and specificity. Each point on the ROC curve gives (1 − specificity, sensitivity) for a given cutoff, so the cutoff that jointly maximizes sensitivity and specificity is given by the point that minimizes the distance to (0,1) (Fig. 6):

```
d = (perf@x.values[[1]] - 0)^2 + (perf@y.values[[1]] - 1)^2
m = which(d == min(d)); print(m)
```

```
[1] 96
```

```
plot(perf); abline(a=0, b=1, col="red");
abline(v= perf@x.values[[1]][m])
```

Although we only did one prediction, you may also pass a matrix of labels and predictions into the `prediction` command, which will make the `x.values` and `y.values` slots in the `performance` object lists as long as the number of predictions given. Thus, we can generalize this operation using `mapply`:

```
mapply(FUN=function(x, y){
  d = (x - 0)^2 + (y-1)^2
  ind = which(d == min(d))
}, perf@x.values, perf@y.values)
```

```
[1] 96
```

This process can be applied when you have predictions from a set of models and want to calculate measures over all predictions or plot overlapping ROC curves.

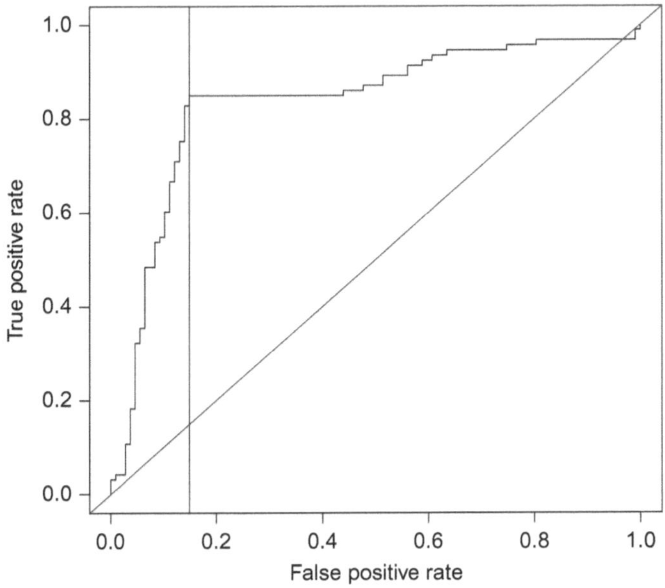

FIGURE 6 Finding the "elbow" of an ROC curve.

In many practical applications, a false positive may be more costly than a false negative, or vice versa. For example, detecting a disease in a patient that does not have the disease may cost a hospital x in dollars, but classifying a patient as nondiseased who has the disease may cost the y dollars. The `cost.fp` and `cost.fn` are optional arguments given to the `performance` command to weight false positives/negatives differently. Overall, the criterion for model/algorithm selection usually depends on the purpose of the analysis and factors external to the data. See Fawcett 2006) for an introduction to ROC analysis.

8 MODELING REPEATED (LONGITUDINAL) BINARY MEASURES

The assumption of independent observations is directly violated when outcomes are repeatedly measured at one time point, are measured longitudinally on the same individual, or clustered by design. These scenarios frequently occur in epidemiological studies.

An important example occurs in estimating treatment efficacy on a binary endpoint (presence/absence of a symptom or a health-related condition) in a randomized controlled trial (RCT). There are two main approaches to modeling longitudinal binary data: modeling the population-average response using GEEs and modeling individual heterogeneity in binary response trajectories using a GLMM. It is generally accepted that both approaches are applicable in problems with longitudinal binary responses, but the major difference is that they estimate different quantities. Moreover, these estimates can be quite different, in contrast to linear mixed models, where estimates tend to be close. The GEE approach estimates differences in the marginal probabilities of response between the different treatment groups in an RCT, whereas the GLMM approach estimates conditional, subject-specific differences in the probabilities of response between the different treatments. Marginal modeling of longitudinal data using GEE has been critiqued in Lindsey and Lambert (1998) (and also by Lee and Nelder, 2004), who highlight some of the conceptual problems with the GEE approach.

We illustrate both GEE and GLMM for repeated binary response with a dataset from Hedeker (2005) (`schiz`) examining the effect of an antipsychotic treatment (one of three drugs: chlorpromazine, fluphenazine, or thoridazine) versus placebo on a binary response relating to their psychotic condition. The treatment group indicator and response variables are denoted as `tx` and `imps79b`, respectively, in the dataset. The time unit was weeks since the initiation of treatment, `week`. A new variable, `sweek`, which is the square root of `week` was created and used as a covariate representing time.

8.1 Generalized Estimating Equations

```
# Schizophrenia Data from Hedeker and Gibbons
# GEE
schiz <- read.table("schizx1.dat.txt", header=FALSE)
names(schiz) <- c("id", "imps79", "imps79b", "imps79o", "int",
                  "tx", "week", "sweek", "txswk")
# imps79 = -9 denotes missing values, so we remove them
schiz <- schiz[schiz$imps79b > -9, ]
dim(schiz)

[1] 1603  9

head(schiz)
    id imps79 imps79b imps79o int tx week sweek txswk
1 1103    5.5       1       4   1  1    0 0.000 0.000
2 1103    3.0       0       2   1  1    1 1.000 1.000
4 1103    2.5       0       2   1  1    3 1.732 1.732
7 1103    4.0       1       2   1  1    6 2.450 2.450
8 1104    6.0       1       4   1  1    0 0.000 0.000
9 1104    3.0       0       2   1  1    1 1.000 1.000
```

We fit a marginal model using GEE using the `geepack` package (Halekoh et al., 2006; Yan and Fine, 2004; Yan, 2002) with an interaction between treatment group indicator and `sweek`. Thus, we are allowing the treatment effect to change with time. Note that data are **assumed** to be sorted by cluster (in this case id) before fitting. Check the number of clusters reported compared to those in the data.

```
# GEE estimation
require(geepack)
schiz = schiz[order(schiz$id, schiz$week), ]
summary(gee1 <- geeglm(imps79b ~ tx * sweek, data=schiz, id=id,
        family=binomial))
Call:
geeglm(formula = imps79b ~ tx * sweek, family = binomial,
       data = schiz, id = id)

Coefficients:
              Estimate  Std.err  Wald  Pr(>|W|)
(Intercept)      3.703    0.490  57.1   4.2e-14   ***
tx              -0.405    0.525   0.6      0.44
sweek           -1.113    0.252  19.5   1.0e-05   ***
tx:sweek        -0.418    0.270   2.4      0.12
---
Signif. codes: 0 '***' 0.001 '**' 0.01 '*' 0.05 '.' 0.1 ' ' 1

Estimated Scale Parameters:
              Estimate  Std.err
(Intercept)      0.914    0.216

Correlation: Structure = independenceNumber of clusters: 437
Maximum cluster size: 5
```

The output is similar to that of `glm()`, but note the estimated scale parameters. The treatment is protective and its effect becomes stronger with time although the interaction term appears to be only marginally significant based on the Wald test. Ideally, we would like to perform a likelihood ratio test (LRT) comparing the models with and without the `tx:sweek` interaction term, but GEE does not accommodate LRTs since GEE estimation is not based on maximizing the log-likelihood, similar to quasi-likelihood methods. This is an important limitation of the GEE approach.

When fitting GEE models, one must specify a correlation structure (independence is the default) using the `corstr`. This correlation structure indicates how one assumes the repeated measures to be related. Now, let us conduct a sensitivity analysis on the correlation structure by comparing the regression coefficients and the corresponding inferences to those from two other correlation structures: exchangeable and AR1 (first-order autoregressive) correlations.

```
# Different covariance structures
summary(gee2 <- geeglm(imps79b ~ tx*sweek, data=schiz, id=id,
family=binomial, corstr="exch"))
```

```
Call:
geeglm(formula = imps79b ~ tx * sweek, family = binomial, data =
      schiz, id = id, corstr = "exch")

Coefficients:

              Estimate  Std.err   Wald  Pr(>|W|)
(Intercept)     3.658    0.485   56.80  4.8e-14    ***
tx             -0.382    0.521    0.54    0.464
sweek          -1.092    0.252   18.85  1.4e-05    ***
tx:sweek       -0.451    0.269    2.80    0.094     .
---
Signif. codes: 0 '***' 0.001 '**' 0.01 '*' 0.05 '.' 0.1 ' ' 1

Estimated Scale Parameters:
              Estimate  Std.err
(Intercept)     0.902    0.203

Correlation: Structure = exchangeable Link = identity

Estimated Correlation Parameters:
       Estimate  Std.err
alpha     0.203   0.0548
Number of clusters:  437  Maximum cluster size: 5

summary(gee3 <- geeglm(imps79b ~ tx*sweek, data=schiz, id=id,
                    family=binomial, corstr="ar1"))

Call:
geeglm(formula = imps79b ~ tx * sweek, family = binomial,
      data = schiz, id = id, corstr = "ar1")

Coefficients:
              Estimate  Std.err   Wald  Pr(>|W|)
(Intercept)     3.717    0.488   57.93  2.7e-14    ***
tx             -0.397    0.524    0.57     0.45
sweek          -1.126    0.249   20.51  5.9e-06    ***
tx:sweek       -0.434    0.267    2.64     0.10

---
Signif. codes: 0 '***' 0.001 '**' 0.01 '*' 0.05 '.' 0.1 ' ' 1

Estimated Scale Parameters:
              Estimate  Std.err
(Intercept)     0.915    0.221

Correlation: Structure = ar1 Link = identity

Estimated Correlation Parameters:
       Estimate  Std.err
alpha     0.332   0.0808

Number of clusters:  437  Maximum cluster size: 5
```

There are some minor quantitative differences as we change the correlation structure, but overall, the coefficient estimates and the inferences are similar. If the results are different, one must make decisions of which structure to choose using tests or information about how the data are collected/generated.

8.2 Generalized Linear Mixed Models

Now, let us look at GLMMs, which are also known as subject-specific or conditional models, for the same dataset. We use the `glmer` function in the `lme4` package (Bates et al., 2014). We first fit a main effects only model. Here, the notation `(1|id)` denotes random intercept.

```
# Generalized linear mixed-effects model estimation
require(lme4)
summary(mod1 <- glmer(imps79b ~ tx + sweek + (1|id), nAGQ=9,
       data=schiz, family=binomial))
```

```
Generalized linear mixed model fit by maximum likelihood (Adaptive
  Gauss-Hermite Quadrature, nAGQ = 9) [glmerMod]
Family: binomial ( logit )
Formula: imps79b ~ tx + sweek + (1 | id)
 Data: schiz

 AIC    BIC   logLik   deviance   df.resid
1266   1288     -629       1258       1599

Scaled residuals:
     Min       1Q   Median      3Q     Max
-16.270   0.024    0.121   0.280   2.984

Random effects:
Groups   Name          Variance   Std.Dev.
id       (Intercept)   4.11       2.03

Number of obs: 1603, groups: id, 437

Fixed effects:
               Estimate   Std. Error   z value   Pr(>|z|)
(Intercept)       6.747        0.534     12.62   < 2e-16    ***
tx               -1.787        0.361     -4.95   7.3e-07    ***
sweek            -2.292        0.176    -13.04   < 2e-16    ***

---
Signif. codes: 0 '***' 0.001 '**' 0.01 '*' 0.05 '.' 0.1 ' ' 1

Correlation of Fixed Effects:
        (Intr)   tx
tx      -0.722
sweek   -0.788   0.235
```

Again, the output format is similar to `glm()` but with additional information about random effects and the correlation of fixed effects. Here, the notation

`(1|id)` denotes random intercept. The option `nAGQ=9` denotes the number of points for adaptive Gauss–Hermite quadrature used in the integration of the likelihood over the distribution of the random effect. Additional quadrature points tend to lead to a better approximation of the likelihood and are more computationally intensive. Observing different results with differing numbers of quadrature points may indicate that the model estimates are local (not global) maxima of the likelihood.

Now, we test whether an interaction between `tx` and `sweek` needs to be introduced. As GLMMs maximize a likelihood, LRTs can be used.

```
summary(mod2 <- glmer(imps79b ~ tx * sweek + (1|id),
nAGQ=9, data=schiz, family=binomial))

Generalized linear mixed model fit by maximum likelihood
   (Adaptive Gauss-Hermite Quadrature, nAGQ = 9) [glmerMod]
 Family: binomial ( logit )
Formula: imps79b ~ tx * sweek + (1 | id)
 Data: schiz

 AIC    BIC   logLik   deviance   df.resid
1260   1287     -625       1250       1598

Scaled residuals:
    Min    1Q      Median   3Q      Max
-14.804   0.030   0.111    0.269   3.392

Random effects:
 Groups   Name          Variance   Std.Dev.
 id       (Intercept)   4.49       2.12
Number of obs: 1603, groups: id, 437

Fixed effects:
                Estimate   Std. Error   z value   Pr(>|z|)
(Intercept)      5.3880      0.6309       8.54    < 2e-16    ***
tx              -0.0219      0.6540      -0.03     0.9733
sweek           -1.4991      0.2907      -5.16    2.5e-07    ***
tx:sweek        -1.0173      0.3345      -3.04     0.0024     **
---
Signif. codes: 0 '***' 0.001 '**' 0.01 '*' 0.05 '.' 0.1 ' ' 1

Correlation of Fixed Effects:
          (Intr)   tx       sweek
tx        -0.782
sweek     -0.829    0.716
tx:sweek   0.555   -0.831   -0.793

anova(mod1, mod2)

Data: schiz
Models:
mod1: imps79b ~ tx + sweek + (1 | id)
mod2: imps79b ~ tx * sweek + (1 | id)
```

```
     Df  AIC  BIC logLik deviance Chisq Chi Df Pr(>Chisq)
mod1  4 1266 1288   -629     1258
mod2  5 1260 1287   -625     1250  8.57      1     0.0034 **
---
Signif. codes: 0 '***' 0.001 '**' 0.01 '*' 0.05 '.' 0.1 ' ' 1
```

The LRT provides evidence that the interaction term is needed. Although GEE and GLMM results are not directly comparable since they estimate different quantities (i.e., marginal vs. conditional effects), we can still do an informal comparison. We see that the GLMM object, `mod2`, shows a strong `time:sweek` interaction when compared to GEE object, `gee3`.

To obtain the CIs for fixed-effect parameters and for the variance components, we could use the `confint` function to compute the CI based on LRT of the profile likelihood.

```
# Generalized linear mixed-effects model estimation
print(confint(mod2, method="profile"))

               2.5 %   97.5 %
.sig01          1.71    2.598
(Intercept)     4.25    6.733
tx             -1.36    1.226
sweek          -2.11   -0.963
tx:sweek       -1.67   -0.347
```

Note that `.sig01` is the standard deviation of the random intercept. The point estimate from `mod2` object is 2.03 and the CIs are `(1.71, 2.60)`. This clearly shows that there is substantial "unexplained" heterogeneity in the probability of response across individuals.

Now, let us look at the impact of the number of quadrature points for evaluating the integral over the random effects distribution, `nAGQ`. In `mod2`, we used nine quadrature points. Let us compare it to the default Laplace approximation, which is `nAGQ=1`.

```
# Impact of number of quadrature points
summary(mod3 <- glmer(imps79b ~ tx * sweek + (1|id),
data=schiz, family=binomial))

Generalized linear mixed model fit by maximum likelihood (Laplace
 Approximation) [glmerMod]
 Family: binomial ( logit )
Formula: imps79b ~ tx * sweek + (1 | id)
  Data: schiz

 AIC    BIC   logLik   deviance   df.resid
1268   1295     -629       1258       1598

Scaled residuals:
    Min      1Q   Median      3Q     Max
-14.189   0.032    0.113   0.274   3.256
```

```
Random effects:
Groups   Name          Variance  Std.Dev.
id       (Intercept)   4.1       2.02

Number of obs: 1603, groups: id, 437

Fixed effects:
             Estimate  Std. Error  z value  Pr(>|z|)
(Intercept)     5.294       0.613     8.64  < 2e-16   ***
tx             -0.016       0.628    -0.03   0.9796
sweek          -1.435       0.281    -5.10  3.4e-07   ***
tx:sweek       -1.039       0.327    -3.17   0.0015   **

---
Signif. codes: 0 '***' 0.001 '**' 0.01 '*' 0.05 '.' 0.1 ' ' 1

Correlation of Fixed Effects:
          (Intr)  tx      sweek
tx        -0.770
sweek     -0.823   0.720
tx:sweek   0.525  -0.828  -0.786

c(logLik(mod2), logLik(mod3))

[1] -625 -629
```

Comparing the log-likelihoods of `mod2` and `mod3`, we can see that the model fit is better—log-likelihood is higher—with `nAGQ=9` points compared to the default `nAGQ=1`. It should be noted that it is possible in `glmer` to specify number of Gauss–Hermite quadrature points only when there is one random effect (i.e., random intercept). When there are two or more random effects (e.g., random intercept and slope), due to greater computational burden, only one Gauss–Hermite quadrature point (i.e., Laplace approximation) is allowed. This is an important limitation of `glmer`.

GLMM has several advantages over GEE for longitudinal responses:

- GLMM provides subject-specific predictions
- Population-averaged effect or marginal effect interpretation of GEE can be conceptually problematic. The marginal trajectories do not correspond to any possible individual. Consequently, it is possible to infer that a treatment is better on average than another treatment, when in fact the converse may be true for each individual (Lindsey and Lambert, 1998).
- Missing data are handled seamlessly in GLMM, provided missing at random assumption is justifiable, whereas GEE methods assume that data are missing completely at random (MCAR), which is a stronger assumption.
- Comparison of nested models with LRT and of nonnested models with AIC or BIC can be done in GLMM, but GEE does not allow such model comparisons.

GEE models are easy to fit and are computationally inexpensive. GLMMs, on the other hand, can be computationally demanding and have convergence issues for binary responses, especially for models with more than a simple random intercept. For example, the following model, with both random intercept and slope, could not be fitted using `lme4::glmer`, although we were able to fit it in SAS with both PROC GLIMMIX and PROC NLMIXED without any convergence problems (results not shown).

```
# Lack of convergence
mod4 <- glmer(imps79b ~ tx * sweek + (sweek|id),
              data=schiz, family=binomial)
```

```
Warning: Model is nearly unidentifiable: very large eigenvalue -
Rescale variables?
```

However, fitting the same model, but with probit link, instead of the default logit link, works!

```
mod5 <- glmer(imps79b ~ tx * sweek + (sweek|id),
data=schiz, family=binomial(link=probit))
print(mod5)
```

```
Generalized linear mixed model fit by maximum likelihood (Laplace
  Approximation) [glmerMod]
 Family: binomial  ( probit )
Formula: imps79b ~ tx * sweek + (sweek | id)
  Data: schiz
    AIC     BIC   logLik   deviance   df.resid
   1272    1309     -629       1258       1596

Random effects:
Groups   Name          Std.Dev.   Corr
id       (Intercept)   1.619
         sweek         0.898      -0.57

Number of obs: 1603, groups: id, 437
Fixed Effects:
(Intercept)       tx     sweek   tx:sweek
      3.406    0.237    -0.736     -1.023
```

Model parameters, however, do not have interpretation as log-odds or log-odds ratios when the model is fitted on probit scale.

9 MODEL SELECTION

Model selection ranks models based on some criteria. In order to choose between various competing models, we need to choose which criteria to use. The AIC (Akaike, 1973) combines the fitted value of the log-likelihood, penalized by the number of degrees of freedom in the model (k):

$$\mathrm{AIC}(k,\beta)=2(k-\mathrm{LL}(\beta)) \tag{9}$$

Better fitting models will have a higher log-likelihood value and lower number of parameters, resulting in a lower value of the AIC. Another commonly used information criterion is Bayes information criterion (Schwarz et al., 1978). For studies where prediction is the goal, misclassification rates, sensitivity, specificity, and accuracy are commonly used criteria. All these can be used for comparing nonnested models; nested models can be compared using traditional inferential techniques, such as LRTs.

We will examine two common methods for model selection: stepwise selection (based on AIC) and penalized regression models.

Stepwise selection starts with an initial model and then attempts to add and/or remove parameters to optimize some sort of goodness-of-fit index, such as the AIC. Forward selection starts at an initial model and sequentially adds covariates until there are no other covariates which meet the selection criteria:

```
stepAIC(glm(outcome=="Good"~1, data=aSAH, family=binomial),
        direction="forward",   scope=list(upper=~age+s100b
        +ndka, lower=~1))

Start: AIC=150
outcome == "Good" ~ 1
          Df   Deviance   AIC
+ s100b    1        124   128
+ age      1        143   147
+ ndka     1        144   148
<none>              148   150

Step: AIC=127.6
outcome == "Good" ~ s100b

          Df   Deviance   AIC
+ ndka     1        120   126
<none>              124   128
+ age      1        122   128

Step: AIC=125.8
outcome == "Good" ~ s100b + ndka

          Df   Deviance   AIC
<none>              120   126
+ age      1        118   126

Call: glm(formula = outcome == "Good" ~ s100b + ndka,
    family = binomial, data = aSAH)

Coefficients:

(Intercept)   s100b     ndka
2.3786        -5.3341   -0.0313

Degrees of Freedom: 112 Total (i.e., Null); 110 Residual
```

```
Null Deviance:   148
Residual Deviance: 120 AIC: 126
```

Backward selection starts with an initial model and sequentially removes terms which do not meet the selection criteria (results not shown):

```
stepAIC(glm(outcome=="Good"~age+s100b+ndka, data=aSAH,
        family=binomial), direction="backward")
```

These can be combined into a sequential process of adding covariates and removing those who no longer meet the selection criteria:

```
stepAIC(glm(outcome=="Good"~age+s100b+ndka+wfns,     data=aSAH,
        family=binomial), scope=list(upper=~age+s100b+ndka
        +wfns, lower=~1))
Warning: glm.fit: fitted probabilities numerically 0 or 1
occurred

Start: AIC=107.7
outcome == "Good" ~ age + s100b + ndka + wfns

Warning: glm.fit: fitted probabilities numerically 0 or 1
occurred
Warning: glm.fit: fitted probabilities numerically 0 or 1
occurred

         Df  Deviance  AIC
- s100b   1      92.8  107
- age     1      93.5  108
<none>           91.7  108
- ndka    1     102.0  116
- wfns    4     118.0  126

Warning: glm.fit: fitted probabilities numerically 0 or 1
occurred

Step: AIC=106.8
outcome == "Good" ~ age + ndka + wfns

Warning: glm.fit: fitted probabilities numerically 0 or 1
occurred
Warning: glm.fit: fitted probabilities numerically 0 or 1
occurred

         Df  Deviance  AIC
<none>           92.8  107
+ s100b   1      91.7  108
- age     1      96.1  108
- ndka    1     103.2  115
- wfns    4     139.7  146

Call: glm(formula = outcome == "Good" ~ age + ndka + wfns, family
     = binomial, data = aSAH)
```

```
Coefficients:
(Intercept)        age       ndka      wfns2      wfns3
     7.0584    -0.0324    -0.0716    -3.7385    -3.2018
      wfns4      wfns5
    -4.3383    -5.9465

Degrees of Freedom: 112 Total (i.e., Null); 106 Residual
Null Deviance:   148
Residual Deviance: 92.8 AIC: 107
```

An additional complication associated with binary outcomes seen here in the warning messages is the issue of large fitted values becoming indistinguishable from 0 or 1 with numerical precision. Another possibility that can also arise is the issue of near or complete separation, where a linear combination of the variables nearly or completely separates the two classes of outcome. Though this may be desirable for the problem, this may lead to an unstable model.

These procedures, like any statistical procedure, should be used with care, as the results can be sensitive to the choice of criteria used to evaluate models, the starting model, and potentially influential data points. While the AIC can be used as an estimate of the asymptotic leave-one-out cross-validation (Stone, 1977) value, selection procedures do not account for the issues of multiple inference, and estimates obtained using these methods may exaggerate the importance of some covariates. For these reasons, shrinkage-based methods of selection and estimation have become increasingly popular.

9.1 Penalized Logistic Regression: The `glmnet` Package

We want to predict, using a GLM, the binary outcome, but want to penalize coefficients that large in magnitude. This penalization is desirable because it helps reduce overfitting to the sample and increase generalizability of the model to new data, and provides a more parsimonious model. The `glmnet` package (Friedman et al., 2010) has functions for fitting the LASSO (Least Absolute Shrinkage and Selection Operator) (Tibshirani, 1996), ridge regression (Hoerl and Kennard, 1970), and elastic nets (convex combinations of ridge regression and the LASSO) (Zou and Hastie, 2005).

The elastic net for logistic regression solves the following optimization (http://web.stanford.edu/~hastie/glmnet/glmnet_alpha.html):

$$\min_{(\beta_0,\beta)\in\mathbb{R}^{p+1}} -\left[\frac{1}{N}\sum_{i=1}^{N} y_i\cdot(\beta_0 + x_i^T\beta) - \log\left(1 + e^{(\beta_0 + x_i^T\beta)}\right)\right] + \lambda\left[(1-\alpha)||\beta||_2^2/2 + \alpha||\beta||_1\right]$$

where N is the number of observations, and λ represents the penalty of a β coefficient being large, and $||\cdot||_p$ represents L_p norm. If $\alpha = 0$, then the optimization reduces to the case of ridge regression; if $\alpha = 1$, the model reduces to the LASSO model. The elastic net corresponds to, as we refer to it, where $0 < \alpha < 1$.

If an elastic net is used, selection of α can be done with cross-validation, similar to the choice of λ, but is commonly set to a fixed value. A range of values of α can also be used to determine how sensitive the model is to the choice of α. The LASSO and ridge regression are more commonly used, however. A nice feature of the LASSO is that it does shrinkage and model selection simultaneously, whereas ridge regression cannot do model selection. See Hastie et al. (2009) for a comprehensive overview of the differences between the LASSO and ridge regression. Weighted regression variants of the LASSO, ridge regression, and elastic net may also be performed, but we will assume equal weighting for each observation.

9.2 Phoneme Data

From the `ElemStatLearn` package (Hastie et al., 2005; Material from the book's webpage and R port and packaging by Halvorsen, K., 2012), we will analyze the phoneme data: each row represents a log-periodogram (256 columns) of a spoken phoneme (distinct unit of sound) along with the true phoneme. We will focus on two of the phonemes: "aa" as the vowel in "dark," and "ao" as the first vowel in "water" (Material from the book's webpage and R port and packaging by Halvorsen, K., 2012). Although there are 256 variables in this dataset that can be investigated, the variables which might best distinguish the two phonemes likely are smaller in number. Thus, the goal of the analysis is to determine which periodograms predict the vowel sounds, reducing the number of periodograms from the 256 given.

We will subset the data into the phonemes of interest and create the simple binary variable *Y* to represent the "ao" phoneme.

```
require(ElemStatLearn)
dataset = ElemStatLearn::phoneme
dataset = dataset[ dataset$g %in% c("ao", "aa"), ]
Y = as.numeric(dataset$g == "ao")
X = as.matrix(dataset[,1:256])
```

9.3 Fitting `glmnet` Models

Below we fit a model using an elastic net (with $\alpha = 0.5$), the LASSO, and ridge regression. Note that `glmnet` does not take formula objects and takes the outcome (`y`) as a factor and `X` as a matrix/`data.frame`.

```
X = scale(X, center=TRUE, scale=TRUE)
require(glmnet)
```

```
lasso.model = glmnet(x=X, y=factor(Y), family="binomial", alpha=1)
ridge.model = glmnet(x=X, y=factor(Y), family="binomial", alpha=0)
enet.model = glmnet(x=X, y=factor(Y), family="binomial",
alpha=0.5)
```

We will examine a subset of the output. By default, `glmnet` runs the penalized regression with 100 values of λ, but here we will present only the first 8.

```
output = cbind(Df = enet.model$df,
'%Dev' = signif(enet.model$dev.ratio, 3),
Lambda = signif(enet.model$lambda, 3))
head(output, 8)
```

```
      Df      %Dev   Lambda
[1,]   0  6.58e-16    0.544
[2,]   6  3.45e-02    0.496
[3,]  10  7.41e-02    0.452
[4,]  12  1.11e-01    0.412
[5,]  17  1.43e-01    0.375
[6,]  19  1.72e-01    0.342
[7,]  21  1.98e-01    0.311
[8,]  21  2.20e-01    0.284
```

The corresponding columns represent the degrees of freedom (nonzero coefficients), percent of deviance explained, and penalty (λ), respectively. Thus, these measures represent a sequence of models—each with a different λ—each with different coefficients. Note the last two entries each have 21 degrees of freedom but different λ and percent deviance—these are two models with the same number of coefficients but have different percent deviances explained. Thus, these are different models; each row/model does not represent a nested model of the row/model preceding it.

9.4 Visualizing the `glmnet` Model

Using `plot`, we can visualize each coefficient by different levels of λ and degrees of freedom (Fig. 7):

```
plot(enet.model, xvar="lambda")
```

Each line represents 1 of the 256 variables. The top axis represents the degrees of freedom and the bottom the values of λ. Information is hard to see in this plot, but we note that sets of covariates are shrunk to zero around $\log(\lambda) = -5$, but some covariates are nonzero even for large values of λ.

Similarly, we can look these coefficients as a function of percent deviance explained:

```
plot(enet.model, xvar="dev")
```

We see in both plots that for models above 30 degrees of freedom, many lines begin to separate from zero. In Fig. 8, the increases in coefficients raise dramatically with modest increases in percent deviance explained.

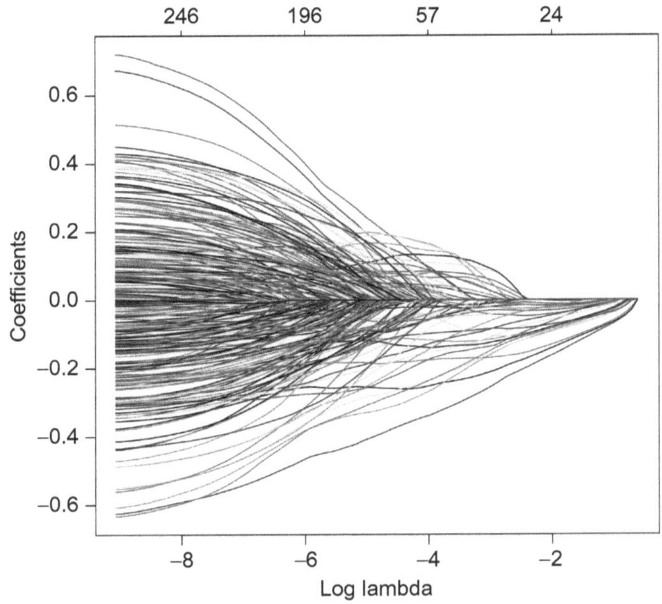

FIGURE 7 Two hundred fifty-six log-periodogram coefficients for each of the 100 λ fit using an elastic net. The top axis represents the degrees of freedom within the model and the bottom axis is the corresponding λ. Each line represents a covariate (log-periodogram).

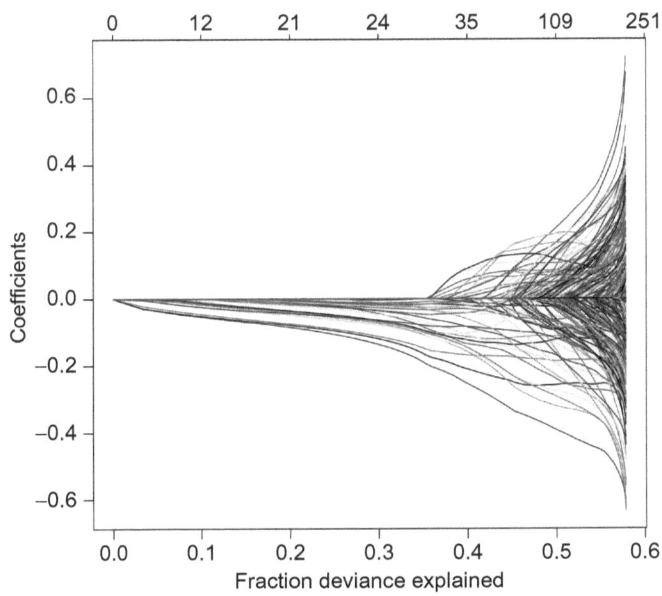

FIGURE 8 Two hundred fifty-six log-periodogram coefficients for each of the 100 lambda fit using an elastic net, compared using fraction deviance explained. The top axis represents the degrees of freedom within the model and the bottom axis is the fraction deviance explained. Each line represents a covariate (log-periodogram).

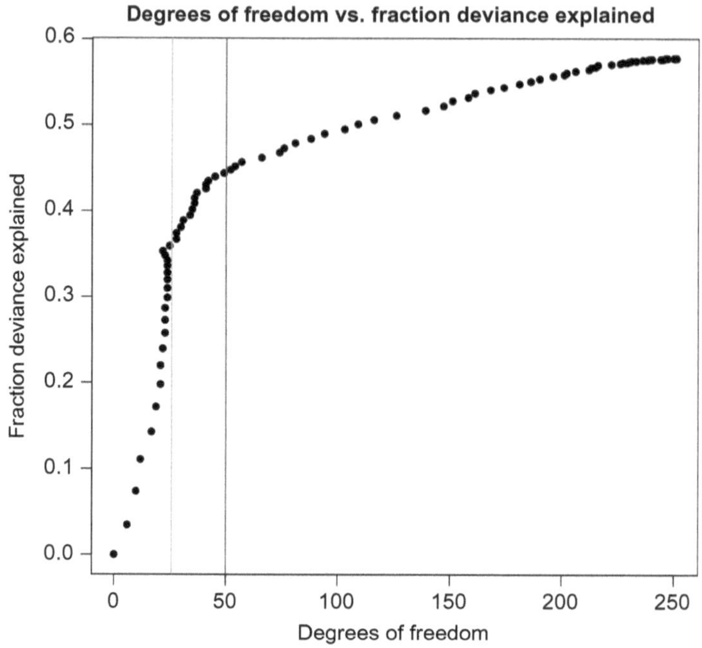

FIGURE 9 Comparison of fraction deviance explained with increasing degrees of freedom.

We can also plot the degrees of freedom directly compared to the deviance:

```
plot(x=output[, "Df"], y=output[, "%Dev"],
    pch=16,
    main="Degrees of Freedom vs. Fraction Deviance Explained",
    xlab="Degrees of Freedom",
    ylab="Fraction Deviance Explained")
abline(v=26, col="green")
abline(v=50, col="red")
```

We see in Fig. 9 that there is a steep increase in deviance explained with ≤ 26 degrees of freedom (green; light gray in the print version), and that after 50 degrees of freedom, there is a leveling off of deviance explained. Figure 9 illustrates that a model with less than 50 nonzero coefficients can result in a model that gives good percent deviance explained compared to much larger models.

9.5 Choosing λ in `glmnet` Using Cross-Validation

Overall though, these plots do not give one end model, which many users want. We implore you to weigh the practical importance of variables in the data, missingness patterns, and interpretability of models along with any output from automated penalization.

In order to choose λ, one option is to use cross-validation. Cross-validation separates the data into training and test datasets, runs a series of values for λ on the training data, estimates the performance of the model based on the test data, and repeats this procedure many times. In many analyses, α is set or varied, but can be chosen using the same concepts as below.

The `cv.glmnet` function incorporates this procedure with the same syntax as `glmnet`. `cv.glmnet` provides additional options for cross-validation, such as the number of folds (`nfolds`) to be used. Note that this procedure may take much longer than the `glmnet` function as it fits many more models.

```
enet.cvmodel.dev = cv.glmnet(x=X, y=factor(Y),
                             family="binomial", alpha=0.5)
```

We plot the results in Fig. 10, which shows a range of λ with similar binomial deviance.

```
plot(enet.cvmodel.dev)
```

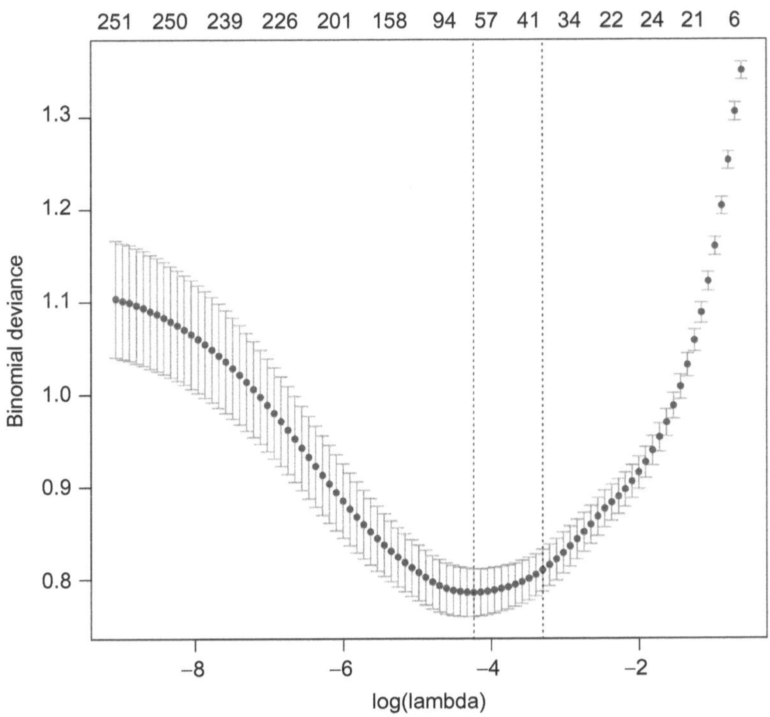

FIGURE 10 Binomial deviance as a function of λ. We see values slightly below -4 and under -3 correspond to similar binomial deviance.

We also denote `type.measure = "class"` so that the penalty used is the misclassification rate compared to the binomial deviance. Use `type.measure= "auc"` to use the AUC cost function.

```
enet.cvmodel = cv.glmnet(x=X, y=factor(Y), family="binomial",
                         alpha=0.5, type.measure = "class")
```

The results of this model (Fig. 11) are similar to those in Fig. 10, but the criterion for choosing λ is different. Figure 11 provides us with the estimated misclassification rate versus the $\log(\lambda)$ and marks a set of λ values that reduce the misclassification rate but are not "overpenalized."

```
plot(enet.cvmodel)
```

We can then extract the "best" λ, chosen by cross-validation. Note that using different cost functions (misclassification rate or deviance) can (and does) result in different λ being chosen.

```
enet.cvmodel$lambda.min
[1] 0.009076
```

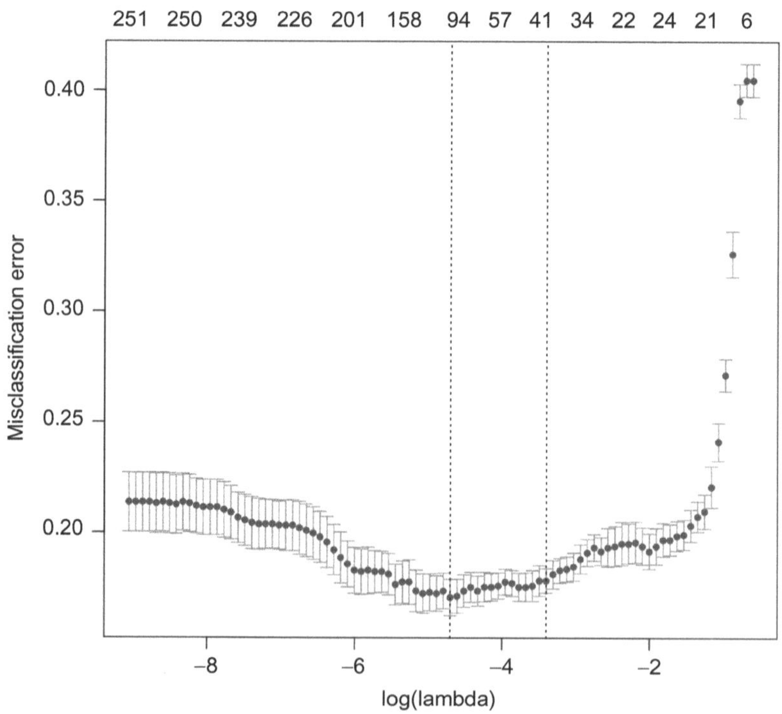

FIGURE 11 Misclassification error as a function of lambda. Similar to the plot for binomial deviance, we see values slightly below − 4 and under − 3 correspond to similar binomial deviance, but these values do not overlap exactly with the previous plot.

```
enet.cvmodel.dev$lambda.min
```

```
[1] 0.01445
```

We can fit the elastic net using this λ (using the misclassification rate cost).

```
enet.model.final = glmnet(x=X, y=factor(Y),
family="binomial", alpha=0.5, lambda = enet.cvmodel$lambda.min)
```

Alternatively, the model has an element `lambda.1se` which gives the λ corresponding to the most regularized model (fewest coefficients) where the error rate is within one standard error of the minimum.

```
enet.cvmodel$lambda.1se
```

```
[1] 0.03339
```

Note that fitting this model with any input λ chosen above does **not** take into account the uncertainty in the selection procedure for λ. Characterizing this uncertainty is an open area of research. Note also that cross-validation chooses subsets randomly and each run may not give the same chosen λ.

10 MACHINE LEARNING METHODS

The machine learning task view (http:/cran.r-project.org/web/views/MachineLearning.html) provides a large collection of machine learning methods for classification and regression. We will focus on 3 in particular: recursive partitioning (`rpart`) (Therneau et al., 2014), random forests (`randomForest`) (Breiman, 2001; Liaw and Wiener, 2002), and gradient boosted modeling (`gbm`) (Ridgeway, G. with contributions from others, 2013). In contrast to many other settings, the goal of using machine learning techniques with binary outcomes many times is prediction rather than estimation or inference. Therefore, we will use misclassification rates as the performance criteria. Again, we will use the `phoneme` data, trying to produce a classifier that has small misclassification error for discriminating the two, previously mentioned, phonemes.

```
require(ElemStatLearn)
dataset = ElemStatLearn::phoneme
dataset = dataset[ dataset$g %in% c("ao", "aa"), ]
dataset$g = factor(dataset$g == "ao")
dataset$speaker = NULL
```

10.1 Splitting the Data in Train and Test Samples

In order to attempt to avoid overfitting, we will subset the data into a training dataset and a test dataset, using 90% of the data for training and 10% for testing:

```
shuffle.rows = sample(nrow(dataset))
train.rows = shuffle.rows[seq(1, floor(nrow(dataset)*.9))]
```

```
train = dataset[ train.rows,]
test = dataset[-train.rows,]
```

This allows us to evaluate and select classifiers on the training data and get estimates of out-of-sample performance using the test data. This should give a better (usually less optimistic) estimate of the classifier performance.

10.2 Recursive Partitioning (`rpart`)

The `rpart` package (Therneau et al., 2014) in our setting does recursive partitioning to create a decision tree that tries cut the explanatory variables to create a prediction. As an observation "goes down the tree," the combination of cut points defines a path that gives a prediction of 0 or 1 for the binary outcome. Let us fit the tree and plot it:

```
require(rpart)
rp = rpart(formula=g ~ ., data = train)
par(xpd = TRUE)
options(scipen=8)
plot(rp)
text(rp, use.n = TRUE, all = TRUE, cex=.5)

rp.prob = predict(rp, newdata = test, type="prob")[,1]
rp.pred = predict(rp, newdata = test, type="class")
```

One appealing aspect of `rpart` is that it results in an interpretable decision tree (Fig. 12). If we observe a new record, we can determine the classification of its outcome by evaluating its covariates to the cutoffs in the tree, determine which bin that observation is in, and then take the majority vote of the outcomes of other observations within that bin.

One detraction from single decision trees such as these is that they may have a higher likelihood for overfitting to the training data than random forests or gbms. We can try to determine the misclassification rate compared to other classifiers using the `test` dataset, which is done in Section 10.5.

10.3 Random Forests

The `randomForest` package (Liaw and Wiener, 2002) fits multiple trees (and thus called a forest). To try to guard against overfitting, random forests bootstrap the data *and* sample the covariates, and fit a decision tree. Thus, many trees are created and single observations and individual covariates have less influence over the overall prediction. Each tree gets a "vote," which is a prediction—0 or 1 in the binary case. You can either extract $P(Y = 1)$ by averaging these predictions or classify the outcome by the category with the highest number of votes. This is one example of an ensembling methods as it ensembles multiple predictors—in this case trees—to create a prediction. We fit a `randomForest` using the same syntax as `rpart`.

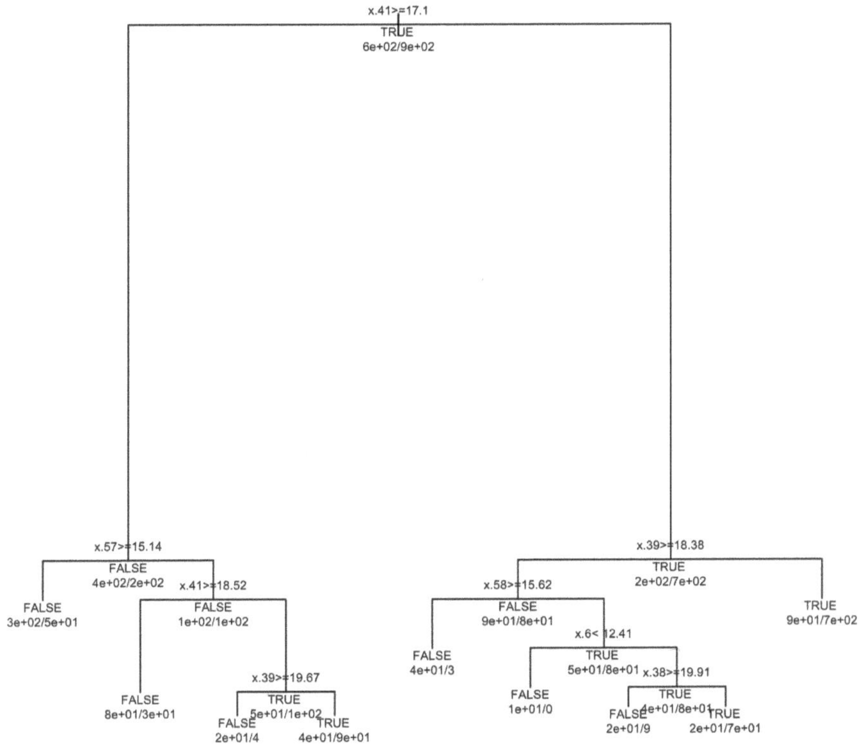

FIGURE 12 Decision tree determined by `rpart`. We see the individual cut points for each log-periodogram and the number of zeroes and the number of ones (given by 0s/1s) in each branch of the tree.

```
require(randomForest)
print({rf = randomForest(formula=g ~ ., data = train)})

Call:
 randomForest(formula = g ~ ., data = train)
               Type of random forest: classification
                     Number of trees: 500
No. of variables tried at each split: 16

    OOB estimate of error rate: 18.25%

Confusion matrix:
        FALSE  TRUE  class.error
FALSE     465   163       0.2596
 TRUE     119   798       0.1298

plot(rf)

rf.prob = predict(rf, newdata=test, type="prob")[, 1]
rf.pred = predict(rf, newdata=test, type="class")
```

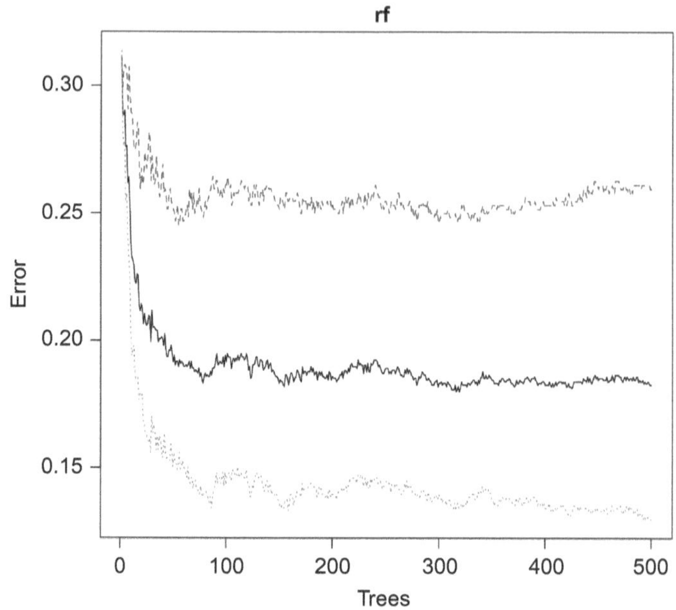

FIGURE 13 Misclassification/error rate as a function of trees grown in the random forest. The black line represents the entire sample, the green line (light gray in the print version) represents the error rate where $Y = 0$ and the red (gray in the print version) line represents the error rate when $Y = 1$.

Figure 13 shows the misclassification rate of the training data overall (black), in the 1's (red; gray in the print version) and the 0's (green; light gray in the print version). We see that the positive outcomes have worse classification, and the overall misclassification rate is around 18%.

10.4 Generalized Boosted Regression Modeling

The `gbm` package (Ridgeway, G. with contributions from others, 2013) is another example of ensembling using generalized boosted regression modeling. The `gbm()` function creates trees, but in a more iterative process than random forests, and subsequently reweights training cases which have been misclassified. The syntax is similar to `glm()`. We needed to convert the outcome to a binary variable rather than a factor.

```
require(gbm)
train$g = as.numeric(train$g)-1
gbmod = gbm( formula=g ~ ., data = train,
distribution="bernoulli")
gb.prob = predict(gbmod, newdata=test, n.trees = 100)
gb.pred = gb.prob > mean(gb.prob)
```

We chose to classify the predicted probabilities by values greater than the mean. In some cases, common thresholds such as $\hat{P}(Y=1)>0.5$ can be used. We can change different thresholds using out `train` dataset to determine a good cutoff.

10.5 Comparison of Results

We can look at the predicted probabilities of each classifier and see how they compare for each group. We will use the `ggplot2` R package (Wickham, 2009) for visualization.

```
library(reshape2); library(ggplot2); library(plyr)
df = data.frame(rf=rf.prob, rp=rp.prob, gbm=gb.prob,
                truth=factor(test$g))
plot(df, pch=16, cex=.5)
```

```
df$id = 1:nrow(df )
long = melt(df, measure.vars = c("rf", "rp", "gbm"))
long$variable = revalue(long$variable, c("rf" = "Random Forest",
                        "rp"="Recursive Partitioning",
                        "gbm"= "Gradient Boosted Modeling"))
ggplot(long, aes(x=variable, colour=truth, y=value))
  + geom_boxplot() + xlab("Classfier") + ylab("Prediction") +
 guides(colour = guide_legend(title = "True Y"))
```

Figure 14 shows that values high on the random forest (`rf`) are negatively correlated with those of the `gbm`, but positively correlated with the `rpart` tree (`rp`). Figure 15 shows good separation between the groups for each classifier, which indicates we can likely determine the "ao" from "aa" sounds with a new periodogram.

Also, we can investigate the overall accuracy of each of the classifiers:

```
accur = function(tab){
 print(paste0("Accuracy = ",
              round(sum(diag(tab))/sum(tab)*
              100, 2), "%"))
 print(tab)
}
accur(table(rp.pred, test$g))
```

```
[1] "Accuracy = 79.07%"

rp.pred  FALSE  TRUE
  FALSE     46    15
   TRUE     21    90
```

```
accur(table(rf.pred, test$g))
```

```
[1] "Accuracy = 83.72%"
```

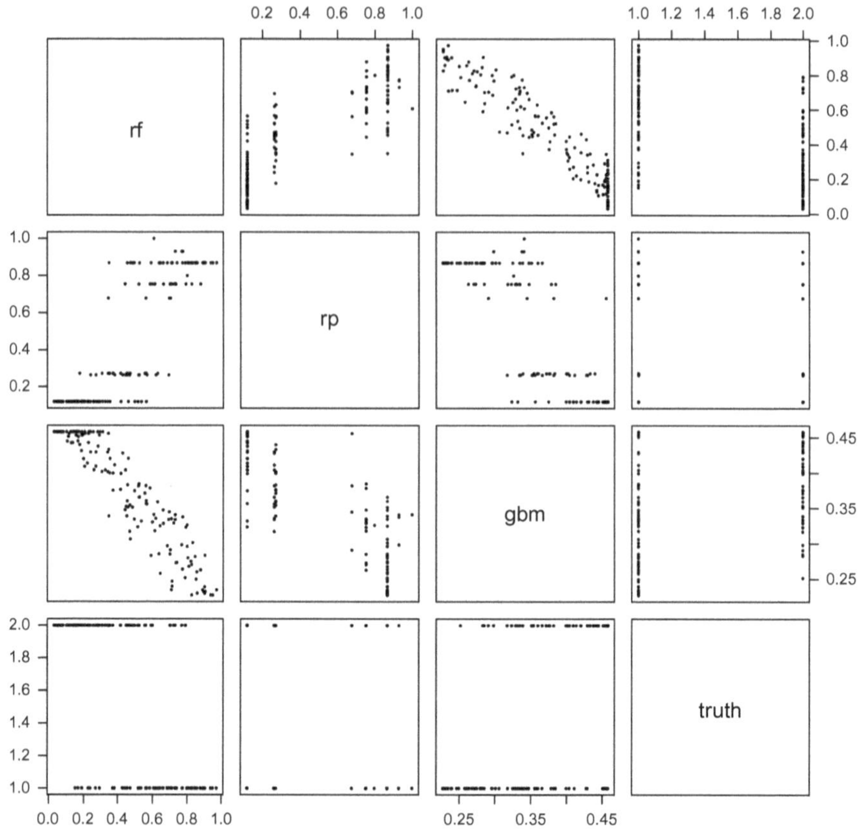

FIGURE 14 Scatterplot matrix of predictions from each classifier and the true outcome for the training dataset.

```
rf.pred   FALSE   TRUE
  FALSE      52     13
   TRUE      15     92
```

```
accur(table(gb.pred, test$g))
```

```
[1] "Accuracy = 77.33%"
```

```
gb.pred   FALSE   TRUE
  FALSE      53     25
   TRUE      14     80
```

The accuracy is comparable across the three classifiers, with the random forest performing the best among them. There are trade-offs between "seeing" the classifier such as the `rpart` model and classification accuracy. In practice, random forests tend to perform reasonably well and are robust to overfitting.

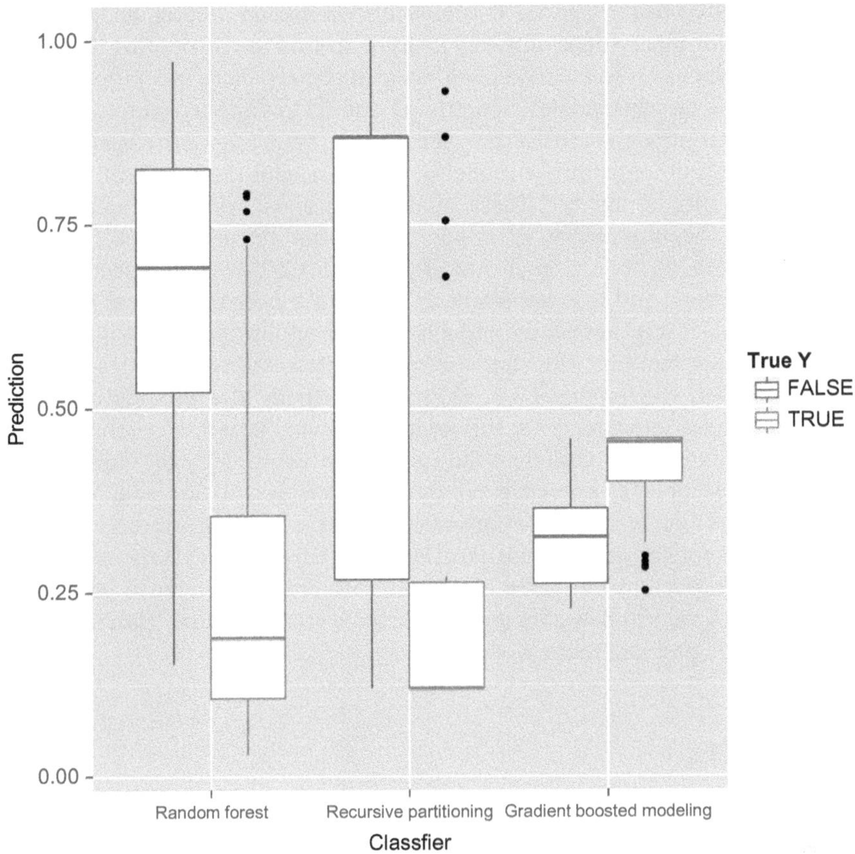

FIGURE 15 Distribution of predictions from each classifier, separated by true value of Y.

No classifier currently provides a standard inferential framework similar to those in the other models presented in this work, such as the variance of the standard deviation of predictions, p-values of covariates, etc.

Overall, there is an entire set of tools for prediction of binary (and categorical) outcomes in machine learning, with trade-offs for interpretability, predictive power, and ability for inference from the algorithm. One must always consider these aspects when selecting an analysis plan.

11 CONCLUDING REMARKS

Regression with binomial outcomes is ubiquitous in practice, and appropriate modeling of binary data depends very strongly upon the nature of the data. Like any other statistical application, proper understanding of the underlying assumptions of each model is important for proper use and interpretation.

R provides a wealth of options for choosing, estimating, testing, and evaluating models for binary data, allowing analysts to find the model that best suits their application. Overdispersion can occur in binary data, and proper inference depends on appropriately identifying and addressing its source. GLMM and marginal models (GEE) are useful for modeling correlated binary responses, but the appropriate model depends on both the nature of the data and the quantity of interest. These models also come with other conceptual difficulties, the importance of which should not be understated, such as Simpsons paradox for marginal models fit with GEE. Model selection is a complex process, and it is important to understand how the choices made in the model selection procedures influence the resulting models. While many researchers are familiar with stepwise model selection methods, their limitations are often underappreciated. Shrinkage methods also provide powerful tools for model selection, with different underlying selection criteria. While these methods can better address the issues of overfitting by providing more parsimonious models, inference on these models is still an area of active research. Machine learning methods also provide a broad set of tools for prediction in binomial data. While the flexibility of these methods avoids the need for an explicit model of the data generation process, inference (e.g., standard errors, *p*-values) based on models obtained from machine learning methods is challenging and more work is needed.

REFERENCES

Akaike, H., 1973. Information theory and an extension of the maximum likelihood principle. In: Second International Symposium on Information Theory, Akademinai Kiado, pp. 267–281.

Bates, D., Maechler, M., Bolker, B., Walker, S., 2014. lme4: linear mixed-effects models using Eigen and S4. R package version 1.1–7.

Belsey, D.A., Kuh, E., Welsch, R.E., 1980. Regression Diagnostics: Identifying Influential Data and Sources of Collinearity. John Wiley, New York, NY.

Breiman, L., 2001. Random forests. Mach. Learn. 45 (1), 5–32.

Breslow, N., 1996. Generalized linear models: checking assumptions and strengthening conclusions. Stat. Appl. 8, 23–41.

Cantoni, E., 2004. Analysis of robust quasi-deviances for generalized linear models. J. Stat. Softw. 10(i04).

Cantoni, E., Ronchetti, E., 2001. Robust inference for generalized linear models. J. Am. Stat. Assoc. 96 (455), 1022–1030.

Cantoni, E., Ronchetti, E., 2006. A robust approach for skewed and heavy-tailed outcomes in the analysis of health care expenditures. J. Health Econ. 25 (2), 198–213.

Canty, A., Ripley, B., 2014. boot: Bootstrap R (S-Plus) functions. R package version 1.2–34.

Cleveland, W.S., Grosse, E., Shyu, M.J., 1992. Local regression models. In: Chambers, J.M., Hastie, T. (Eds.), Statistical Models in S. Chapman and Hall, New York, pp. 309–376.

Collett, D., 2003. Modelling Binary Data. Chapman & Hall/CRC, London, UK.

Cook, R.D., 1977. Detection of influential observation in linear regression. Technometrics 19, 15–18.

Cook, R.D., Weisberg, S., 1980. Characterizations of an empirical influence function for detecting influential cases in regression. Technometrics 22 (4), 495–508.

Davison, A., Hinkley, D., 1997. Bootstrap Methods and Their Application, vol. 193. Cambridge University Press, Cambridge, UK.

DeLong, E.R., DeLong, D.M., Clarke-Pearson, D.L., 1988. Comparing the areas under two or more correlated receiver operating characteristic curves: a nonparametric approach. Biometrics. 44, 837–845.

Faraway, J.J., 2005. Extending the Linear Model with R: Generalized Linear, Mixed Effects and Nonparametric Regression Models. CRC Press, Boca Raton, FL,

Fawcett, T., 2006. An introduction to ROC analysis. Pattern Recogn. Lett. 27 (8), 861–874.

Fox, J., Weisberg, S., 2011. An R Companion to Applied Regression, second ed. Sage, Thousand Oaks CA.

Friedman, J., Hastie, T., Tibshirani, R., 2010. Regularization paths for generalized linear models via coordinate descent. J. Stat. Softw. 33 (1), 1.

Halekoh, U., Højsgaard, S., Yan, J., 2006. The R package geepack for generalized estimating equations. J. Stat. Softw. 15 (2), 1–11.

Hastie, T., Tibshirani, R., Friedman, J., Franklin, J., 2005. The elements of statistical learning: data mining, inference and prediction. Math. Intell. 27 (2), 83–85.

Hastie, T., Tibshirani, R., Friedman, J., Hastie, T., Friedman, J., Tibshirani, R., 2009. The Elements of Statistical Learning, vol. 2(1). Springer, New York, NY.

Hedeker, D., 2005. Encyclopedia of Statistics in Behavioral Sciences. In: Everitt, B., Howell, D. (Eds.), Wiley (Chapter Generalized linear mixed models), New York, NY.

Heritier, S., Cantoni, E., Copt, S., Victoria-Feser, M.P., 2009. Robust Methods in Biostatistics, vol. 825. John Wiley & Sons, West Sussex, UK.

Hoerl, A.E., Kennard, R.W., 1970. Ridge regression: biased estimation for nonorthogonal problems. Technometrics 12 (1), 55–67.

Hosmer, D.W., Lemeshow, S., Sturdivant, R.X., 2000. Introduction to the Logistic Regression Model. Wiley Online Library, Hoboken, NJ.

Hothorn, T., Bretz, F., Westfall, P., 2008. Simultaneous inference in general parametric models. Biometric. J. 50 (3), 346–363.

Huber, P.J., 1967. The behavior of maximum likelihood estimates under nonstandard conditions. In: Proceedings of the Fifth Berkeley Symposium on Mathematical Statistics and Probability vol.1, (1), pp. 221–233.

Lee, Y., Nelder, J., 2004. Conditional and marginal models: another view. Stat. Sci. 19, 219–238.

Liaw, A., Wiener, M., 2002. Classification and regression by randomForest. R News 2 (3), 18–22.

Lindsey, J., Lambert, P., 1998. On the appropriateness of marginal models for repeated measurements in clinical trials. Stat. Med. 17, 447–469.

Material from the book's webpage and R port and packaging by Halvorsen, K., 2012. ElemStatLearn: Data sets, functions and examples from the book: "The Elements of Statistical Learning, Data Mining, Inference, and Prediction" by Trevor Hastie, Robert Tibshirani and Jerome Friedman. R package version 2012.04–0.

McCullagh, P., Nelder, J.A., 1989. Generalized linear models. Chapman and Hall, New York, NY.

Nelder, J.A., Wedderburn, R.W.M., 1972. Generalized linear models. J. R. Stat. Soc. A: Gen. 135 (3), 370–384.

Ridgeway, G., with contributions from others, 2013. gbm: generalized boosted regression models. R package version 2.1.

Robin, X., Turck, N., Hainard, A., Tiberti, N., Lisacek, F., Sanchez, J.C., Müller, M., 2011. pROC: an open-source package for R and S+ to analyze and compare ROC curves. BMC Bioinformatics 12 (1), 77.

Rovers, M.M., Glasziou, P., Appelman, C.L., Burke, P., McCormick, D.P., Damoiseaux, R.A., Gaboury, I., Little, P., Hoes, A.W., 2006. Antibiotics for acute otitis media: a meta-analysis with individual patient data. Lancet 368 (9545), 1429–1435.

Schwarz, G., et al., 1978. Estimating the dimension of a model. Ann. Stat. 6 (2), 461–464.

Sing, T., Sander, O., Beerenwinkel, N., Lengauer, T., 2005. ROCR: visualizing classifier performance in R. Bioinformatics 21 (20), 3940–3941.

Stone, M., 1977. An asymptotic equivalence of choice of model by cross-validation and Akaike's criterion. J. R. Stat. Soc. B: Methodolog., 39, 44–47.

Therneau, T., Atkinson, B., Ripley, B., 2014. rpart: recursive partitioning and regression trees. R package version 4.1–8.

Tibshirani, R., 1996. Regression shrinkage and selection via the lasso. J. R. Stat. Soc. B: Methodolog., 58, 267–288.

Valdora, M., Yohai, V.J., 2014. Robust estimators for generalized linear models. J. Stat. Plan. Inference 146, 31–48.

Wang, J., Zamar, R., Marazzi, A., Yohai, V., Salibian-Barrera, M., Maronna, R., Zivot, E., Rocke, D., Martin, D., Maechler, M., Konis., K., 2014. robust: robust library. R package version 0.4–16.

White, H., 1980. A heteroskedasticity-consistent covariance matrix estimator and a direct test for heteroskedasticity. Economet.: J. Economet. Soc., 48, 817–838.

Wickham, H., 2009. ggplot2: Elegant Graphics for Data Analysis. Springer, New York, NY.

Yan, J., 2002. Geepack: yet another package for generalized estimating equations. R news 2 (3), 12–14.

Yan, J., Fine, J., 2004. Estimating equations for association structures. Stat. Med. 23 (6), 859–874.

Zou, H., Hastie, T., 2005. Regularization and variable selection via the elastic net. J. R. Stat. Soc. B: Stat. Methodol. 67 (2), 301–320.

Chapter 8

Computing Tolerance Intervals and Regions Using R

Derek S. Young[1]
Department of Statistics, University of Kentucky, Lexington, Kentucky, USA
[1]*Corresponding author: e-mail: derek.young@uky.edu*

ABSTRACT

Tolerance intervals provide bounds on a specified proportion of the sampled population (P) with a given confidence level (γ). While they are, perhaps, less known than confidence and prediction intervals, there are some applications where tolerance intervals are commonly used, such as in quality control, setting engineering tolerances, and environmental monitoring. We present a general introduction to the topic followed by overviews of how to calculate tolerance intervals for some continuous and discrete distributions, nonparametric tolerance limits, tolerance intervals for regression settings, and multivariate normal tolerance regions. Calculating tolerance intervals and regions under these as well as other settings can be accomplished using the R package `tolerance` (Young, 2010). For the settings in our discussion, we present real-data examples and demonstrate how to calculate tolerance intervals and regions using the `tolerance` package.

Keywords: Discrete quality assessment, Multivariate normal, Nonparametrics, Order statistics, Regression, `tolerance` Package

1 INTRODUCTION

Statistical intervals provide a way to quantify the uncertainty about a scalar quantity of a population based on a random sample of data. For example, confidence intervals are used to obtain bounds for an unknown scalar population parameter of interest, such as the mean or variance. Prediction intervals provide bounds on one or more future observations from the sampled population. The multivariate analogues to these intervals are confidence regions and prediction regions, respectively. A third type of interval is a *tolerance interval*. Tolerance intervals provide bounds to capture at least a specified proportion (P) of the sampled population with a given confidence level (γ). The quantity P is referred to as the *content* of the tolerance interval, while the confidence level γ reflects the sampling variability. The chosen levels

Handbook of Statistics, Vol. 32. http://dx.doi.org/10.1016/B978-0-444-63431-3.00008-5

are typically reflected by writing that one seeks to calculate a "P/γ tolerance interval." Moreover, the levels of P and γ are between 0 and 1 and typically chosen from the set $\{0.90, 0.95, 0.99\}$.

Tolerance intervals have been studied quite extensively in the literature, with some of the earliest works being Wilks (1941, 1942) and Wald (1943). Jílek (1981) and Jílek and Ackermann (1989) provide extensive bibliographies on tolerance intervals and regions, while a thorough literature review is available in Patel (1986). A few texts also treat the topic of tolerance intervals. Guttman (1970) and Krishnamoorthy and Mathew (2009) are texts devoted to the theoretical development and application of tolerance intervals and regions. The text by Hahn and Meeker (1991) also handles the topic, but in the broader context of statistical intervals.

Closed-form solutions to tolerance intervals exist for very few settings. As such, approximations and numerical methods often need to be employed. Computing tolerance intervals and regions in `R` (R Development Core Team, 2014) is quite easy with the `tolerance` add-on package (Young, 2010). The `tolerance` package has functions for computing tolerance intervals for numerous univariate parametric distributions, the nonparametric setting, and some regression settings. It also has functions for constructing multivariate tolerance regions. After installing the package, one can type the following for a listing of all functions:

```
> help(package="tolerance")
```

The focus of this chapter is to demonstrate some of these functions applied to real datasets. But first, we provide formal definitions of one- and two-sided tolerance intervals.

1.1 Formal Definition

In this section, we provide a basic definition of one- and two-sided tolerance intervals. A more rigorous development of the definitions can be found in Krishnamoorthy and Mathew (2009).

Consider the random variable X with x the corresponding realization. Let $\mathbf{X} = (X_1, X_2, \ldots, X_n)$ denote a random sample of size n from the distribution $F_X(\cdot)$. A P/γ two-sided tolerance interval, say $(L(\mathbf{X}), U(\mathbf{X}))$, depends on the random sample $\mathbf{X}$, and satisfies the following:

$$P_{\mathbf{X}}(P_X[L(\mathbf{X}) \leq X \leq U(\mathbf{X}) | \mathbf{X}] \geq P) = \gamma. \tag{1}$$

The quantities $L(\mathbf{X})$ and $U(\mathbf{X})$ are referred to as *tolerance limits*. Limits for one-sided upper and lower tolerance intervals satisfy similar expressions, namely

$$P_{\mathbf{X}}(P_X[X \leq U(\mathbf{X}) | \mathbf{X}] \geq P) = \gamma \tag{2}$$

and

$$P_{\mathbf{X}}(P_X[L(\mathbf{X}) \leq X | \mathbf{X}] \geq P) = \gamma, \tag{3}$$

respectively. Note that the quantities to determine one-sided tolerance limits are equivalent to confidence intervals on the respective quantiles.

For the two-sided tolerance interval defined in Eq. (1), note that many such limits could be constructed from a given sample. Thus, one may desire to compute a central (or equal-tailed) tolerance interval. Let q_P denote the Pth quantile of $F_X(x)$. A P/γ equal-tailed tolerance interval, written as $(L_e(\mathbf{X}), U_e(\mathbf{X}))$, satisfies the following:

$$P_{\mathbf{X}}\left(L_e(\mathbf{X}) \leq q_{(1-P)/2} \text{ and } q_{(1+P)/2} \leq U_e(\mathbf{X})\right) = \gamma. \tag{4}$$

As noted in Krishnamoorthy and Mathew (2009), equal-tailed tolerance intervals are often used to calculate reference limits, which are limits used by clinicians to identify the expected measurement range from a reference population. See Harris and Boyd (1995), for a detailed treatment of reference limits and regions.

2 TOLERANCE INTERVALS FOR CONTINUOUS DISTRIBUTIONS

We first use the general definitions of Section 1.1 to construct tolerance intervals for three common continuous distributions: normal, exponential, and Weibull.

2.1 Tolerance Intervals for the Normal Distribution

A normal random variable X has the cumulative distribution function:

$$F_X(x;\mu,\sigma) = \int_{t=-\infty}^{x} \left(2\pi\sigma^2\right)^{-1/2} \mathrm{e}^{-((t-\mu)^2)/2\sigma^2}\, \mathrm{d}t, \tag{5}$$

where $-\infty < x < +\infty$, $-\infty < \mu < +\infty$ is the mean, and $\sigma > 0$ is the standard deviation. Normal tolerance intervals require estimates of both parameters ($\hat{\mu}$ and $\hat{\sigma}$). Typically, the sample mean $\overline{X}$ and the sample standard deviation S are used.

The formulas for one-sided lower and upper normal tolerance limits are as follows:

$$L(X) = \overline{X} - k_1 S \quad \text{and} \quad U(X) = \overline{X} + k_1 S, \tag{6}$$

respectively. In the above, the value k_1 is called a *k-factor*, and is calculated as

$$k_1 = \frac{1}{\sqrt{n}} t_{n-1;\gamma}\left(\sqrt{n} z_P\right), \tag{7}$$

where $t_{d;q}(\delta)$ is the qth percentile of a noncentral t-distribution with d degrees of freedom and noncentrality parameter δ, and z_q is the qth percentile of the standard normal distribution. A two-sided normal tolerance interval can be found by

$$\overline{X} \pm k_2 S. \tag{8}$$

k_2 is the solution to the integral equation

$$\sqrt{\frac{2n}{\pi}} \int_0^\infty P\left(\chi_m^2 > \frac{m\chi_{1;P}^2(z^2)}{k_2^2}\right) \mathrm{e}^{-(1/2)nz^2}\,\mathrm{d}z = \gamma, \tag{9}$$

where $m = n - 1$, χ_f^2 is the chi-squared random variable with f degrees of freedom, and $\chi_{f;q}^2(\delta)$ is the noncentral chi-squared distribution with f degrees of freedom and noncentrality parameter δ. Note that m will not always necessarily equal $n - 1$, such as in the case of constructing tolerance intervals in a one-way analysis of variance setup.

Example 1. Let us consider the example taken from chapter 2 of Krishnamoorthy and Mathew (2009). This example involves a production machine that is set to fill a liter of milk in plastic bags. The accuracy of the machine is checked by taking a sample of $n = 20$ bags at the end of a production shift and then measuring the actual amount of milk in each plastic bag. The data are given in Table 1.

One- and two-sided tolerance intervals for the normal setting can be calculated using the `normtol.int` function. First, suppose that we are interested in a 0.95/0.95 one-sided upper tolerance limit for the milk data. This can be calculated as follows:

```
> x <- c(0.968, 0.982, 1.030, 1.003, 1.046, 1.020,
+         0.997, 1.010, 1.027, 1.010, 0.973, 1.000,
+         1.044, 0.995, 1.020, 0.993, 0.984, 0.981,
+         0.997, 0.992)

> normtol.int(x=x, alpha=0.05, P=0.95, side=1)
  alpha    P  x.bar 1-sided.lower 1-sided.upper
1  0.05 0.95 1.0036     0.9506455      1.056555
```

TABLE 1 Measured Amount of Milk (in Liters) in Plastic Bags

0.968	0.982	1.030	1.003	1.046
1.020	0.997	1.010	1.027	1.010
0.973	1.000	1.044	0.995	1.020
0.993	0.984	0.981	0.997	0.992

The function takes various arguments, of which four are illustrated above. `x` is a vector of data, `alpha` controls the confidence level (i.e., $\alpha = 1 - \gamma$), `P` is the content level, and `side` is used to specify which type of tolerance interval to calculate (i.e., one- or two-sided). Hence, a one-sided 0.95/0.95 upper tolerance limit is 1.0566, which means we can be 95% confident that no more than 95% of the bags are filled with 1.0566 L of milk.

Next, suppose we want to calculate a 0.95/0.95 two-sided tolerance interval. As shown in Eq. (8), we need to find k_2 through solving an integral equation. This can be somewhat computationally burdensome and many approximations exist in the literature. Some of these are available to use in the `normtol.int` function and can be set using the `method` argument. Below, we show the results using approximations due to Howe (1969) and Weissberg and Beatty (1960) as well as for the exact method:

```
> normtol.int(x=x, alpha=0.05, P=0.95, side=2, method="HE")
  alpha    P  x.bar 2-sided.lower 2-sided.upper
1  0.05 0.95 1.0036     0.9425343      1.064666

> normtol.int(x=x, alpha=0.05, P=0.95, side=2, method="WBE")
  alpha    P  x.bar 2-sided.lower 2-sided.upper
1  0.05 0.95 1.0036     0.9427822      1.064418

> out.norm <- normtol.int(x=x, alpha=0.05, P=0.95, side=2,
+                          method="EXACT", m=100)

> out.norm
  alpha    P  x.bar 2-sided.lower 2-sided.upper
1  0.05 0.95 1.0036      0.942593      1.064607
```

Notice for the exact method that an additional argument `m` is required. This is used to specify the number of subintervals for the underlying `integrate` function. By default, `m=50`, but ideally the larger the number of subintervals the more accurate the result. As can be seen above, both approximations are fairly close to the result using the exact method.

There is also some plotting capabilities with the `tolerance` package. The `plottol` function allows one to plot the univariate data on a control chart and/or a histogram along with the calculated tolerance limits. This is controlled through the `plot.type` option. Below, we demonstrate how to obtain both the control chart and histogram for the `out.norm` output from above:

```
> plottol(out.norm, x=x, plot.type="both")
```

These plots are given in Fig. 1. Note that the default title of the plots is "$(100 \times \gamma)\%/(100 \times P)\%$," which is flipped from the P/γ convention used in this chapter.

One final note is that due to the relationship between the log-normal and the normal distributions, we can also easily calculate tolerance intervals for

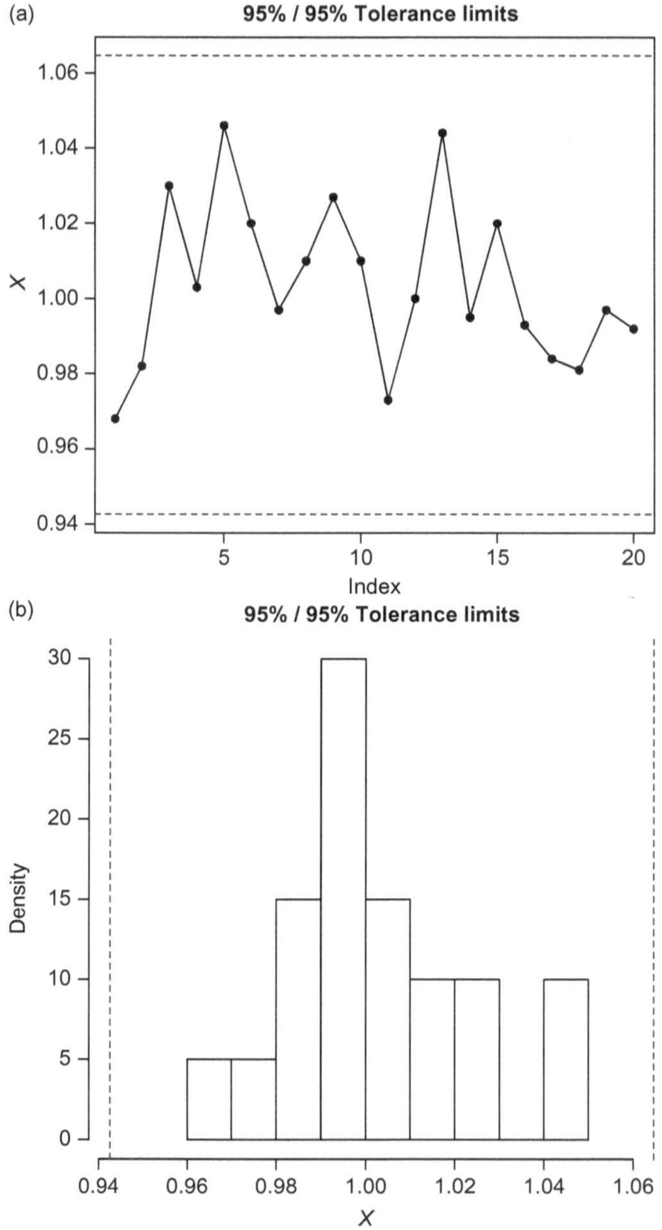

FIGURE 1 (a) Control chart and (b) histogram of the milk dataset with 0.95/0.95 two-sided tolerance limits overlaid.

log-normal data. This is controlled through the argument `log.norm`. Type `?normtol.int` for more information.

2.2 Tolerance Intervals for the Exponential Distribution

An exponential random variable X has cumulative distribution function

$$F_X(x;\lambda) = 1 - e^{-\lambda x}, \tag{10}$$

where $x > 0$ and $\lambda > 0$ is a scale (or rate) parameter. Lifetime data can often be appropriately modeled using an exponential distribution. Tolerance intervals for such data can be very informative. More discussion on tolerance intervals for exponentially distributed data can be found in Blischke and Murthy (2000).

Let $\hat{\lambda}$ be the maximum likelihood estimate of λ. The formulas for the one-sided lower and upper P/γ tolerance limits for exponentially distributed data are as follows:

$$L(X) = \frac{2n\hat{\lambda}\ln(1-P)}{\chi^2_{2n;\gamma}} \quad \text{and} \quad U(X) = \frac{2n\hat{\lambda}\ln(P)}{\chi^2_{2n;\gamma}}, \tag{11}$$

respectively. Approximate tolerance limits for the two-sided setting are found by replacing γ by $(1 + \gamma)/2$ and P by $(1 + P)/2$ in the above formulas.

Example 2. We next consider the data presented in Bhaumik and Gibbons (2006), which is given in Table 2. These data are vinyl chloride concentration measurements collected from clean upgradient monitoring wells. While Bhaumik and Gibbons (2006) assessed these data using a gamma distribution, we use an exponential distribution for our demonstration. We are interested in finding a 0.99/0.95 one-sided upper tolerance limit for the vinyl chloride concentration in the monitoring wells. This can be calculated as follows:

TABLE 2 Vinyl Chloride Concentrations (μg/L) from Monitoring Wells

5.1	2.4	0.4	0.5	2.5	0.1	6.8
1.2	0.5	0.6	5.3	2.3	1.8	1.2
1.3	1.1	0.9	3.2	1.0	0.9	0.4
0.6	8.0	0.4	2.7	0.2	2.0	0.2
0.5	0.8	2.0	2.9	0.1	4.0	

```
> x <- c(5.1, 2.4, 0.4, 0.5, 2.5, 0.1, 6.8,
+        1.2, 0.5, 0.6, 5.3, 2.3, 1.8, 1.2,
+        1.3, 1.1, 0.9, 3.2, 1.0, 0.9, 0.4,
+        0.6, 8.0, 0.4, 2.7, 0.2, 2.0, 0.2,
+        0.5, 0.8, 2.0, 2.9, 0.1, 4.0)

> exptol.int(x=x, alpha=0.05, P=0.99, side=1)
  alpha    P lambda.hat 1-sided.lower 1-sided.upper
1  0.05 0.99   1.879412    0.01455445      11.76605
```

Thus, we are 95% confident that at least 99% of the vinyl chloride concentration measurements are below 11.7661 μg/L. Note that this limit is well above the maximum value in the dataset, which is 8.0 μg/L. This is not unexpected since we are attempting to project for a percentile that is relatively far out in the tail of a right-skewed distribution. We also note that the `exptol.int` function can handle type II censored data through the `type.2` argument. Type `?exptol.int` for more information.

2.3 Tolerance Intervals for the Weibull Distribution

A random variable X has a Weibull distribution if it has cumulative distribution function

$$F_X(x; \theta, \beta) = 1 - e^{-(x/\theta)^\beta}, \tag{12}$$

where $x > 0$, with shape parameter $\beta > 0$ and scale parameter $\theta > 0$. Like the exponential distribution, the Weibull distribution is often used for lifetime data as well as other engineering applications. Let $\hat{\beta}$ and $\hat{\theta}$ be the maximum likelihood estimates of the shape and scale parameters, respectively. Using the methodology of Bain and Engelhardt (1981), formulas for one-sided lower and upper P/γ tolerance limits for Weibull data are as follows:

$$\begin{aligned} L(X) &= \exp\left\{ \ln(\hat{\theta}) - \frac{t_{n-1;\gamma}(-\sqrt{n}\lambda_{1-P})}{\hat{\beta}\sqrt{n-1}} \right\} \text{ and} \\ U(X) &= \exp\left\{ \ln(\hat{\theta}) - \frac{t_{n-1;1-\gamma}(-\sqrt{n}\lambda_{P})}{\hat{\beta}\sqrt{n-1}} \right\}, \end{aligned} \tag{13}$$

respectively, where $\lambda_P = \ln(-\ln(1-P))$. Just like for the exponential setting, approximate tolerance limits for the two-sided Weibull setting are found by replacing γ by $(1 + \gamma)/2$ and P by $(1 + P)/2$ in the above formulas.

Example 3. We next consider the data presented in Grubbs (1971) on the failure mileages of $n = 19$ military carriers. The data are presented in Table 3. Suppose that we are interested in a 0.90/0.90 one-sided upper

TABLE 3 Failure Mileages of Military Carriers

162	200	271	302	393	508
539	629	706	777	884	1008
1101	1182	1463	1603	1984	2355
2880					

tolerance limit for the failure mileage of all such military carriers. This can be calculated using the `exttol.int` function as follows:

```
> x <- c(162, 200, 271, 302, 393, 508,
+        539, 629, 706, 777, 884, 1008,
+        1101, 1182, 1463, 1603, 1984, 2355,
+        2880)
```

```
> exttol.int(x=x, alpha=0.10, P=0.90, side=1, dist="Weibull")
  alpha P shape.1 shape.2 1-sided.lower 1-sided.upper
1 0.1 0.9 1.408074 1100.593   124.9571   2729.129
```

Thus, we are 90% confident that 90% of all such military carriers will have a failure mileage no greater than approximately 2729.129 miles. In the above, the maximum likelihood estimates for the Weibull distribution (given by `shape.1` and `shape.2`) are found using a Newton–Raphson algorithm that is initialized by the corresponding method of moments estimates.

Note in the above code the use of the argument `method="Weibull"`. The `exttol.int` function also provides tolerance limits for the Gumbel distribution for the minimum and the Gumbel distribution for the maximum due to their relationship with the Weibull distribution. More details on how these extreme value tolerance limits are calculated can be found in Young (2010) or by typing `?exttol.int`.

3 TOLERANCE INTERVALS FOR DISCRETE DISTRIBUTIONS

Suppose we have a discrete random variable X, which has the distribution $F_X(x; n, \theta)$ that depends only on a single parameter θ and a "size" quantity n (e.g., sample size or number of necessary trials). Moreover, suppose that the distribution is stochastically nondecreasing in θ. Hahn and Chandra (1981) presented a method for constructing equal-tailed tolerance intervals for discrete distributions, which can be summarized in two primary steps:

1. Based on the observed value x of the random variable X, construct a two-sided $100 \times \gamma\%$ confidence interval for θ, $(\theta_{L;\alpha}(x,n),\theta_{U;\alpha}(x,n))$.

2. For a future size m, find the minimum integer $U(x)$ and the maximum integer $L(x)$ such that

$$1 - F(L(x); \theta_{L;\alpha}(x,n), m) \geq \frac{1+P}{2} \tag{14}$$

and

$$F(U(x); \theta_{U;\alpha}(x,n), m) \geq \frac{1+P}{2}, \tag{15}$$

respectively.

The above can be modified to obtain one-sided tolerance limits by replacing the confidence interval in Step 1 by the one-sided $100 \times \gamma\%$ confidence bound for θ and then replacing $(1 + P)/2$ in Step 2 by P.

The above method is for a two-sided setting that results in equal-tailed tolerance intervals because we are controlling the percentages in both tails. However, one can control the percentage in the center of the distribution by using a slightly different criterion for Step 2. Namely,

2a. For a future size m, find integers $U(x) > L(x)$ such that

$$F(U(x); \theta_{U;\alpha}(x,n), m) - F(L(x); \theta_{L;\alpha}(x,n), m) \geq P. \tag{16}$$

Note that the coverage probabilities of the resulting tolerance intervals depend on if the objective is to control the percentages in both tails or to control the central percentage. However, Krishnamoorthy et al. (2011) point out that one typically seeks to control the central percentage of the sampled population since it is often unnecessary to control the tail percentages in practical applications (e.g., discrete quality assessment).

The choice of confidence interval in Step 1 impacts the coverage probabilities and expected widths of the resulting tolerance interval. Coverage properties of one-sided tolerance limits are typically similar to those of the confidence intervals used to construct them, while coverage properties of two-sided tolerance intervals often require further analysis. Krishnamoorthy et al. (2011) present some coverage results for binomial and Poisson tolerance intervals, and Young (2014) presents an extensive coverage study on negative binomial tolerance intervals.

3.1 Tolerance Intervals for the Binomial Distribution

A binomial random variable X has the cumulative distribution function

$$F_X(x; n, \pi) = \sum_{i=0}^{x} \binom{n}{i} \pi^i (1-\pi)^{n-i}, \tag{17}$$

for $x = 0, \ldots, n$ and $0 \leq \pi \leq 1$. Binomial tolerance intervals are often used where X is considered the number of defective (or acceptable) units in a random sample of size n with proportion π of such units in the population. To apply the discrete tolerance interval algorithm discussed above, we must first compute a $100 \times \gamma\%$ confidence interval for π. The problem of constructing a confidence interval for a binomial proportion is quite complex due to the lattice structure of the binomial distribution; see Brown et al. (2001) for a thorough discussion of this problem. For example, the exact confidence interval for π developed by Clopper and Pearson (1934) has the following limits:

$$\pi_L = B_{m,n-m+1;(1-\gamma)/2} \text{ and } \pi_U = B_{m+1,n-m;(1+\gamma)/2}, \tag{18}$$

where $B_{a,b;q}$ denotes the qth quantile of a beta distribution with shape parameters a and b.

Example 4. Let us consider the wafer data analyzed in Krishnamoorthy and Mathew (2009). This example concerns analyzing the locations of 50 chips in each of 30 wafers. A total of $x = 347$ chips were found on these $n = 1500$ locations; i.e., $\hat{\pi} = 347/1500 = 0.2313$ estimates the proportion of defective chips in a wafer. Suppose, we want a 0.99/0.95 equal-tailed tolerance interval for the number of defective chips in a wafer. To find this, we use the `bintol.int` function as follows:

```
> bintol.int(x=347, n=1500, m=50, alpha=0.05, P=0.99, method="CP",
+            side=2)
  alpha    P     p.hat 2-sided.lower 2-sided.upper
1  0.05 0.99 0.2313333             4            21
```

Thus, we are 95% confident that at least 99% of the wafers contain between 4 and 21 defective chips. In the code given above, note that `method="CP"` is used to specify that the exact method of Clopper and Pearson (1934) be used. Currently, 12 possible ways for constructing confidence intervals for a binomial proportion are available. Type `?bintol.int` for more information.

3.2 Tolerance Intervals for the Poisson Distribution

A Poisson random variable X has the cumulative distribution function

$$F_X(x; n, \lambda) = \sum_{i=0}^{x} \frac{e^{-\lambda}\lambda^i}{i!}, \tag{19}$$

for $x = 0,1,\ldots$ and where $\lambda > 0$ is a mean (or rate) parameter. Poisson tolerance intervals are often constructed for $S = \sum_{i=1}^{n} X_i$, where the $X_1,\ldots,X_n$ are an *iid* sample from a Poisson distribution with mean λ. Note that S is Poisson distributed with mean $n\lambda$. Like the binomial setting, there are many different ways to calculate confidence bounds for λ. For example, the exact confidence interval for λ, which is sometimes referred to as the tabular method and is attributable to Garwood (1936), has the following limits:

$$\lambda_L = \frac{1}{2n}\chi^2_{2s;\frac{1-\gamma}{2}} \text{ and } \lambda_U = \frac{1}{2n}\chi^2_{2(s+1);\frac{1+\gamma}{2}}, \quad (20)$$

where s is the observed value of S and $\chi^2_{f;q}$ is the qth quantile of a chi-squared distribution with f degrees of freedom.

Example 5. Wang and Tsung (2009) analyze a subset of data on surface defects on $n = 21$ rectangular steel plates. The total number of defects is $s = 35$. Hence, $\hat{\lambda} = 35/21 = 1.6667$ is an estimate for the average number of defects per steel plate. Suppose, we want a 0.95/0.95 equal-tailed tolerance interval for the number of surface defects on a future sample of $m = 30$ steel plates. To find this, we use the `poistol.int` function as follows:

```
> poistol.int(x=35, n=21, m=30, alpha=0.05, P=0.95,
+               method="TAB", side=2)
  alpha    P lambda.hat 2-sided.lower 2-sided.upper
1  0.05 0.95   1.666667            24            86
```

Thus, we are 95% confident that 95% of all future samples of size 30 steel plates will have between 24 and 86 defects. Note that like the binomial tolerance interval code, there is an option for `method`. Currently, eight possible ways for constructing confidence intervals for a Poisson distribution rate are available. Type `?poistol.int` for more information.

3.3 Tolerance Intervals for the Negative Binomial Distribution

A negative binomial random variable X has the cumulative distribution function

$$F_X(x;n,v) = \sum_{k=0}^{x} \binom{k+n-1}{k} v^n (1-v)^k, \quad (21)$$

for $x = 0,1,\ldots$ and where $v > 0$ is the probability of success within a given trial. An exact confidence interval for v that is given in Casella and Berger (2001) has the following limits:

$$v_L = \frac{n}{n+(x+1)F_{2(x+1),2n;(1+\gamma)/2}} \quad \text{and} \quad v_U = \frac{nF_{2n,2x;(1+\gamma)/2}}{nF_{2n,2x;(1+\gamma)/2}+x}, \quad (22)$$

where $F_{f_1,f_2;q}$ is the qth quantile of the F distribution with f_1 and f_2 degrees of freedom.

Example 6. Kikuchi (1987) presented results from a study assessing the maternal congenital heart disease status of mothers who gave birth to infants with a birth weight ≥ 2500 g. Pregnant mothers were recruited under an inverse sampling scheme until $n = 5$ mothers with the disease were observed.

There were $x = 141$ mothers observed with no maternal congenital heart disease. Suppose, we want a 0.90/0.90 equal-tailed tolerance interval for the total number of mothers necessary to recruit if it is desired to observe $m = 7$ mothers with the disease. To find this, we use the `negbintol.int` function as follows:

```
> negbintol.int(x=141, n=5, m=7, alpha=0.10, P=0.90,
+                method="CB", side=2)
  alpha   P     pi.hat 2-sided.lower 2-sided.upper
1   0.1 0.9 0.03424658            48           862
```

In the above, the tolerance limits are for the total number of failures before m successes are attained. Thus, we are 90% confident that 90% of all future studies requiring $m = 7$ mothers with the disease to be observed, will require 55 to 869 subjects; i.e., we added 7 to the limits. There are also eight different options for the `method` argument. Details on these different methods can be found in Young (2014) or by typing `?negbintol.int`.

4 NONPARAMETRIC TOLERANCE INTERVALS

When no distributional assumptions are made about the data, then one can construct nonparametric tolerance intervals based on order statistics. All that is assumed about the random sample $X_1, \ldots, X_n$ is that the underlying distribution function $F_X(\cdot)$ is a continuous, non-decreasing, probability distribution. In the following discussion, let $X_{(1)} < X_{(2)} < \cdots < X_{(n)}$ denote the corresponding order statistics of our random sample. Note that we write the ordering with strict inequality since $P_X\{X_i = X_j\} = 0$, $i \neq j$ for a continuous distribution. However, in practice the recorded data might have some equal values due to rounding or measurement error.

Suppose $B \sim \text{Bin}(n, P)$ (i.e., a binomial distribution with sample size n and probability of success P) and let k be the smallest integer for which $P(B \leq k - 1) \geq \gamma$. Then $X_{(k)}$ and $X_{(n-k+1)}$ are P/γ nonparametric one-sided lower and upper tolerance limits, respectively, for the underlying continuous distribution. The existence of such tolerance limits based on order statistics requires the sample size to satisfy the condition that

$$n \geq \frac{\ln(1-\gamma)}{\ln(P)}. \tag{23}$$

For a P/γ nonparametric two-sided tolerance interval, let $k = s - r$, where $1 \leq r < s \leq n$. Then $[X_{(r)}, X_{(s)}]$ is a P/γ two-sided tolerance interval for any such r and s. Choosing $s = n - r + 1$ yields the interval $[X_{(r)}, X_{(n-r+1)}]$, which is called the *truncated sample range* (Wilks, 1941). Note that square brackets are used since the limits determined by the order statistics are included in the interval. For such a two-sided tolerance interval to exist, n should satisfy the condition

$$(n-1)P^n - nP^{n-1} + 1 \geq \gamma. \tag{24}$$

Krishnamoorthy and Mathew (2009) note that a shorter two-sided nonparametric tolerance interval based on work in Hahn and Meeker (1991) can sometimes be constructed if we ignore the above truncated sample range convention. In general, the above method for constructing nonparametric tolerance intervals is known to be conservative. However, some recent methodologies have attempted to address this problem; cf. Frey (2010) and Young and Mathew (2014).

Example 7. As shown in Young and Mathew (2014), the `distfree.est` function can be used to calculate the minimum sample size necessary for the existence of one- and two-sided tolerance intervals as presented in the above formulas. The function takes four arguments:

- `n`: the necessary sample size to cover a proportion P of the population with confidence γ.
- `alpha`: this is equal to $1 - \gamma$.
- `P`: the proportion of the population to be covered.
- `side`: whether a one- or two-sided tolerance interval is assumed (determined by `side = 1` or `side = 2`, respectively).

The user needs to specify two of the three quantities: `n`, `alpha`, and `P`. The function will then solve for the third unspecified quantity. Thus, for determining `n` we only need to specify values for `alpha` and `P`. Also, any of these values can take on a vector of values to solve for multiple cases. For example, suppose we want the minimum sample sizes for one-sided tolerance limits under all combinations of $\gamma \in \{0.80, 0.90, 0.95, 0.99\}$ and $P \in \{0.50, 0.75, 0.80, 0.90, 0.95, 0.99\}$:

```
> distfree.est(alpha=c(0.20, 0.10, 0.05, 0.01),
+                 P=c(0.50, 0.75, 0.80, 0.90, 0.95, 0.99), side=1)
     0.5 0.75 0.8 0.9 0.95 0.99
0.2    3    6   8  16   32  161
0.1    4    9  11  22   45  230
0.05   5   11  14  29   59  299
0.01   7   17  21  44   90  459
```

The above quantities are reported in Young and Mathew (2014), but they also match those reported in Table 8.1 of Krishnamoorthy and Mathew (2009). Note that using `side=2` gives us the minimum sample sizes for the corresponding two-sided tolerance intervals:

```
> distfree.est(alpha=c(0.20, 0.10, 0.05, 0.01),
+                 P=c(0.50, 0.75, 0.80, 0.90, 0.95, 0.99), side=2)
     0.5 0.75 0.8 0.9 0.95 0.99
0.2    5   11  14  29   59  299
0.1    7   15  18  38   77  388
0.05   8   18  22  46   93  473
0.01  11   24  31  64  130  662
```

Example 8. Sokal and Rohlf (2012) presented a dataset on sulfur dioxide (SO_2) content for $n = 41$ U.S. cities averaged over the years 1969–1971. These data can be accessed from the `gamlss.data` package (Stasinopoulos and Rigby, 2014). We do not assume any specific distribution and, thus, will be interested in calculating an 0.80/0.90 two-sided nonparametric tolerance interval on these SO_2 levels. This can be accomplished using the `nptol.int` function as follows:

```
> library(gamlss.data)
> attach(usair)
> out <- nptol.int(x1,alpha=0.10,P=.80,side=2, method="WILKS")
> out
  alpha P 2-sided.lower 2-sided.upper
1 0.1 0.8          45.7          70.3
```

Hence, we are 90% confident that at least 80% of the SO_2 levels are between 45.7 $\mu g/m^3$ and 70.3 $\mu g/m^3$. Note in the above that we have specified `method="WILKS"` for the truncated sample range. Three other options are currently available. First is the shorter interval that was mentioned above. This can be specified by `method="HM"`, which refers to the work in Hahn and Meeker (1991). The second additional option is `method="WALD"`, which refers to the work of Wald (1943) and will produce all possible two-sided nonparametric tolerance intervals that meet the specified P/γ level using the condition

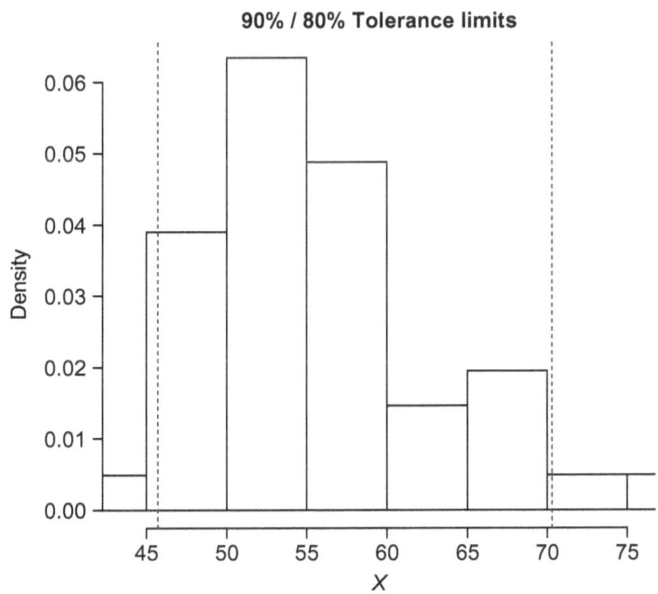

FIGURE 2 The air pollution data of Sokal and Rohlf (2012) with 0.80/0.90 two-sided tolerance limits overlaid.

in Eq. (24). The third additional option is `method="YM"`, which refers to the work of `youngmat` and will produce tolerance limits based on interpolated or extrapolated order statistics. Type `?nptol.int` for more information.

Figure 2 is a histogram of these data along with the calculated 0.80/0.90 tolerance limits. As in the normal example from earlier, we can apply the `plottol` as follows:

```
> plottol(out, x=x1, plot.type="hist")
```

Note that we specified `plot.type="hist"` to generate only the histogram.

5 REGRESSION TOLERANCE INTERVALS

Thus far, we have illustrated tolerance intervals for some standard univariate distributions. We now turn our attention to the regression setting. For the regression setting, we can either construct pointwise tolerance intervals or simultaneous tolerance intervals. The current functionality of the `tolerance` package only produces the former; hence, we will only demonstrate pointwise regression tolerance intervals in this section.

Wallis (1951) was the first paper to thoroughly develop the problem of tolerance intervals in the linear regression setting. Lieberman and Miller (1963) and Wilson (1967) both developed approaches for estimating simultaneous tolerance intervals in a regression setting. Mee et al. (1991) extended the use of simultaneous tolerance intervals for constructing calibration intervals. Carroll and Ruppert (1991) considered the issues of weighting and transformations in constructing regression tolerance intervals. Finally, Young (2013) extended the Wallis (1951) procedure for the nonlinear regression setting as well as developed an approach for nonparametric regression tolerance intervals. These procedures found in Young (2013) will be discussed below.

5.1 Linear Regression Tolerance Intervals

We first discuss tolerance intervals for the linear regression model

$$\mathbf{Y} = \mathbf{X}\beta + \epsilon, \tag{25}$$

where $\mathbf{Y} = (Y_1,\ldots, Y_n)^{\mathrm{T}}$ is an n-dimensional vector of responses, $\mathbf{X}$ is a known $n \times p$ matrix of rank p with the first column being 1's for an intercept term, β is an unknown $p \times 1$ vector of regression parameters, and ϵ is a vector of *iid* normally distributed error terms with mean 0 and variance σ^2. Moreover, let $\hat{\beta}$ be the ordinary least squares estimate, S^2 be the residual mean square error based on $n - p$ degrees of freedom, and $\hat{\mathbf{Y}} = \mathbf{X}\hat{\beta}$ be the fitted values.

Next, consider a vector of future predictor values of interest, say $\underline{X} = \underline{X}_h$. Let $\hat{Y}(\underline{X}_h)$ be the fitted value from the estimated regression model. Then P/γ lower and upper regression tolerance bounds are given by

$$L(\underline{X}_h) = \hat{Y}(\underline{X}_h) - k_{l,h}S \text{ and } U(\underline{X}_h) = \hat{Y}(\underline{X}_h) + k_{l,h}S, \tag{26}$$

respectively. In the above, $l \in \{1,2\}$ is the index for specifying one- or two-sided tolerance intervals. Formulas for these k-factors are as follows:

$$k_{1,h} = \frac{t_{n-p;\gamma}\left(\sqrt{n_h^*} z_P\right)}{\sqrt{n_h^*}} \tag{27}$$

and

$$k_{2,h} = \sqrt{\frac{(n-p)\chi^2_{1;P}(1/n_h^*)}{\chi^2_{n-p;1-\gamma}}}. \tag{28}$$

The quantity

$$n_h^* = (\underline{X}_h^{\mathrm{T}}(\mathbf{X}^{\mathrm{T}}\mathbf{X})^{-1}\underline{X}_h)^{-1} \tag{29}$$

is sometimes referred to as the *effective number of observations* (Wallis, 1951).

Example 9. Let us look at one of the widely used Old Faithful Geyser datasets. The dataset consists of the waiting time between eruptions and the duration of the eruption (both in minutes) for the Old Faithful Geyser in Yellowstone National Park, located in Wyoming, USA. The data were collected continuously between August 1 and 15, 1985. The data were first reported in Azzalini and Bowman (1990) and have been analyzed using various statistical techniques, like mixture models and clustering algorithms. We use a simple linear regression model to analyze a modified version of the original data found in the `datasets` package. There are $n = 299$ observations in the original dataset, but only $n = 272$ in the modified version.

Suppose we want 0.95/0.95 pointwise two-sided tolerance intervals for the duration times of Old Faithful at each of the different observed eruption times. Moreover, suppose we are interested in tolerance intervals at the specific times of 65 and 70 min. To find these, we use the `regtol.int` function as follows:

```
> library(datasets)
> data(faithful)
> attach(faithful)
> out <- regtol.int(reg=lm(eruptions ~ waiting), new.x=cbind
+                    (c(65, 70)), side=2, alpha=0.05, P=0.95)
```

```
> out[is.na(out[,3]),]
    alpha   P  y    y.hat 2-sided.lower 2-sided.upper
98  0.05 0.95 NA 3.041801      1.991828      4.091773
109 0.05 0.95 NA 3.419940      2.370322      4.469559
```

The last three arguments of the `regtol.int` function (i.e., `side`, `alpha`, and `P`) are all the same as in previous functions. The first argument `reg` is the regression function that one wishes to fit. It is an object of class `lm`, so one can also fit tolerance intervals for multiple linear regression models (i.e., more than one predictor). The second argument is `new.x`, which consists of any new predictor values for which one wants to compute tolerance intervals. Note that this must be a matrix or data frame with dimensions matching the number of predictors that appear in the specified model. Finally, we only present the tolerance intervals for the two new predictor values that we specified. This is accomplished by subsetting the output on the `NA`s that occur in the third column. The third column, `y`, corresponds to the observed values of the dataset, which is why these values are `NA` for any new levels of the predictors. Moreover, the output is a data frame that is sorted by the predicted value column, `y.hat`. The first row of the above output is for the waiting time of 60 min. Thus the interpretation is that we are 95% confident that 95% of the eruption duration times for a waiting time of 60 min will be between approximately 2.0 and 4.1 min.

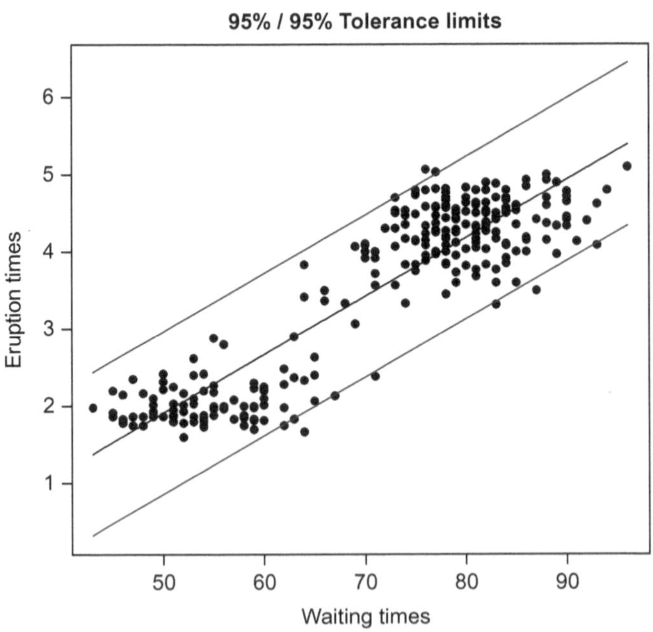

FIGURE 3 The Old Faithful Geyser data with 0.95/0.95 pointwise two-sided tolerance intervals.

We are also able to use the plottol function to plot the regression data and the pointwise tolerance intervals:

```
> plottol(out, x=cbind(1, waiting), y=eruptions, side="two",
+         x.lab="Waiting Times", y.lab="Eruption Times")
```

Note that the x argument must be the full matrix of predictors, including a column of 1's if the intercept is fit. The y argument is simply the data used as the response in the regression model. The plot using the above function is given in Fig. 3.

5.2 Nonlinear Regression Tolerance Intervals

Young (2013) developed an approach for nonlinear regression tolerance intervals that closely follows the linear regression setting. Consider the nonlinear regression model

$$\mathbf{Y} = f(\mathbf{X}, \beta) + \epsilon, \tag{30}$$

where $\mathbf{Y}$ and $\mathbf{X}$ are defined as in the linear regression setting, but now β is a q-dimensional vector of regression parameters, $f(\cdot)$ is a parametric nonlinear regression function, and ϵ is a vector of *iid* error terms, which may or may not be normal. Denote the maximum likelihood estimate of β by $\hat{\beta}$, which can be obtained using one of the iterative nonlinear optimization algorithms available in the nls function. Moreover, let S^2 be an estimate of the error variance based on $n - q$ degrees of freedom and $\hat{\mathbf{Y}} = f(\mathbf{X}, \hat{\beta})$ be the fitted values.

Next, define the $n \times q$ pseudo-design matrix $\mathbf{D}$ to have ith row equal to the gradient vector evaluated at the maximum likelihood estimate $\hat{\beta}$:

$$\underline{D}_i = \nabla f(\underline{x}_i, \hat{\beta}) = \left. \frac{\partial f(\underline{x}_i, \beta)}{\partial \beta^{\mathrm{T}}} \right|_{\beta = \hat{\beta}}. \tag{31}$$

As before, let $\hat{Y}(\underline{X}_h)$ be the fitted values from the estimated nonlinear regression model at a new set of predictor values. Then, approximate nonlinear regression intervals can be found by replacing the degrees of freedom for the linear regression k-factors by $n - q$ and also by replacing the effective number of observations by

$$n_h^* = (\underline{D}_h^{\mathrm{T}} (\mathbf{D}^{\mathrm{T}} \mathbf{D})^{-1} \underline{D}_h)^{-1}. \tag{32}$$

Example 10. We next look at an example involving biochemical oxygen demand (BOD) in a sample of water. BOD is the amount of dissolved oxygen needed by an organism in a body of water at a given temperature over a specified amount of time to break down organic material that is present. We analyze a small BOD dataset of size $n = 6$ from Marske (1967) using a nonlinear regression where BOD (in mg/L) is the response Y and time (in days) is the predictor X. We fit the following first-order nonlinear regression model:

$$Y = \beta_0(1 - \exp\{-\exp\{\beta_1\}X\}). \tag{33}$$

Suppose we want 0.95/0.90 pointwise two-sided tolerance intervals for the BOD at each of the different observed times when evaluating the quality of a certain water sample. Moreover, suppose we are interested in tolerance intervals at the specific time of 6 days. To find these, we use the `regtol.int` function as follows:

```
> data(BOD)

> attach(BOD)

> formula <- as.formula(demand ~ A*(1-exp(-exp(lrc)*Time)))

> out.nl <- nlregtol.int(formula=formula,
+                          xy.data=BOD[,2:1], x.new=cbind(6),
+                          side=2, alpha=0.10, P=0.95,
+                          start=c(A=20, lrc=log(.35)))
> out.nl
  alpha    P    y.hat    y 2-sided.lower 2-sided.upper
1  0.1 0.95  7.88745  8.3     -3.273941      19.04884
2  0.1 0.95 12.52498 10.3      1.290310      23.75965
3  0.1 0.95 17.79749 15.6      6.882233      28.71275
4  0.1 0.95 16.85487 16.0      6.261148      27.44859
5  0.1 0.95 15.25167 19.0      4.523460      25.97989
6  0.1 0.95 18.67758 19.8      6.816094      30.53906
7  0.1 0.95 18.35172   NA      6.945754      29.75768

> plottol(out.nl, x=Time, y=demand, side="two",
+          x.lab="Time (Days)", y.lab="Demand (mg/L)")
```

For the first observation, the lower bound is about −3.27 mg/L, which is not physically possible. So one would need to truncate this specific value at 0. Note in the above that any predictions for new levels of the predictors appear at the bottom of the outputted data frame. In this case, for a given level of 6 days, we are 90% confident that 95% of the BOD measurements will be between about 6.9 and 29.8 mg/L. For more about passing starting values for the underlying `nls` routine, see `?nlregtol.int` or `?nls`. We have also included a plot of the fitted nonlinear regression model with the 0.95/0.90 pointwise two-sided tolerance intervals in Fig. 4.

5.3 Nonparametric Regression Tolerance Intervals

Young (2013) also developed an approach for nonparametric regression tolerance intervals. For this approach, the methodology of Section 4 is applied to

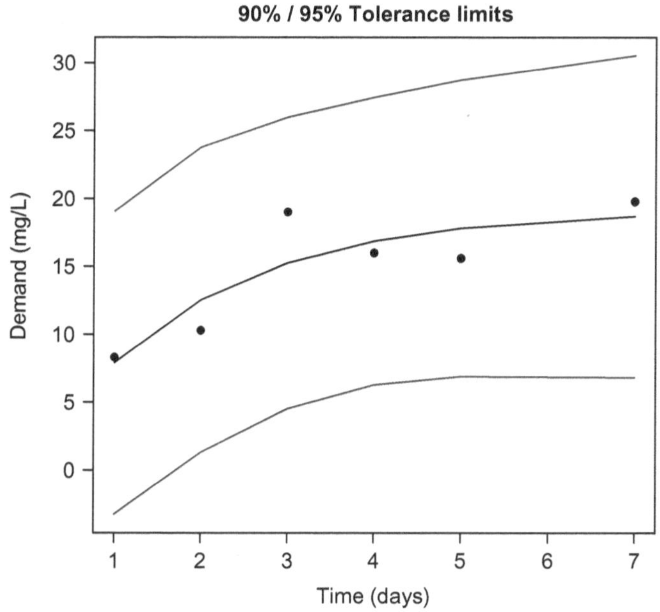

FIGURE 4 The BOD data with 0.95/0.90 pointwise two-sided tolerance intervals.

the raw residuals from a nonparametric regression fit. Consider the nonparametric regression model

$$\mathbf{Y} = m(\mathbf{X}) + \epsilon, \tag{34}$$

where $m(\cdot)$ is a regression function to be estimated and ϵ is an error term whose elements are assumed to be drawn from an unspecified continuous distribution. As noted in Young (2013), there is no need to assume constant error variance since it is often the case that $\text{Var}(\epsilon_i) = \sigma^2(\underline{x}_i)$, where $\sigma^2(\cdot)$ is some continuous, bounded function. Since the functional form of $m(\cdot)$ is unknown, we can use nonparametric approaches like kernel regression, smoothing splines, or LOESS (Cleveland et al., 1988) for estimation. We denote the fitted values from the nonparametric regression fit by $\hat{\mathbf{Y}} = \hat{m}(\mathbf{X})$.

One also needs to determine the degree of smoothness of the nonparametric regression fit. This is controlled through some user-specified quantity, e.g., bandwidth for kernel regression, tuning parameter for smoothing splines, and span for LOESS. As emphasized in Young (2013), the validity of the resulting tolerance limit will depend upon the correctness of the estimated functional form $\hat{m}(\cdot)$ through determining an appropriately chosen smoothing parameter. This can be accomplished by using, for example, a generalized or local cross-validation criterion (see Hastie et al., 2009; Cummins et al., ,2001, respectively). Closely tied to the appropriateness of the chosen fit is the inherent bias in nonparametric regression fits, which will impact the nominal level of the specified confidence level for a statistical interval (Ruppert et al., 2003). Thus, the nonparametric regression tolerance intervals is only approximate, but the additional coverage study in Young (2013) demonstrates that it does have good performance relative to the nominal level.

We will now apply the method of Section 4 to the raw residuals from a nonparametric fit to construct tolerance limits. Let $e_{(1)},\ldots,e_{(n)}$ be the ordered residuals from the fitted nonparametric regression equation and $e_{(r)}$ and $e_{(n-r+1)}$ (i.e., the limits for the truncated sample range) be the P/γ lower and upper nonparametric tolerance bounds for the residuals. Then approximate P/γ nonparametric regression tolerance bounds can be calculated by

$$L=\hat{Y}(\underline{X}_h)+e_{(r)} \text{ and } U=\hat{Y}(\underline{X}_h)+e_{(n-r+1)}, \tag{35}$$

where $\hat{Y}(\underline{X}_h)$ is the fitted value from the estimated regression model for a vector of future predictor values of interest $\underline{X}_h$. One-sided tolerance bounds can be similarly constructed. Young (2013) also notes that in the case of nonconstant error variance, one can use the scaled residuals $e_i^*=e_i/\hat{\sigma}^2(x_i)$, where $\hat{\sigma}^2(\cdot)$ is an estimated variance function. This would require one to use the ordered scaled residuals $e_{(1)}^*,\ldots,e_{(n)}^*$ in the above approach.

Example 11. We analyze an occupational dataset where the prestige ranking for $n = 102$ occupations is modeled as a function of the corresponding average income (in Canadian dollars) as reported from the 1971 Canadian Census. The methodology for determining the prestige rankings is from Pineo and Porter (1967). The data are available in the `car` package by Fox and Weisberg (2011).

Suppose we wish to identify occupations with high or low levels of prestige relative to a given annual salary. We accomplish this by constructing 0.90/0.95 pointwise tolerance intervals for a nonparametric regression fit and then by identifying those observations that fall outside of the limits. As in Young (2013), we use a LOESS fit with a span of 0.70. We first construct this fit, save the fitted values, and then use the `npregtol.int` function as follows:

```
> library(car)
> data(Prestige); attach(Prestige)
> out.np <- loess(prestige~income, span=0.7)
> y.hat <- out.np$fit
> out.tol <- npregtol.int(x=income, y=prestige, y.hat=y.hat,
+                         side=2, alpha=0.05, P=0.90)
> plottol(out.tol, x=income, y=prestige, y.hat=y.hat,
+         x.lab="Average Income (Canadian Dollars)",
+         y.lab="Prestige of Canadian Occupations",side="two")
```

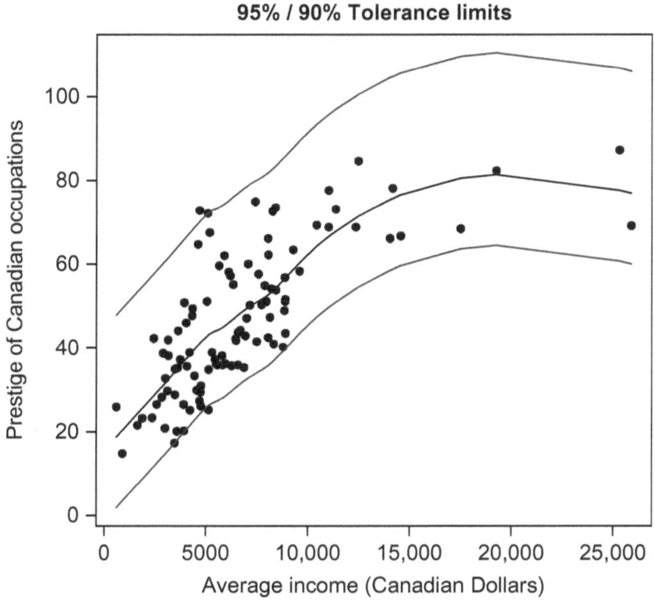

FIGURE 5 The Canadian occupational prestige dataset with 0.90/0.95 pointwise two-sided tolerance intervals.

The data and the corresponding 0.90/0.95 tolerance intervals are given in Fig. 5. Note that around the average income of $5000 Canadian dollars are two occupations that fall outside the limits. These can be extracted as follows:

```
> out.tol2 <- out.tol[order(out.tol[,4]),]
> low.ind <- which((out.tol2[,4]<out.tol2[,6]))
> up.ind <- which((out.tol2[,4]>out.tol2[,7]))
> rownames(Prestige[order(Prestige[,4]),])[low.ind]
[1] "slaughterers.1"
> rownames(Prestige[order(Prestige[,4]),])[up.ind]
[1] "ministers"
```

Thus, using the 0.90/0.95 tolerance intervals as (subjective) thresholds, one could view that slaughterers (i.e., meat butchers) have a lower level of prestige, given their average income while ministers have a higher level of prestige given their average income.

6 MULTIVARIATE TOLERANCE REGIONS

So far, the settings discussed have been inherently univariate procedures to obtain one- or two-sided tolerance intervals. For multivariate data, we can construct a tolerance region, which is akin to constructing confidence and prediction regions. However, there is considerably less literature on tolerance regions for multivariate data and the majority of the literature focuses on the multivariate normal setting. Some work has been developed for nonparametric multivariate tolerance regions (cf. Di Bucchianico et al., 2001; Li and Liu, 2008), but such procedures can be quite computationally expensive, especially for larger dimensions. The focus of this section will be solely on the multivariate normal setting.

Let $\mathbf{X}_1, \ldots, \mathbf{X}_n$ be an *iid* sample from a d-dimensional multivariate normal distribution, which has density

$$f_{\mathbf{X}}(\mathbf{x};\mu,\Sigma) = \frac{1}{(2\pi)^{p/2}|\Sigma|^{1/2}} \exp\left\{-(1/2)\left\|\Sigma^{-1/2}(\mathbf{x}-\mu)\right\|^2\right\}, \tag{36}$$

where $\mu \in \mathbb{R}^d$, Σ is a $d \times d$ symmetric, positive-definite matrix, and $\|\mathbf{A}\|^2 = \mathbf{A}^{\mathrm{T}}\mathbf{A}$ for any matrix $\mathbf{A}$. Multivariate normal tolerance regions are constructed using the sample mean vector and sample variance–covariance matrix, which are given by

$$\overline{\mathbf{X}} = \frac{1}{n}\sum_{i=1}^{n}\mathbf{X}_i \quad \text{and} \quad \mathbf{S} = \frac{1}{n-1}\sum_{i=1}^{n}\left\|(\mathbf{X}_i - \overline{\mathbf{X}})^{\mathrm{T}}\right\|^2, \tag{37}$$

respectively. Formally, a multivariate normal tolerance region of dimension d is given by

$$\left\{\mathbf{X} : \left\|\mathbf{S}^{-1/2}(\mathbf{X}-\overline{\mathbf{X}})\right\|^2 \leq c\right\}, \tag{38}$$

where the tolerance factor c satisfies

$$P_{\overline{\mathbf{X}},\mathbf{S}}\left[P_{\mathbf{X}}\left\{\left\|\mathbf{S}^{-1/2}(\mathbf{X}-\overline{\mathbf{X}})\right\|^2 \leq c \middle| \overline{\mathbf{X}},\mathbf{S}\right\} \geq P\right] = \gamma. \tag{39}$$

As shown in Krishnamoorthy and Mathew (2009), c does not depend on the unknown parameters μ and Σ. Therefore, the tolerance factor c can be estimated using Monte Carlo simulation.

Several Monte Carlo procedures for determining c have been developed in the literature. Krishnamoorthy and Mathew (1999) provide a good overview and thorough coverage study of these approximation procedures, while the latter approach developed in Krishnamoorthy and Mondal (2006) appears to be superior in terms of coverages. Many of these different procedures for determining c are available in the `mvtol.region` function, however, we will only apply the one due to Krishnamoorthy and Mondal (2006) for our examples. But first, we outline the algorithm presented in their work:

Algorithm: For given n, d, P, and γ, compute the following:

1. Generate $Q_j^2 \sim \chi^2/n$, $j = 1,\ldots,d$, and $\mathbf{V} \sim W_d\left((1/(n-1))\mathbf{I}_d, n-1\right)$, where $W_d(\mathbf{H}, f)$ is the d-dimensional Wishart distribution with f degrees of freedom and scale matrix $\mathbf{H}$.
2. Compute the eigenvalues, $\lambda_1,\ldots,\lambda_d$ of $\mathbf{V}$.
3. Compute the following quantities:

$$c_j = \sum_{i=1}^{d} \frac{1 + jQ_i^2}{l_i^h}, \quad h = 1,2,3 \text{ and } a = \frac{c_2^3}{c_3^2}. \tag{40}$$

4. Set $C = \left(\frac{c_3}{c_2}\left(\chi^2_{a;P} - a\right) + c_1\right)$.
5. Repeat Steps 1-4 B times for some large value of B.
6. Take the 100γth percentile of the simulated C values as an estimate of c.

Example 12. To demonstrate how to calculate c, we will analyze the classic Iris dataset of Fisher (1936). We first subset the data to only look at those flowers belonging to the species *Iris virginica*, which is assigned the name `iris2` below. This leaves us with $n = 50$ tetravariate observations, where the four measurements are sepal length, sepal width, petal length, and petal width. All measurements are in centimeters. For illustrative purposes, we begin by only assessing the petal length and petal width measurements, which we treat as bivariate normal. Below is the code for obtaining the tolerance factor c for a 0.90/0.90 bivariate tolerance region:

```
> iris2 <- subset(iris,Species=="virginica")[,1:4]
> set.seed(100)
> out1 <- mvtol.region(x=iris2[,3:4], alpha=0.10, P=0.90, B=5000,
+                      method="KM")
> out1
         0.1
0.9 6.090448
```

For reproducibility, we set a seed since the function `mvtol.region` uses the Monte Carlo algorithm discussed earlier. There are two arguments in the `mvtol.region` function worth highlighting. First is the `B` argument, which controls how many iterations to use for the Monte Carlo approximation. Second is the `method` argument, which allows the user to select which Monte Carlo method to use. As stated earlier, we use the algorithm due to Krishnamoorthy and Mondal (2006), which corresponds to `method="KM"`. Currently, the `method` argument can take on eight different methods, most of which are discussed in Krishnamoorthy and Mathew (1999).

Using the tolerance factor determined above, we can then proceed to plot the resulting bivariate tolerance region:

```
> plottol(out1, x=as.matrix(iris2[,3:4]), x.lab="Petal Length",
+          y.lab="Petal Width")
```

Note that the x argument must take an object that is of the class matrix, which is why the as.matrix function is applied to the data subset. The resulting plot is given in Fig. 6a.

Example 13. For illustrating a trivariate normal tolerance region, we now consider the sepal width measurement along with the two petal measurements. The computation to determine c for a 0.90/0.90 trivariate tolerance region is analogous to the previous example:

```
> set.seed(100)
> out2 <- mvtol.region(x=iris2[,2:4], alpha=0.10, P=0.90, B=5000,
+                       method="KM")
> out2
        0.1
0.9 8.23237
```

However, plotting of the three-dimensional tolerance region is done using the ellipse3d and plot3d functions in the rgl package (Adler and Murdoch, 2014). The only additional argument that appears below is the (optional) z.lab argument to provide a label for the third axis.

```
> plottol(out1, x=as.matrix(iris2[,2:4]), x.lab="Petal Length",
+          y.lab="Petal Width", z.lab="Sepal Width")
```

Since the resulting figure is interactive, Fig. 6b is simply a screenshot of the resulting tolerance region.

One can easily construct a tolerance region for all four measurements on the *I. virginica* flowers. However, there are currently no data visualization tools in the tolerance package to display summaries of multivariate normal tolerance regions for dimensions greater than 3.

7 FINAL REMARKS

In this chapter, we have given an overview of some tolerance interval and tolerance region calculations, along with applications using the tolerance package in R. There are numerous important topics not covered here, such as computing tolerance intervals for random and mixed effects models, multivariate regression models, or when using a Bayesian framework. For more details on the development of additional topics like these, we refer to the text by Krishnamoorthy and Mathew (2009).

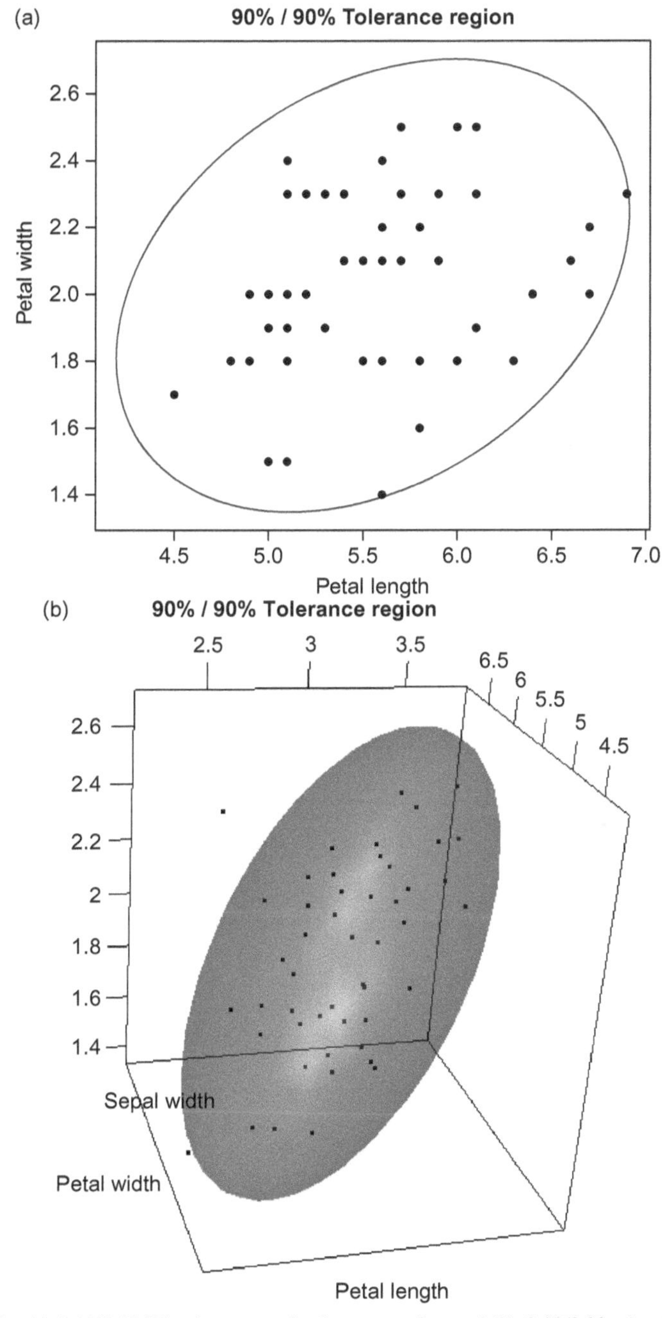

FIGURE 6 (a) 0.90/0.90 Bivariate normal tolerance region and (b) 0.90/0.90 trivariate normal tolerance region using subsets of the *Iris* data.

We have also demonstrated only some of the capabilities of the `tolerance` package. Additional capabilities of the packages include

- functions for calculating tolerance intervals for many other univariate distributions, such as the Cauchy (`?cautol.int`), hypergeometric (`?hypertol.int`), and uniform (`?uniftol.int`) distributions;
- fiducial-based tolerance intervals for functions of binomial (`?fidbintol.int`), negative binomial (`?fidnegbintol.int`), and Poisson distributions (`?fidpoistol.int`), as developed in Mathew and Young (2013); and
- a function for constructing tolerance regions for the multivariate multiple linear regression setting (`?mvregtol.region`).

While the documentation of the `tolerance` package is complete for all of the included functions, some additional demonstrations are also provided in Young (2010). We also note that the `tolerance` package is continuing to grow and there have been some requests made by end users incorporated into the package. So, one can expect additional capabilities in future versions of the package.

REFERENCES

Adler, D., Murdoch, D., 2014. rgl: 3D Visualization Device System (OpenGL). R Package Version 0.93.996. http://CRAN.R-project.org/package=rgl.

Azzalini, A., Bowman, A.W., 1990. A look at some data on the Old Faithful geyser. J. R. Stat. Soc. C Appl. Stat. 39, 357–365.

Bain, L.J., Engelhardt, M., 1981. Simple approximate distributional results for confidence and tolerance limits for the Weibull distribution based on maximum likelihood. Technometrics 23, 15–20.

Bhaumik, D.K., Gibbons, R.D., 2006. One-sided approximate prediction intervals for at least p of m observations from a gamma population at each of r locations. Technometrics 48, 112–119.

Blischke, W.R., Murthy, D.N.P., 2000. Reliability: Modeling, Prediction, and Optimization. John Wiley & Sons, New York, NY.

Brown, L.D., Cai, T.T., DasGupta, A., 2001. Interval estimation for a binomial proportion. Stat. Sci. 16, 101–133.

Carroll, R.T., Ruppert, D., 1991. Prediction and tolerance intervals with transformation and/or weighting. Technometrics 33, 197–210.

Casella, G., Berger, R.L., 2001. Statistical Inference, second ed. Duxbury, Belmont, CA.

Cleveland, W.S., Devlin, S.J., Grosse, E., 1988. Regression by local fitting. J. Econ. 37, 87–114.

Clopper, C.J., Pearson, E.S., 1934. The use of confidence or fiducial limits illustrated in the case of the binomial. Biometrika 26, 404–413.

Cummins, D.J., Filloon, T.G., Nychka, D., 2001. Confidence intervals for nonparametric curve estimates: towards more uniform pointwise convergence. J. Am. Stat. Assoc. 96, 233–246.

Di Bucchianico, A., Einmahl, J.H.J., Mushkudiani, N.A., 2001. Smallest nonparametric tolerance regions. Ann. Stat. 29, 1320–1343.

Fisher, R.A., 1936. The use of multiple measurements in taxonomic problems. Ann. Eugenics 7, 179–188.

Fox, J., Weisberg, S., 2011. An R Companion to Applied Regression, second ed. Sage, Thousand Oaks, CA. http://socserv.socsci.mcmaster.ca/jfox/Books/Companion.

Frey, J., 2010. Data-driven nonparametric tolerance sets. J. Nonparametr. Stat. 22, 169–180.

Garwood, F., 1936. Fiducial limits for the Poisson distribution. Biometrika 28, 437–442.

Grubbs, F.E., 1971. Approximate fiducial bounds for the reliability of a series system for which each component has an exponential time-to-fail distribution. Technometrics 13, 865–871.

Guttman, I., 1970. Statistical Tolerance Regions: Classical and Bayesian. Charles Griffin and Company, London.

Hahn, G.J., Chandra, R., 1981. Tolerance intervals for Poisson and binomial random variables. J. Qual. Technol. 13, 100–110.

Hahn, G.J., Meeker, W.Q., 1991. Statistical Intervals: A Guide for Practitioners. Wiley-Interscience, New York, NY.

Harris, E.K., Boyd, J.C., 1995. Statistical Bases of Reference Values in Laboaratory Medicine. Marcel-Dekker, New York, NY.

Hastie, T., Tibshirani, R., Friedman, J., 2009. The Elements of Statistical Learning: Data Mining, Inference, and Prediction, second ed. Springer, New York, NY.

Howe, W.G., 1969. Two-sided tolerance limits for normal populations - some improvements. J. Am. Stat. Assoc. 64, 610–620.

Jílek, M., 1981. A bibliography of statistical intervals. Math. Oper. Stat. Ser. Stat. 12, 441–456.

Jílek, M., Ackermann, A., 1989. A bibliography of statistical tolerance regions, II. Statistics 20, 165–172.

Kikuchi, D.A., 1987. Inverse sampling in case control studies involving a rare exposure. Biom. J. 29, 243–246.

Krishnamoorthy, K., Mathew, T., 1999. Comparison of approximation methods for computing tolerance factors for a multivariate normal population. Technometrics 41, 234–249.

Krishnamoorthy, K., Mathew, T., 2009. Statistical Tolerance Regions: Theory, Applications, and Computation. Wiley, Hoboken, NJ.

Krishnamoorthy, K., Mondal, S., 2006. Improved tolerance factors for multivariate normal distributions. Commun. Stat. Simul. Comput. 35, 461–478.

Krishnamoorthy, K., Xia, Y., Xie, F., 2011. A simple approximate procedure for constructing binomial and Poisson tolerance intervals. Commun. Stat. Theory Methods 40, 2443–2458.

Li, J., Liu, R.Y., 2008. Multivariate spacings based on data depth: I. Construction of nonparametric multivariate tolerance regions. Ann. Stat. 36, 1299–1323.

Lieberman, G.J., Miller, R.G., 1963. Simultaneous tolerance intervals in regression. Biometrika 50, 155–168.

Marske, D., 1967. An Implementation of Matchmaking Analysis in Condor. (Master's thesis) University of Wisconsin-Madison.

Mathew, T., Young, D.S., 2013. Fiducial-based tolerance intervals for some discrete distributions. Comput. Stat. Data Anal. 61, 38–49.

Mee, R.W., Eberhardt, K.R., Reeve, C.P., 1991. Calibration and simultaneous tolerance intervals for regression. Technometrics 33, 211–219.

Patel, J.K., 1986. Tolerance limits - a review. Commun. Stat. Theory Methods 15, 2719–2762.

Pineo, P.C., Porter, J., 1967. Occupational prestige in Canada. Can. Rev. Sociol. Anthropol. 4, 24–40.

R Development Core Team, 2014. R: A Language and Environment for Statistical Computing. R Foundation for Statistical Computing, Vienna, Austria. ISBN: 3-900051-07-0. http://www.R-project.org/.

Ruppert, D., Wand, M.P., Carroll, R.J., 2003. Semiparametric Regression. Cambridge University Press, Cambridge, UK.

Sokal, R.R., Rohlf, F.J., 2012. Biometry, fourth ed. W. H. Freeman and Company, New York.

Stasinopoulos, M., Rigby, B., 2014. gamlss.data: GAMLSS Data. R Package Version 4.2–7. http://CRAN.R-project.org/package=gamlss.data.

Wald, A., 1943. An extension of Wilks' method for setting tolerance limits. Ann. Math. Stat. 14, 45–55.

Wallis, W.A., 1951. Tolerance intervals for linear regression. In: Neyman, J. (Ed.), Second Berkeley Symposium on Mathematical Statistics and Probability. University of California Press, Berkely, CA, pp. 43–51.

Wang, H., Tsung, F., 2009. Tolerance intervals with improved coverage probabilities for binomial and Poisson variables. Technometrics 51, 25–33.

Weissberg, A., Beatty, G.H., 1960. Tables of tolerance-limit factors for normal distributions. Technometrics 2, 483–500.

Wilks, S.S., 1941. Determination of sample sizes for setting tolerance limits. Ann. Math. Stat. 12, 91–96.

Wilks, S.S., 1942. Statistical prediction with special reference to the problem of tolerance limits. Ann. Math. Stat. 13, 400–409.

Wilson, A.L., 1967. An approach to simultaneous tolerance intervals in regression. Ann. Math. Stat. 38, 1536–1540.

Young, D.S., 2010. tolerance: an R package for estimating tolerance intervals. J. Stat. Softw. 36, 1–39. http://www.jstatsoft.org/v36/i05/.

Young, D.S., 2013. Regression tolerance intervals. Commun. Stat. Simul. Comput. 42, 2040–2055.

Young, D.S., 2014. A procedure for approximate negative binomial tolerance intervals. J. Stat. Comput. Simul. 84, 438–450.

Young, D.S., Mathew, T., 2014. Improved nonparametric tolerance intervals based on interpolated and extrapolated order statistics. J. Nonparametr. Stat. 26, 415–432.

Chapter 9

Modeling the Probability of Second Cancer in Controlled Clinical Trials

Kao-Tai Tsai[1] and Karl E. Peace[1]
Jiann-Ping Hsu College of Public Health, Georgia Southern University, Georgia, USA
[1]Corresponding author: e-mail: Tsai0123@yahoo.com; kepeace@georgiasouthern.edu

ABSTRACT
Due to the advancement of medical technologies and cancer care, the long-term survival of cancer patients has been substantially increased. For some patients, increases in survival have been offset by the long-term late effects of cancer and its treatments. One of the most life-threatening sequelae is the diagnosis of a new malignant cancer. The number of patients with multiple primary cancers is growing with second new cancer now representing approximately 16% of all cancers reported to the SEER Program of National Cancer Institute. Second malignant cancers reflect not only the late effects of the original cancers and therapies but also the influence of shared etiologic factors, genetic susceptibility, environmental exposures, cancer drug exposures, and older age, etc. In this research, we attempt to outline a framework to model the patient-level probability of the occurrence of new cancer malignancies using patient's demographics, medical history and concomitant medications, clinical laboratory tests, drug efficacy, and adverse event data—which are all commonly collected in clinical trials.

Keywords: Second malignant cancer, Repeated measures, Control clinical trials

1 INTRODUCTION

Recent medical research has substantially advanced medical technologies and cancer care. As a consequence, the long-term survival of cancer patients has been substantially increased. For example, as of January 2008, there were approximately 11.9 million cancer survivors in the United States (about 4% of the population) and the 5-year relative survival rate for all cancers combined has increased steadily to reach 66.1% for patients diagnosed from 1999 to 2006.

For some patients, increases in survival have been offset by the long-term late effects of cancer and its treatments. One of the most life-threatening sequelae is the diagnosis of a new malignant cancer. The number of patients

Handbook of Statistics, Vol. 32. http://dx.doi.org/10.1016/B978-0-444-63431-3.00009-7

with multiple primary cancers is growing with second new cancer now representing approximately 16% of all cancers reported to the SEER Program of National Cancer Institute.

Second cancers reflect not only the late effects of therapies but also the influence of shared etiologic factors (in particular, tobacco and excessive alcohol intake), genetic susceptibility, environmental exposures, host effects, and combinations of factors, including gene–environment interactions. Risks for selected second new cancers are also modified by age at exposure and attained age.

During the course of a clinical trial, huge amounts of data are collected. These usually include patient demographic data, medical history and concomitant medications, and efficacy and adverse event data of the medical intervention, in addition to extra data collected during the follow-up phase if feasible. It would be a great advantage for the researcher to investigate these data and identify the risk factors, which can potentially lead to the second new cancer.

In this article, we attempt to outline a framework to model the patient-level probability of the occurrence of new cancer malignancies using demographic, medical history and concomitant medication data, efficacy and adverse event, and clinical laboratory test data of the medical intervention. Such data are commonly collected in clinical trials during the active treatment and follow-up phases.

2 DIFFICULTIES IN SECOND CANCER RESEARCH

Journal articles about clinical trials are typically based on the active controlled trial phase in addition to short-term follow-up with small to moderate sample sizes (with few exceptions), which is inadequate to capture the incidence of second cancer due to its medium to long latency. In addition, most cancer trials involve combinations of multiple drugs. Patients crossing over from one treatment to another when their drug does not show efficacy leads to an inadequate control group for fair comparisons. Furthermore, the time between the end of controlled trial phase and the occurrence of second malignancies can be years, during which many uncontrolled factors are operative and make proper attribution of the causal-effect to be very difficult if not impossible.

On the other hand, most of the reports of systematic reviews are retrospective with mixtures of patient populations. In addition, under-reporting of patient-level detail incidence data by clinicians makes the tracking of new cancers even more challenging.

3 CURRENT KNOWLEDGE OF SECOND MALIGNANCY

Cancer survivors constitute about 3.5% of the U.S. population, but second malignancies among this high-risk group account for approximately 16% of all cancer

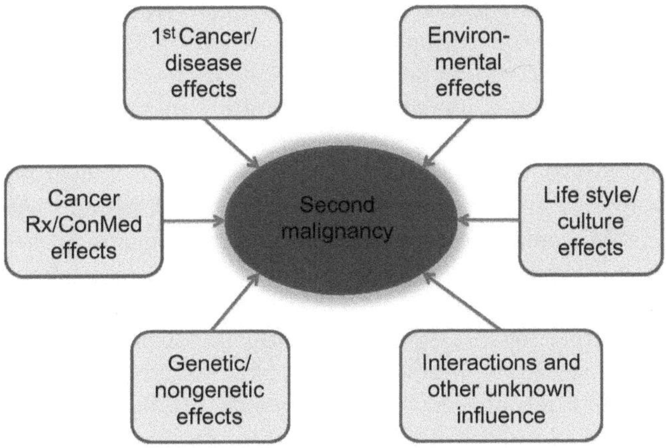

FIGURE 1 Known factors affecting second malignancy occurrence.

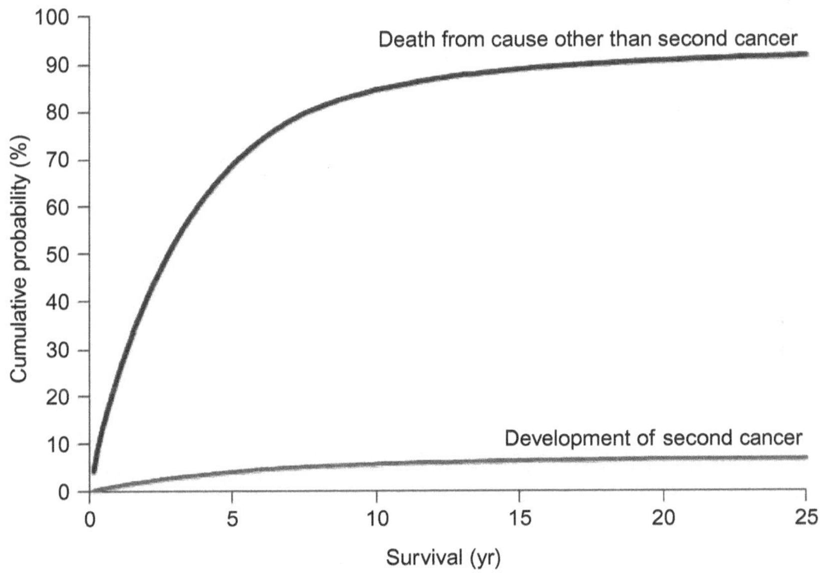

FIGURE 2 Second malignancy and death.

incidence (Travis et al., 2006). Based on the NCI SEER database, compared to the general population, cancer survivors have a 14% increased risk of developing a new cancer (Curtis, 2006). In the following, we discuss a few potential causal-effect factors commonly cited in cancer research. Given the great extent of complexities, they still remain very active research areas (Figs. 1 and 2).

Drug Effects: Ironically, medicines or medical procedures for treating current diseases may also cause new cancers later. For example, although radiotherapy is a component of the standard of care for many primary malignancies, it may also be complicit in the development of second cancer (Featherstone et al., 2005). In addition, the long term adverse effects of some popular cytogenetic and immune therapies are still not well studied. The effects of these treatments and their combinations can only become clearer after broader exposure of larger patient populations with longer durations.

Disease Effects: In a Swedish study, based on 5652 patients with multiple myeloma (MM) precursor disease (MGUS) and IgG/IgA, an eightfold increased risk of developing myelodysplastic syndrome or acute myeloid leukemia (MDS/AML) was observed (Mailankody et al., 2011). These observations support a role for disease related factors in MDS/AML following MM (Mailankody et al., 2011). Similar findings exist for other types of malignancies with certain kinds of biomarkers and genetic profiles. Therefore, a better understanding of underlying molecular mechanisms across disease subgroups and risk of a second malignancy may form the basis for targeted therapies to minimize risks of second malignancies.

Genetic Effects: It has been estimated that genetic variations may account for up to 95% of the variability in drug disposition and effects. Polymorphisms in genes encoding drug-metabolizing enzymes, DNA repair pathways, drug transporters, and targets may also contribute to an individual's susceptibility for subsequent malignancies (Evans et al., 2003). Non-genetic factors that may modulate treatment effects include age, race, organ function, concomitant therapy, drug interactions, and the disease itself.

Environmental Effects: Studies indicate that exposure to ionizing radiation at younger ages and higher doses increase the risk of developing cancer (Landgren et al., 2011). Chronic antigen stimulation from prior autoimmune, infectious, inflammatory, allergic disorders, and immune dysregulation may play a role in the pathogenesis of some specific types of cancers (Kristinsson et al., 2011). In addition, socioeconomic status has been shown to influence cancer survival, suggesting that life-style factors in these disorders are of importance.

Behavior Effects: The commonly proposed behavioral risk factors (e.g., tobacco, alcohol, and diet) and obesity are among the various types of behavioral effects that can potentially affect the occurrence of cancers (Kyle et al., 2007). Large scale epidemiological studies exist to provide potential causal effect of behaviors and prevalence of cancers and other health effects.

4 CLINICAL TRIAL DATABASE

Controlled clinical trials usually collect huge amounts of data, especially for large scale trials—which can enroll hundreds or thousands of patients. These data may include demographics: age, gender, disease stage, performance score, baseline cytogenetic status, medical history and previous treatments, and data from the experimental treatment: efficacy, adverse effects, clinical laboratory data, etc. Efficacy data from cancer trials include variables such as disease progression, time-to-progression, overall survival, censoring status, etc. Additional data from the follow-up phase may also be collected, if that is part of the trial design.

With such a broad database, it may be informative for physicians to explore the database in assessing possible causal effect between disease and treatment. Since not all the variables will be relevant to the objectives of an analysis, it is crucial to select the proper variables to be included in the statistical model as covariates.

4.1 Laboratory Test Data Analysis

In clinical trials, clinical laboratory data are usually collected at visits throughout the trial and are therefore longitudinal in nature. Thus repeated measure methodologies that take into account the serial correlation between data points should be utilized.

As a general practice to better understand the data, it is usually very informative to examine the data distributions graphically prior to formal statistical modeling. With the graphical capability in R, this can be easily performed for each clinical laboratory parameter to investigate whether a treatment effect exists. For example, after examining all clinical laboratory tests data in the illustrative data used for this article, a few parameters were found to exhibit interesting patterns, which can potentially be attributed to the status of second malignancies.

The data for clinical laboratory test for lymphocytes is plotted here across the various visits of the study. Figure 3 is a boxplot of lymphocyte test values plotted between the visits of the study and the status of malignancy. The green (gray in the print version) boxes are for the subjects without second malignancy and the red (dark gray in the print version) for those with second malignancy.

Even though this plot shows all the lymphocyte values, it is simply too crowded to reveal any distinctive patterns. Therefore, it is preferable to use the line plot as shown in Fig. 4, which shows the lines of median values for each visit by malignancy status. It clearly shows the different patterns of lymphocytes progression in time between the two malignancy groups.

Examining the patterns of all the clinical laboratory tests data, their correlations, and discussions with substantive subject experts are crucial steps to properly selecting the clinically relevant factors for the formal statistical model.

For statistical modeling of repeated measures longitudinal data, a common approach is to use a mixed effects model to identify significant parameters. The specification of the mixed effects model and the parameter estimations can be found in many monographs. In R, one can use the function *glm* for various kinds of data distributions.

Specifically, in our illustrative data analysis, we denote the clinical laboratory data, at time points $\{t_{ij} : j = 1,\cdots,n_i\}$, for subject i by

$$y_{ij} = \{y_i(t_{ij}), j = 1,\cdots,n_i\}$$

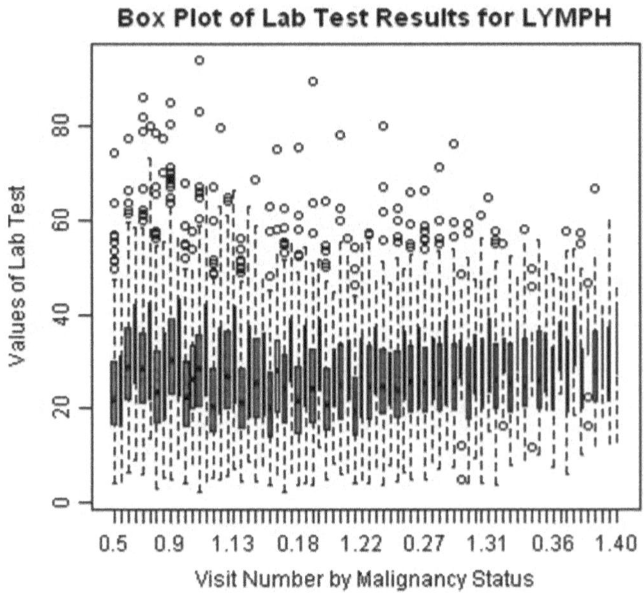

FIGURE 3 Boxplot of lymphocyte values.

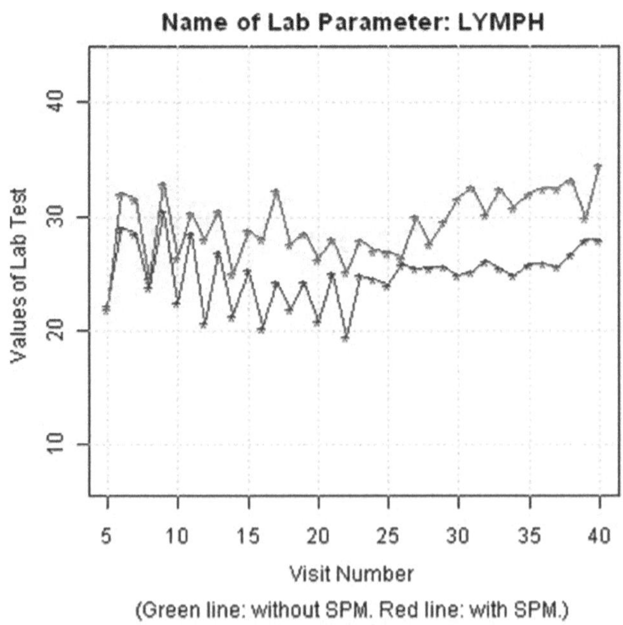

FIGURE 4 Line plot of lymphocyte values. (Green (dark gray in the print version) line: Without SPM. Red (gray in the print version) line: With SPM.)

and estimate the laboratory tests effects by assuming a linear mixed effects model for the repeated measures:

$$\begin{aligned} y_i(t) &= m_i(t) + \epsilon_i(t) \\ &= x_i'(t)\beta + z_i'(t)b_i + \epsilon_i(t), \epsilon_i(t) \sim N(0,\sigma^2); \end{aligned} \tag{1}$$

where β is the vector of the unknown fixed effects parameters, b_i is the vector of random effect parameters and $b_i \sim N(0,\Sigma)$, $x_i(t)$ is the design matrix for the fixed effect, $z_i(t)$ is the design matrix for the random effect, and $\epsilon_i(t)$ is the measurement error independent of b_i.

In addition to the linear model, we check the goodness-of-fit using graphical tools instead of just examining the goodness-of-fit statistics. Graphical methods can sometimes suggest proper transformation of covariates and can provide visual inspection of how well models fit the data. For example, the *mmps* function in R package *car* provides the marginal modeling plot (not shown here) for each variable and compares the fit of a linear model with the fit of a nonparametric model so that any substantial deviation with potential needs for variable transformation can be easily detected.

4.2 Medical History and Concomitant Medicines

In Phase II or Phase III clinical trials, participants are usually diagnosed with the disease that the experimental treatments are designed to show treatment benefit. Many participants may even have been through several previous treatments with no or little success. Therefore, these participants carry their medical history of the disease and previous treatments into the new trial. This kind of background history can often obscure the effect of the experimental treatment; therefore, it is critical for clinicians to understand the potential contaminations of treatment effect arising from these background factors.

The medical history may include previous treatments, previous medications, and life style factors, etc. Such data may sometimes be imprecise with possibly many missing values. This can pose substantial data analysis challenges. Therefore, careful analyses are needed to examine these variables to gain a good understanding of the role these variables may have on treatment effect.

In the illustrative example in this article, the experimental drug was designed to treat MM, a human blood disease. After combing through all the medical history data, the following variables were found to have substantial association with the second malignancy: aortic valve stenosis, basal cell carcinoma, benign prostatic hyperplasia, chronic bronchitis, cardiac failure, cholecystitis, chronic obstructive pulmonary disease, epistaxis, herpes zoster, hypoaesthesia, infection, menopause, peripheral oedema, oesophagitis, periarthritis, polyneuropathy, prostate cancer, sciatica, upper respiratory tract infection, vitamin B12 deficiency, and vitamin D deficiency.

These medical history items were selected using a simple χ^2 test of significance in the relationship to the occurrence of the malignancy. The corresponding graphs were also examined to investigate any potential influential outliers.

Similar to the analysis of medical history, the following concomitant medicines (denoted by ATC codes) were found to be significant in the occurrence of second malignancies: A06AD (lLaxatives), C09DA (angiotensin II antagonists and diuretics), C10AB (lipid modifying agents, plain), D06BB (antivirals), J01DB and J01FF (antibacterials), N05CD (benzodiazepine derivatives), N05CF (cyclopyrrolones), and R03AK (adrenergics and other anti-asthmatics).

Several graphical methods can be applied to demonstrate association between two sets of categorical variables. The most commonly used are bioplot (Gabriel, 1971), correspondence analysis, and multidimensional scaling, etc. These functions are all available in R packages. For example, the biplot of the association between concomitant medicine and status of second malignancy is shown in Fig. 5. In this figure, patients were divided into four categories by their status of second malignancy (present/absent) and status of concomitant medicine usage (present/absent). They were placed at the four tips of the two crossing straight lines (in clockwise order from first quadrant: no malignancy but with concomitant medicines, had malignancy and had concomitant medicines, no malignancy and no concomitant medicines, had malignancy but no concomitant medicines). The concomitant medications were placed in various places in the plot. By projecting these concomitant medicines into the straight lines, one can measure the degree of their association with these four categories.

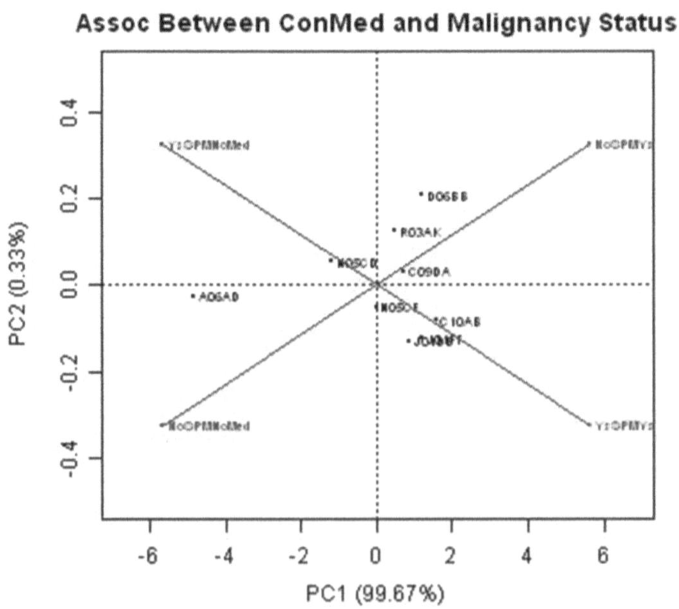

FIGURE 5 Assoc between ConMed and status of malignancy.

4.3 Efficacy Data Consideration

In clinical trials, drug efficacy and adverse events are two primary concerns. Investigators want to maximize efficacy while minimizing drug safety issues simultaneously. Therefore, to gain a more comprehensive picture of the treatment effect, one needs to consider these two aspects of drug effects together in data analyses.

In our illustrative example, time to disease progression is one of the key efficacy variables. Therefore, it is desirable to investigate whether hazard of disease progression is a contributing factor of second malignancy. Several methods are available in the literature for the estimation of disease progression hazard. We used Mueller and Wang 1994) for our estimation in this article. The results using other methods general produced the same results.

More specifically, we let (T_i^*, C_i, T_i) be the vector of (failure time, censor time, observation time) for the i-th subject ($i = 1, \cdots, n$). The hazard estimate can be obtained by smoothing the increments of the Nelson–Aalen estimator

$$\Lambda_n(t) = \sum_{i=1}^{n} \delta_{[i]} I_{(X_{(i)} \leq t)} / (n - i + 1) \tag{2}$$

of the cumulative hazard function $\Lambda(t)$, where $\delta_{[i]}$ is the censoring indicator of $X_{(i)}$. Using the kernel method, one can estimate the kernel hazard rate function by

$$\Lambda_n(t) = \sum_{i=1}^{n} \{\delta_{[i]} / (n - i + 1)\} (1/h) K((t - X_{(i)})/h). \tag{3}$$

The estimated cumulative hazard of all patients is shown in Fig. 6 and the *Q–Q* plot of the cumulative hazards for the patients with or without second malignancy is shown in Fig. 7. Two phenomena can be easily observed: the longer the patients were treated, the higher the hazard, and the hazard for the patients with second malignancy is higher than that of the patients who did not have second malignancy. Therefore, the graphical difference suggests the inclusion of hazard of disease progression as one of the important factors in the final statistical model.

5 INTEGRATED ANALYSIS

After preliminary analyses of the medical history, concomitant medicines, efficacy and drug adverse event data, clinical laboratory tests data, and possibly other externally related data, one can incorporate the significant variables found from each analysis to construct a more comprehensive model.

In our illustrative example, we use the following mixed effects model to aggregate these variables and to estimate the possible effect of each covariate on the occurrence of second malignancy; the R code with the function *glmer* is shown below

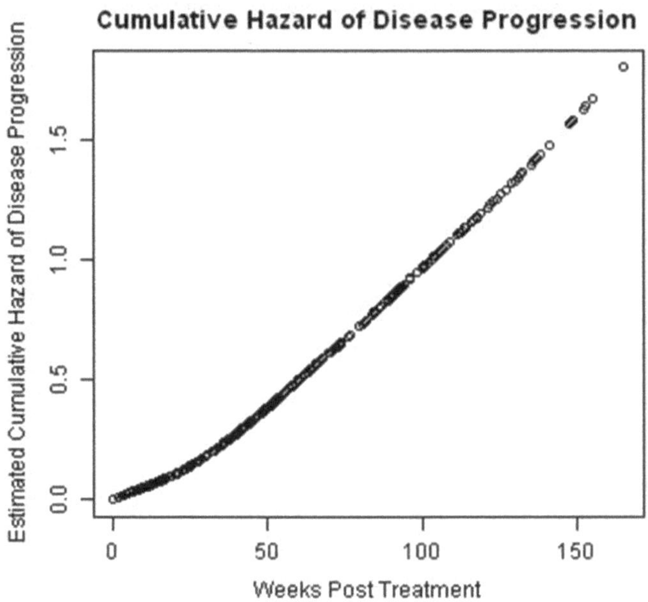

FIGURE 6 Est cumulative hazards.

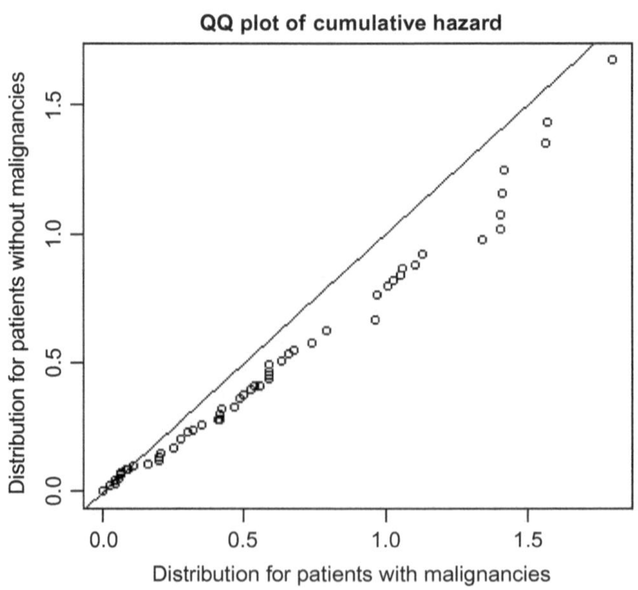

FIGURE 7 *QQ* plot of Est hazards.

```
glmer(SPM ~ lymp + nuet + age + pscore + pfscumhaz + trt +
sex + stage + creat + sigmedhist + conmed + (1|trt),
data=data1, na.action=na.omit, family=binomial)
```

The results from the *glmer* need to be carefully examined to further select the statistically significant and biologically or clinically meaningful variables to avoid colinearity among these variables and to arrive at a parsimonious model.

In addition to parametric modeling shown above, one can also use model-based recursive partitioning as proposed by Breiman et al. 1984) and Hothorn et al. 2006). Recursive partitioning models have the advantage to partition the entire patient population into subgroups with more homogeneous characteristics and can better estimate their respective probabilities of having a second malignancy. This advantage is generally not easy to be distilled in the parametric model especially with a model having several variables. In addition, one can also use the R functions, such as *tree* (in *Tree* package) and *ctree* (in *Party* package), to provide a graphical presentation of the results which can be a very useful tool to communicate with nonstatisticians. An example is shown in Fig. 8.

In Fig. 8, the recursive partitioning tree splits the patients into two groups by the most significant covariate (concomitant medicine use in this example). The right-branch is further split by the values of lymphocytes, medical history, and

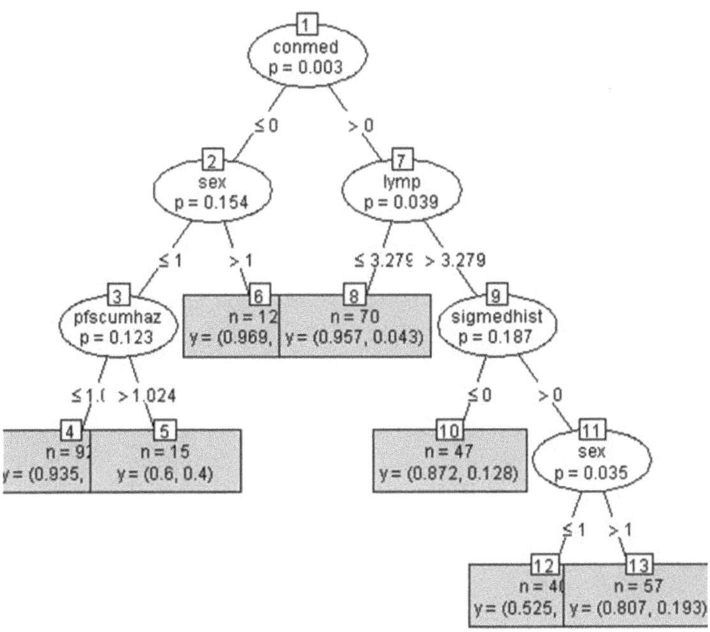

FIGURE 8 Recursive partitioning tree.

sex. Similar splits are performed for the left branch with sex and hazard of disease progression. With a more homogeneous subset of the patients, the boxes provide the sample size therein and the probabilities of having second malignancies. Therefore, the recursive partitioning tree provides a clearer picture for readers to understand how the covariates affect the risk of the second malignancies. The depth of the partitioning tree can be controlled by the analysts depending on the chosen level of significance. It is important to balance the information exhibited and the complexity of the tree. An overly large tree can produce confusion for readers and be practically useless.

In parametric modeling, such as *glmer* above, one way to rank the relative importance among the model covariates is by their nominal significance level. The R package *randomForest* (Breiman, 2001) provides another way to rank the importance of covariates by generating many recursive-partitioning trees (hence the *Forest*) and the covariates that are statistically significant more often in the forest are considered more important. There are a couple of ways to measure the significance. One approach is to use the reduction of mean squared errors, and an alternative is to use the Gini index. Given the estimated class probabilities $\{p(j|t)|\, j = 1,\cdots,J\}$ of a node t in a classification tree, the Gini index is a measure of node impurity defined as

$$\text{Gini Index}(t) \equiv 1 - \sum_j p^2(j|t). \tag{4}$$

The Gini index can be interpreted as the empirical variance of a given classification (Light and Margolin, 1971). The results from these two methods may not always be the same; therefore, it is important to also consider other factors such as clinical meaningfulness to make the final choice. An example is shown in Figs. 9 and 10. Even though they are somewhat different, both of these two figures indicate that the values of lymphocytes and neutrophils are most important among all of these covariates.

The fitted values, namely, the probability of having a second malignancy, from the *glmer* model above are categorized by the status of second malignancy and are plotted in Fig. 11 using boxplot. The group of patients without second malignancy, as shown in the left shorter box, clearly has much smaller fitted values than the one on the right of the graph, which is for patients with second malignancy.

The misclassification rates for various cutoff probabilities are also shown in Fig. 12. The curve running from lower-left corner to the upper-right corner of the plot represents the misclassification rate for the patients with second malignancy. The curve running from upper-left corner to the lower-right corner of the plot represents the misclassification rate for the patients without second malignancy. The curve in between provides the overall misclassification rates. The overall misclassification rate is about 13% if the cutoff probabilities are near 0.2 or greater.

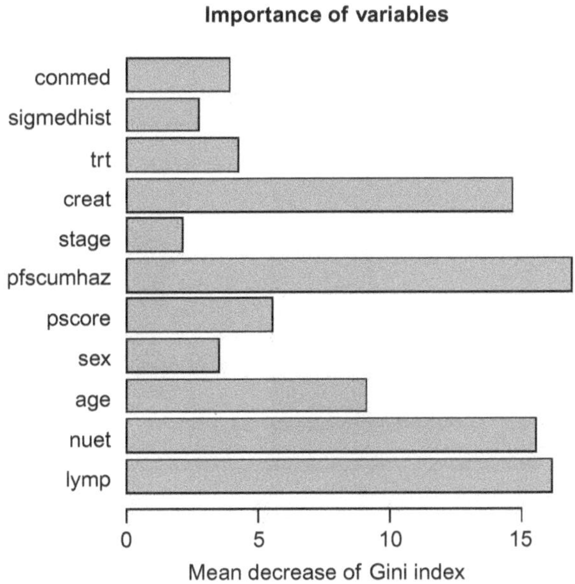

FIGURE 9 Importance of variables.

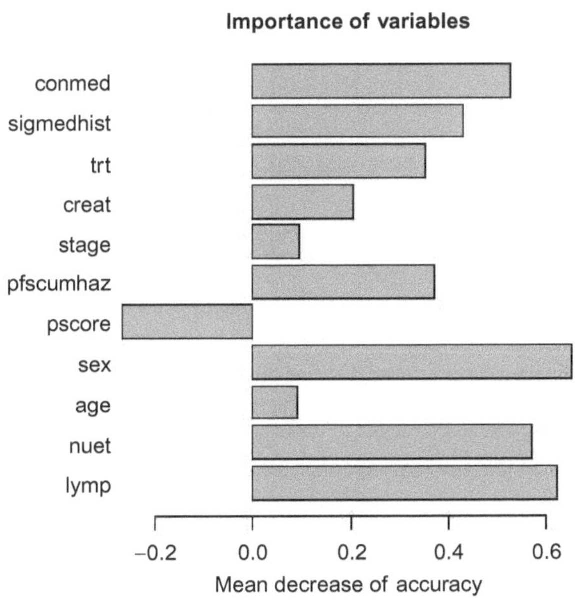

FIGURE 10 Importance of variables.

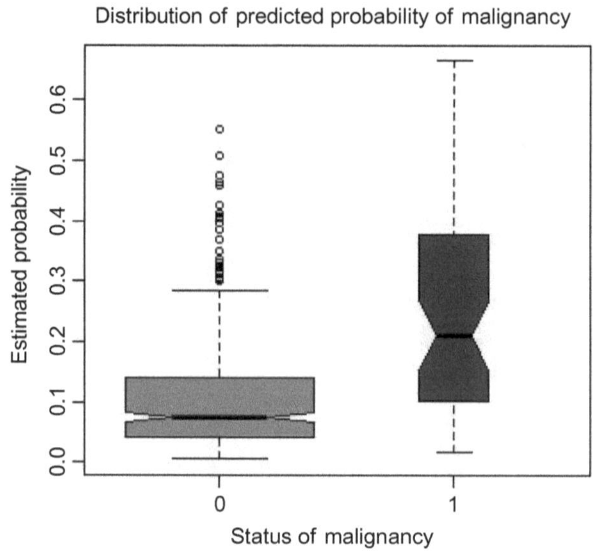

FIGURE 11 Predicted probabilities of malignancies.

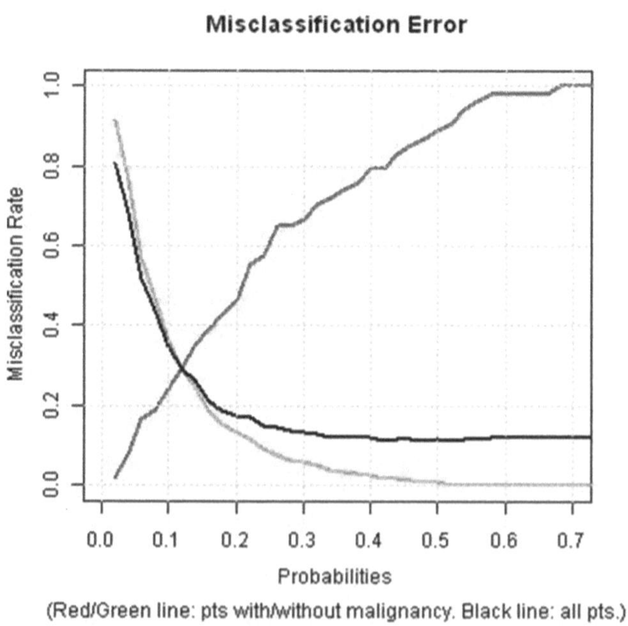

FIGURE 12 Misclassification error of malignancy prediction.

6 ASSESSING MODEL ADEQUACY

After fitting a model to data, as a general practice, one assesses goodness-of-fit to ensure that the data can be adequately explained by the proposed model. This can be especially more challenging for categorical responses than for continuous responses when fitted using the generalized linear model due to the discrete nature of the responses.

Instead of just examining the significance of the goodness-of-fit statistic, Landwehr et al. 1984) proposed a graphical method to assess the model adequacy by creating pseudo measures and uses these pseudo measures to evaluate the goodness-of-fit. More specifically, according to their proposed method, the following steps were taken to assess the adequacy of the model:

1. partition the N observations into K non-overlapping clusters with N_k observations in each;
2. form the $N \times K$ matrix $\mathbf{Z} = (z_{ik})_{N \times K}$ with $z_{ik} = 1$, if ith observation is in the kth cluster and 0 otherwise;
3. compute the local estimate $\hat{p}_l$ by fitting

$$\text{logit}(p_l) = \mathbf{Z}\gamma + \mathbf{X}\beta + \epsilon,$$

where **X** if the vector of covariates.

4. use the model fitted above to compute the local deviance contribution of each observation, $d(\hat{p}_{l,jk}, y_{jk})$, and sum up the deviances within each group and denote it by D_{lk} (Fig. 13);

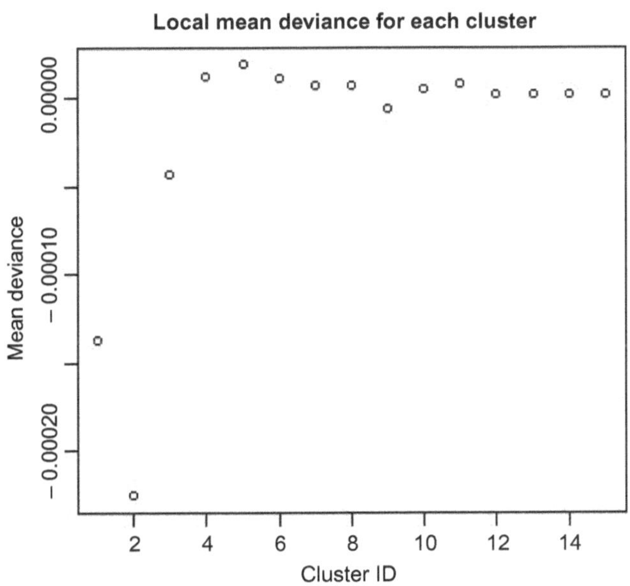

FIGURE 13 Goodness-of-fit of GLM model.

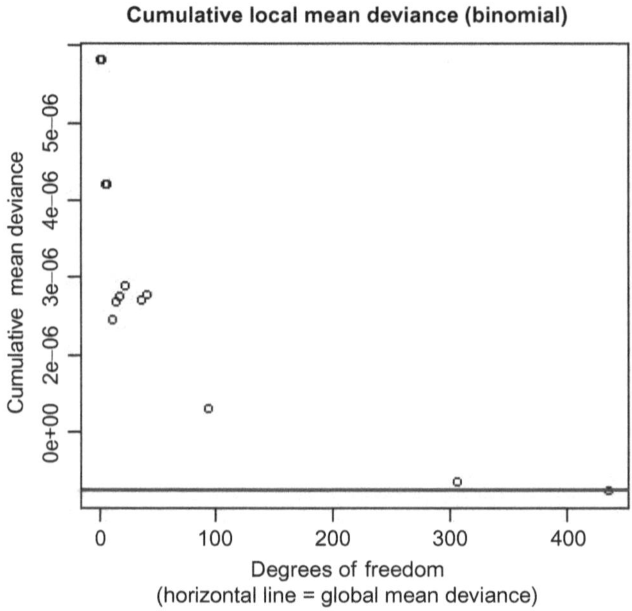

FIGURE 14 Goodness-of-fit of GLM model.

5. compute running estimates of approximate pure error

$$D_L(t) = \sum_{k=1}^{t} D_{lk} / \sum_{k=1}^{t} (N_k - 1).$$

The values $D_L(t)$ represent the local mean deviance calculated from the tightest t clusters;

6. plot $D_L(t)$ against its degrees of freedom $df(t) = \sum_{k=1}^{t} (N_k - 1)$, for $t = 1, \cdots, k$ and superimpose the plot with the line of global mean deviance, $Y = D_L(K)$. Substantial deviation of the line representing global mean deviance from the points of $D_L(t)$ for larger values of t indicates systematic lack-of-fit (Fig. 14).

Depending on the adequacy of the model, one may need to revise the model by including other covariates or further extend the linear model to other generalized additive models such as

$$\mathbf{y} = g(\mu^{\mathbf{b}}) + \epsilon, \tag{5}$$

with

$$g(\mu^{\mathbf{b}}) = \mathbf{X}\beta + \sum_{j=1}^{k} f_j(\mathbf{x}_{\mathbf{j}i1}, \cdots, \mathbf{x}_{\mathbf{j}ij}) + \mathbf{Zb}, \tag{6}$$

and

$$\mathbf{b} \sim N(0, \psi_\theta), \quad \epsilon \sim N(0, \boldsymbol{\Lambda}\sigma^2), \tag{7}$$

where $g(\cdot)$ is link function, $\mu^{\mathbf{b}} \equiv \mathbf{E}(\mathbf{y}|\mathbf{b})$, $\mathbf{y}|\mathbf{b} \sim$ exponential family, and f_j is a smooth/non-smooth function of the columns of $\mathbf{X}$. The error term ϵ may have several variance components to take care of the possible random effects from multiple factors in the data.

7 SUMMARY

The occurrence of second malignant cancers poses a unique challenge in medical practice and research. With advanced medical research, which produces new or better medical technologies and more effective medications for various diseases, human life prolongs and, hence, increases the possibility to acquire diseases which relate to factors such as older age, exposure of cancer-causing substances, environmental hazard, or personal life-styles, etc. Many of these factors are correlated among themselves and even confounded with other unknown factors uncontrollable by researchers.

Research on second malignancy based on controlled clinical trial data usually is inadequate due to the durations of the active controlled trials, the medium to long latency of most second malignancies, and insufficient duration of patient follow-up after finishing the active controlled treatment.

Aggregated results from epidemiological research provides some general indications of the potential causes of diseases. However most studies are usually unable to ascertain more specific underlying causes of second malignancies, not to mention the prediction of disease at the patient-level unless the studies were specifically designed for that purpose.

In this article, we outline a general approach to analyze commonly collected clinical trial data and long-term follow-up data if it is available, with the objective to identify the covariates which can potentially contribute to the occurrence of a second malignancy. Using data from well-designed clinical trials that was carefully collected, we have demonstrated satisfactory prediction of the occurrence of second malignancies for patients in the database. The analysis performed and the statistical model derived in this article can be refined with more good quality data from other clinical trials. Therefore, collaborations among researchers, data providers, and regulatory agencies are essential for further advanced research and discovery of new medicines or technologies.

The data analysis and statistical modeling can be quite computationally intensive, especially if more data are available for incorporation. Modern *Big Data* technologies in data processing and integration, data storage, and modeling analytics related to properly handling of big data may become useful. The analyses conducted in this article were all based on the platform and programs provided in the various R packages. These packages enable researchers to visually examine the data distributions, which is essential to understand the data (but usually not routinely conducted) and to perform advanced statistical analysis and modeling with a high level of efficiency and effectiveness.

REFERENCES

Breiman, L., Friedman, J.H., Olshen, R.A., Stone, C.J., 1984. Classification and Regression Trees. Wadsworth.

Breiman, L., 2001. Random forests. Machine Learning 45 (1), 5–32.

Curtis, R.E., Freedman, D.M., Ron, E., Ries, L.A.G., Hacker, D.G., Edwards, B.K., Tucker, M.A., Fraumeni Jr., J.F. (Eds.), 2006. New Malignancies Among Cancer Survivors: SEER Cancer Registries, 1973–2000. NIH Publ, Bethesda, MD.

Evans, W.E., McLeod, H.L., 2003. Pharmacogenomics - drug disposition, drug targets, and side effects. N. Engl. J. Med. 348, 538–549.

Featherstone, C., Delaney, G., Jacob, S., Barton, M., 2005. Estimating the optimal utilization rates of radiotherapy for hematologic malignancies from a review of the evidence. Cancer 103, 393–401.

Gabriel, K.R., 1971. The biplot graphical display of matrices with application to principal component analysis. Biometrika 58, 453–467.

Hothorn, T., Hornik, K., Zeileis, A., 2006. Unbiased recursive partitioning: a conditional inference framework. Journal of Computational and Graphical Statistics 15 (3), 651–674.

Kristinsson, S.Y., Björkholm, M., Hultcrantz, M., Derolf, Å.R., Landgren, O., Goldin, L.R., 2011. Chronic antigen stimulation from prior autoimmune, infectious, inflammatory, allergic disorders and immune dysregulation may play a role in pathogenesis of both MM and AML/MDS. J. Clin. Oncol. 2897–2903.

Kyle, R.A., Vincent Rajkumar, S., 2007. Epidemiology of the plasma-cell disorders. Clin. Haematol. 20, 637–664.

Landgren, O., Thomas, A., Mailankody, S., 2011. Myeloma and second primary cancers. N. Engl. J. Med. 365 (23), 2241–2242.

Landwehr, J.M., Pregibon, D., Shoemaker, A.C., 1984. Graphical methods for assessing logistic regression models. JASA 79, 61–71.

Light, R.J., Margolin, B.H., 1971. An analysis of variance for categorical data. JASA 66, 534–544.

Mailankody, S., Pfeiffer, R.M., Kristinsson, S.Y., Korde, N., Bjorkholm, M., Goldin, L.R., Turesson, I., Landgren, O., 2011. Risk of acute myeloid leukemia and myelodysplastic syndromes after multiple myeloma and its precursor disease (MGUS). Blood 118 (15), 4086–4092.

Mueller, H.G., Wang, J.L., 1994. Hazard rates estimation under random censoring with varying kernels and bandwidths. Biometrics 50, 61–76.

Travis, L.B., Rabkin, C.S., Brown, L.M., Allan, J.M., Alter, B.P., Ambrosone, C.B., Begg, C.B., Caporaso, N., Chanock, S., DeMichele, A., Figg, W.D., Gospodarowicz, M.K., Hall, E.J., Hisada, M., Inskip, P., Kleinerman, R., Little, J.B., Malkin, D., Ng, A.K., Offit, K., Pui, C.-H., Robison, L.L., Rothman, N., Shields, P.G., Strong, L., Taniguchi, T., Tucker, M.A., Greene, M.H., 2006. Cancer survivorship - genetic susceptibility and second primary cancers: research strategies and recommendations. J. Natl. Canc. Inst. 98 (1), 15–25.

Chapter 10

Bayesian Networks

Marepalli B. Rao*[,1] and C.R. Rao[†,‡]
*Department of Environmental Health, University of Cincinnati, Cincinnati, Ohio, USA
†Research Professor, University of Buffalo, Buffalo, New York, USA
‡CRRAO AIMSCS, Hyderabad, India
[1]Corresponding author: e-mail: Marepalli.rao@uc.edu

ABSTRACT
A Bayesian network model depicts interrelationships in the form of conditional distributions for a collection of random variables. The model is described in terms of a directed acyclic graph in which the nodes are random variables and the directed arcs spell out the structure of conditional distributions. In this chapter, a rudimentary introduction is provided to describe network models. The R package "bnlearn" is invoked to fit a network model using the data on the underlying random variables. Best network model is identified using the "hill-climbing algorithm" via Bayesian Information criterion score. A Bayesian network model is akin to a structural equation model.

Keywords: Directed Acyclic Graphs, Joint distributions, Conditional distributions, Marginal distributions, Model selection, Hill-climbing algorithm

1 INTRODUCTION

In Bayesian networks, we deal with a number of interrelated (random) variables. We explore how the joint distribution of the variables can be described by exploiting what we know about their natural interrelationships via conditional distributions. We use graph theory to explain their interrelationship. If data are available on the random variables, we fit a Bayesian network model which describes their relationship in a succinct way. Bayesian networks are a marriage between probability theory and graphs. One of the main goals in Bayesian networks is prediction. Bayes theorem plays a crucial part in this connection.

The subject has applications in a variety of fields. A well-known example is the Quick Reference Model (QMR)-DT, a decision-theoretic reformulation of the QMR. Here, the variables are diseases and symptoms. The probabilities are conditionals of symptoms given diseases. The Bayesian network would

Handbook of Statistics, Vol. 32. http://dx.doi.org/10.1016/B978-0-444-63431-3.00010-3

invert these conditionals to get diseases given symptoms. See Jordan (1998). There are a number of evocative books and articles on this subject. Notable among these are Lauritzen (1996), Edwards (2000), Jensen (2001), Neapolitan (2004), Pearl (1988), Almond (1995), Cowell et al. (1999), Castillo et al. (1997), Nagarajan et al. (2013), Samatova et al. (2013), Hojsgaard et al. (2012), and Heckerman (1995), among others. The purpose of this chapter is to lay out core ideas behind Bayesian networks and use of R.

2 JOINT AND CONDITIONAL DISTRIBUTIONS

Suppose we want to understand interrelationship between the following five binary random variables in a target population.

For a subject in the population, let

H: History of Smoking	Yes or no
B: Bronchitis	Present or absent
L: Lung Cancer	Present or absent
F: Fatigue	Present or absent
C: Chest X-ray	Positive or negative

The interrelationship between these variables can be explained by describing the joint distribution of these variables. We need 32 (2^5) pieces of information in the form of joint probabilities such as

$$\Pr(H = \text{Yes}, B = \text{Present}, L = \text{Present}, F = \text{Present}, C = \text{Positive});$$
$$\Pr(H = \text{Yes}, B = \text{Present}, L = \text{Present}, F = \text{Present}, C = \text{Negative});$$
etc.

Actually, we need 31 pieces of information because the sum of these probabilities is unity. Suppose we have 10 binary random variables. The joint distribution of these variables requires 2^{10}-1 (1023) pieces of information. Imagine the sample size required in order to estimate the joint probabilities.

In order to simplify the construction of the joint distribution, one is encouraged to study carefully the random variables on hand and then build a Bayesian network. Before we work on this idea, let us spend some time on writing joint probabilities in different ways using the idea of conditional distribution.

For any three events A, B, and C, the joint probability can be factorized via conditional probabilities in several ways.

$$\begin{aligned}\Pr(A \cap B \cap C) &= \Pr(A)\Pr(B|A)\Pr(C|A \cap B)\\ &= \Pr(B)\Pr(A|B)\Pr(C|B \cap A)\\ &\text{etc.}\end{aligned}$$

In the context of three binary random variables X, Y, and Z each taking values 0 or 1, we need to spell out all the eight probabilities in order to describe their joint distribution:

$$\begin{aligned}
&\Pr(X=0, Y=0, Z=0);\\
&\Pr(X=0, Y=0, Z=1);\\
&\Pr(X=0, Y=1, Z=0);\\
&\Pr(X=0, Y=1, Z=1);\\
&\Pr(X=1, Y=0, Z=0);\\
&\Pr(X=1, Y=0, Z=1);\\
&\Pr(X=1, Y=1, Z=0);\\
&\text{and}\\
&\Pr(X=1, Y=1, Z=1).
\end{aligned}$$

These numbers should be nonnegative and add up to unity. It is enough to spell out seven of these probabilities and the eighth one can be obtained by subtraction.

There is another way to get the joint distribution.

1. Spell the distribution of X, i.e., $\Pr(X=0)$ and $\Pr(X=1)$.
2. Spell the conditional distribution of $Y|X$, i.e., the conditional distribution of $Y|X=0$ and the conditional distribution of $Y|X=1$ ($\Pr(Y=0|X=0)$ and $\Pr(Y=1|X=0)$ and $\Pr(Y=0|X=1)$ and $\Pr(Y=1|X=1)$).
3. Spell the conditional distribution of $Z|X$, Y, i.e., the conditional distributions of $Z|X=0$, $Y=0$; $Z|X=0$, $Y=1$; $Z|X=1$, $Y=0$; and $Z|X=1$, $Y=1$. Equivalently,

$$\begin{aligned}
&\Pr(Z=0|X=0, Y=0)\,\text{and}\\
&\Pr(Z=1|X=0, Y=0);\\
&\Pr(Z=0|X=0, Y=1)\,\text{and}\\
&\Pr(Z=1|X=0, Y=1);\\
&\Pr(Z=0|X=1, Y=0)\,\text{and}\\
&\Pr(Z=1|X=1, Y=0);\\
&\text{and}\\
&\Pr(Z=0|X=1, Y=1)\,\text{and}\\
&\Pr(Z=1|X=1, Y=1).
\end{aligned}$$

Even though the conditional distribution specification sounds too many pieces of information, we need only seven pieces of information here! Once we have all these distributions on hand, we can construct the joint distribution of X, Y, and Z. For example,

$$\Pr(X=0, Y=0, Z=0) = \Pr(X=0)^*\Pr(Y=0|X=0)^*\Pr(Z=0|X=0, Y=0)$$

When it comes to conditional distributions, there is no need to follow the order stipulated above. We could follow a different path: marginal distribution of Y, conditional distribution of $Z|Y$, and conditional distribution of $X|Y$, Z, for example.

Let us look at the example of the variables H, B, L, F, and C described earlier. The following network lets us build their joint distribution using the conditionals that flow from the network:

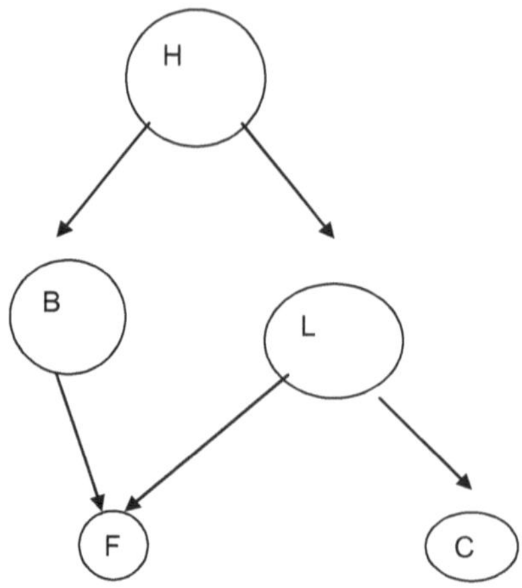

The network is a graph. The nodes are the random variables. The edges represent direct influences. We built the graph from natural considerations.

1. There is an edge from H to L because History of Smoking has a direct influence on Lung Cancer.
2. There is an edge from H to B because History of Smoking has a direct influence on Bronchitis.
3. There is an edge from L to C because the presence of Lung Cancer has a direct influence on the result of Chest X-ray. (This particular Chest X-ray is designed to detect the presence or absence of Lung Cancer.)
4. There is no edge from H to C because a History of Smoking has an influence on the result of Chest X-ray only through its influence on the presence of Lung Cancer.
5. There is an edge from B to F because Bronchitis has a direct influence on Fatigue.
6. There is an edge from L to F because Lung Cancer has a direct influence on Fatigue.

One way to construct Bayesian networks is by creating edges that represent direct influences. We use our understanding and knowledge to assess which variable influences which variable directly.

Where do we go from here? We need some more terminology.

1. H is a root node. It has no parents.
2. B has one parent H.

3. L has one parent H.
4. C has one parent L.
5. F has two parents B and L.

The Bayesian network model as exemplified by its graph is that the conditional distribution of any random variable in the network given all other random variables preceding the variable depends only on its parents.

In order to build the joint distribution of the five variables on hand based on the network presented, we need the following:

1. The marginal distribution of H.
2. The conditional distribution of B | H.
3. The conditional distribution of L | H.
4. The conditional distribution of F | B, L.
5. The conditional distribution of C | L.

This amounts to 11 pieces of information. We can work out the joint distribution of H, B, L, F, and C! This seemingly limited information is good enough to compute the entire joint distribution and hence any marginal probability and conditional probability.

The standard factorization of the following joint probability is given by

$$\begin{aligned}
&\Pr(\mathrm{H}=\mathrm{Yes}, \mathrm{B}=\mathrm{Present}, \mathrm{L}=\mathrm{Present}, \mathrm{F}=\mathrm{Present}, \mathrm{C}=\mathrm{Positive}) \\
&\quad = \Pr(\mathrm{H}=\mathrm{Yes}) \\
&\quad * \Pr(\mathrm{B}=\mathrm{Present} \mid \mathrm{H}=\mathrm{Yes}) \\
&\quad * \Pr(\mathrm{L}=\mathrm{Present} \mid \mathrm{H}=\mathrm{Yes}, \mathrm{B}=\mathrm{Present}) \\
&\quad * \Pr(\mathrm{F}=\mathrm{Present} \mid \mathrm{H}=\mathrm{Yes}, \mathrm{B}=\mathrm{Present}, \mathrm{L}=\mathrm{Present}) \\
&\quad * \Pr(\mathrm{C}=\mathrm{Positive} \mid \mathrm{H}=\mathrm{Yes}, \mathrm{B}=\mathrm{Present}, \mathrm{L}=\mathrm{Present}, \mathrm{F}=\mathrm{Present})
\end{aligned}$$

The network model simplifies the calculations to:

$$\begin{aligned}
&\Pr(\mathrm{H}=\mathrm{Yes}, \mathrm{B}=\mathrm{Present}, \mathrm{L}=\mathrm{Present}, \mathrm{F}=\mathrm{Present}, \mathrm{C}=\mathrm{Positive}) \\
&\quad = \Pr(\mathrm{H}=\mathrm{Yes}) \\
&\quad * \Pr(\mathrm{B}=\mathrm{Present} \mid \mathrm{H}=\mathrm{Yes}) \\
&\quad * \Pr(\mathrm{L}=\mathrm{Present} \mid \mathrm{H}=\mathrm{Yes}) \\
&\quad * \Pr(\mathrm{F}=\mathrm{Present} \mid \mathrm{B}=\mathrm{Present}, \mathrm{L}=\mathrm{Present}) \\
&\quad * \Pr(\mathrm{C}=\mathrm{Positive} \mid \mathrm{L}=\mathrm{Present}).
\end{aligned}$$

This is how we obtain every other joint probability. The unrestricted joint distribution needs 31 pieces of information, i.e., knowledge of 31 probabilities. The Bayesian network model requires only 11 pieces of information.

There are two ways to go about building a Bayesian network for a given set of random variables:

1. Utilize our personal knowledge about the variables to build a network. Estimate the needed distributions in the network using the data collected. Test goodness of fit.

2. Use a Bayesian network learning algorithm on the data collected to come out with a well-fitted network.

How do we use the given network model? Suppose we are looking at a subject from the population. We want to speculate whether or not the subject has Lung Cancer. Suppose we are to be provided with information on H, B, F, and C about the subject. If the model holds, we simply say that we do not need information on B, F, and C. His chances of Lung Cancer depend only on his History of Smoking. We need only the conditional distribution L | H.

Suppose we are provided with information only on C about the subject. Then we need to work out the conditional distribution of L | C from the joint distribution. There are simple ways of calculating the conditional distribution using the Bayes theorem.

3 GENERALITIES AND ISSUES

Let us get into generalities. Look at a network model postulated for the variables on hand. Two issues arise:

1. In practice, we may not know all conditional distributions in a network model. How do we get them?
2. How do we know that the Network model postulated is a good summary of the population?

We can answer these two questions in one stroke. These issues fall under the realm of statistical inference. Resort to sampling. Take a random sample of decent size n from the population. Believe in the model. Estimate all those conditional distributions stemming from the network using the data. Issue 1 is solved. Using these estimated conditional distributions, build the joint distribution of the random variables. Do a chi-squared test of goodness of fit.

As an illustration, look at the network model postulated in Section 2. Note the observed frequency O of the event, for example,

$$\{H = \text{Yes}, B = \text{Present}, L = \text{Present}, F = \text{Present}, C = \text{Positive}\}.$$

How many cases O are there in the sample meeting these requirements? Calculate the expected frequency E as per the model, i.e.,

$$E = n^* \Pr(H = \text{Yes})^* \Pr(B = \text{Present} | H = \text{Yes})^* \Pr(L = \text{Present} | H = \text{Yes})$$
$$^* \Pr(F = \text{Present} | B = \text{Present}, L = \text{Present})^* \Pr(C = \text{Positive} | L = \text{Present}).$$

Do this for every possible joint event. The chi-squared statistic is given by $\chi^2 = \sum \frac{(O-E)^2}{E}$. Compare this value with an appropriate critical value called from the theoretical chi-squared distribution.

Some examples of Bayesian networks and their interpretation:

Example 1

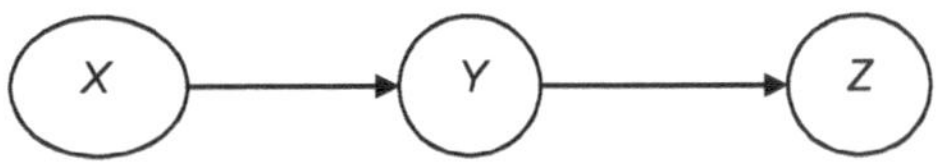

Interpretation of the model:

The joint distribution of X, Y, and Z is the product of the marginal distribution of X, conditional distribution of Y given X, and conditional distribution of Z given Y. The variable Z has two predecessors. Only one is a parent. The conditional distribution of Z given X and Y is the same as the conditional distribution of Z given Y. This means: given Y, and Z are conditionally independent.

Example: Suppose each of A, B, and C is binary with the following distribution:

X	Y	Z	Pr
0	0	0	0.025
0	0	1	0.025
0	1	0	0.18
1	0	0	0.175
1	1	0	0.06
1	0	1	0.175
0	1	1	0.27
1	1	1	0.09

From the joint distribution, the following distributions can be worked out.

Marginal distribution of X:

X	0	1
Pr	0.5	0.5

Conditional distribution of $Y|X=0$:

Y	0	1
Pr	0.1	0.9

Conditional distribution of $Y|X=1$:

Y	0	1
Pr	0.7	0.3

Conditional distribution of $Z|Y=0$:

Z	0	1
Pr	0.5	0.5

Conditional distribution of $Z|Y=1$:

Z	0	1
Pr	0.4	0.6

Conditional distribution of $Z|X=0$, $Y=0$:

Z	0	1
Pr	0.5	0.5

Conditional distribution of $Z|X=1$, $Y=0$:

Z	0	1
Pr	0.5	0.5

Conditional distribution of $Z|X=0$, $Y=1$:

Z	0	1
Pr	0.4	0.6

Conditional distribution of $Z|X=0$, $Y=0$:

Z	0	1
Pr	0.4	0.6

Thus, we see that the conditional distribution of $Z|X, Y$ is the same as the conditional distribution of $Z|Y$. This leads to the network model $X \rightarrow Y \rightarrow Z$.

Warning: One can write joint probabilities in a different style. For example, one can write

$$\Pr(X=0, Y=0, Z=0) = \Pr(Z=0)^{*}\Pr(X=0|Z=0)^{*}\Pr(Y=0|Z=0, Y=0)$$
$$=0.44^{*}(0.205/0.44)^{*}(5/41)=0.025$$

The conditional probabilities do not simplify.

Suppose the random variables A, B, and C have a trivariate normal distribution with means 0, 0, and 0 and dispersion matrix

$$\sum = \begin{pmatrix} 1 & \rho_{AB} & \rho_{AC} \\ \rho_{AB} & 1 & \rho_{BC} \\ \rho_{AC} & \rho_{BC} & 1 \end{pmatrix}$$

One can show that the random variables have the network model described above if and only if $\rho_{AB} \times \rho_{BC} = \rho_{AC}$.

Example 2

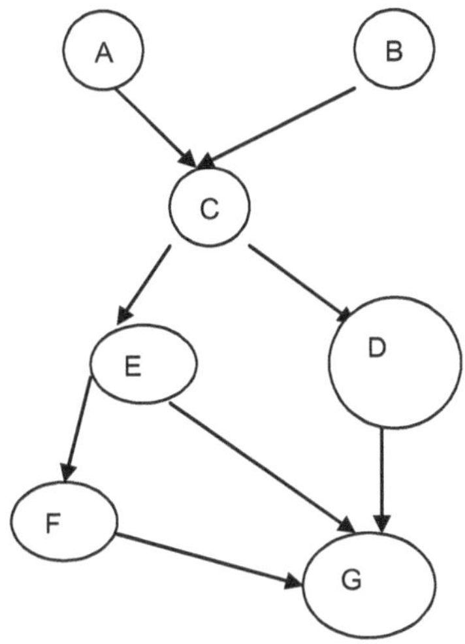

What kind of information do we need to build the joint distribution of the random variables A, B, C, D, E, F, and G as per this model?

1. The distribution of A.
2. The distribution of B.
3. The conditional distribution of $C \mid A, B$.
4. The conditional distribution of $D \mid C$.
5. The conditional distribution of $E \mid C$.
6. The conditional distribution of $F \mid E$.
7. The conditional distribution of $G \mid D, E, F$.

4 GRAPH THEORY

A graph has two components: a set V whose elements are called vertices or nodes and a set E whose elements are called edges. The set E consists of some ordered pairs (A, B) with A and B in V. The ordered pair (A, B) is interpreted as a directed edge from the vertex A to the vertex B. We can denote the ordered pair by $A \rightarrow B$. As an example, suppose

$$V = \{X, Y, Z, U, V\}$$

and

$$E = \{(U, X), (U, Y), (V, X), (V, Y), (X, Z), (Y, Z)\}$$

In the directed graph $G = (V, E)$, there are directed edges from U to X, U to Y, V to X, V to Y, X to Z, and Y to Z. In order to understand a specific graph, it is better to draw a picture. A pictorial representation is the same as the one given above.

A path in a directed graph is any sequence of the type $A_1 \to A_2 \to \cdots \to A_n$. For example, in the graph described above, $U \to X \to Z$ is a path from the node U to the node Z passing through the node X.

A path $A_1 \to A_2 \to \cdots \to A_n$ is said to be cycle if $A_1 = A_n$, i.e., the initial node and the terminal node are the same. In the graph described above, there are no cycles.

A directed graph $G = (V, E)$ is said to be acyclic if there are no cycles in it. For such a graph we use the acronym DAG.

Only DAGs are allowed in Bayesian networks.

Could the following be a Bayesian network?

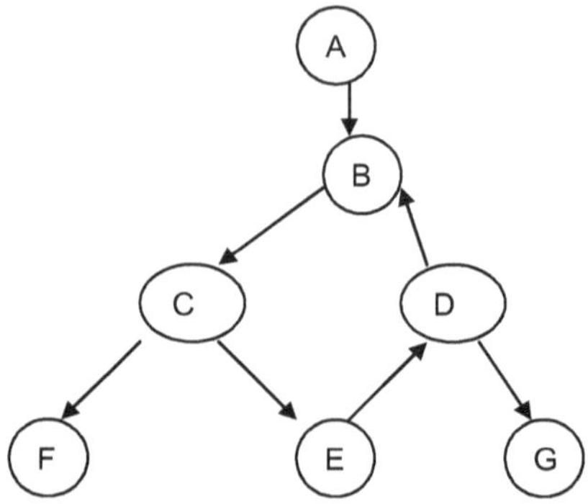

No. There is a cycle: $B \to C \to E \to D \to B$. This cannot be a model. Why we do not allow such a graphical model? There will be complications in building the joint distribution of the variables. The joint distribution cannot be determined uniquely from the conditional distributions specified by the network. In some sections of the literature, a Bayesian network is called a graphical model.

Look at the following graphical model:

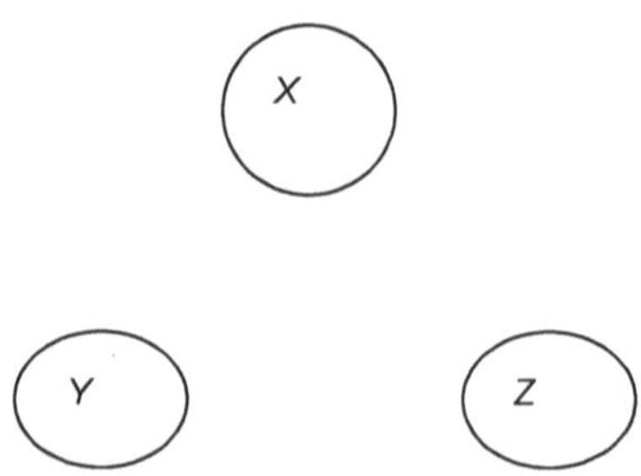

There are no directed edges. This is still a graph. The model is that all the random variables X, Y, and Z are mutually independent. The joint distribution of X, Y, and Z factorizes into the distribution of X, distribution of Y, and distribution of Z.

$$\Pr(X=x,Y=y,Z=z) = \Pr(X=x)*\Pr(Y=y)*\Pr(Z=z)$$

This is the simplest model of all possible graphical models for three random variables X, Y, and Z. Look at the following graphical model.

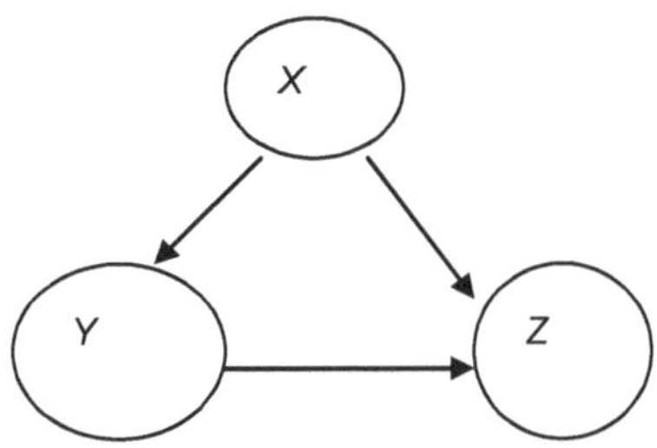

This is the other extreme. We need:

1. the distribution of X;
2. the conditional distribution of $Y|X$;
3. the conditional distribution of $Z|X$, Y.

There is no reduction in the number of pieces of information required.

5 A CASE STUDY

Wisconsin High School Data: College Plans

Sewell and Shah (1968), two sociologists, investigated factors that influence the intention of high school students to attend college. They measured the following variables for 10,418 Wisconsin high school seniors:

Gender: male (M) or female (F)
Socioeconomic status (SES): low (L); lower middle (LM); upper middle (UM); high (H)
Intelligent quotient (IQ): low (L), lower middle (LM), upper middle (UM); high (H)
Parental Encouragement (PE): low (L); high (H)
College Plans (CP): yes (Y); no (N)

The goal is to understand the (possibly causal) relationships among these variables. Find a model which adequately summarizes the data.

The whole data can be arranged in the form of a $2 \times 4 \times 4 \times 2 \times 2$ contingency table. The data have been analyzed using log-linear models. Our goal is to try a Bayesian network model. We have a total of 128 configurations. No configuration has a zero frequency. The data are summarized below:

Sex	SES	IQ	PE	CP	Frequency
M	L	L	L	Y	4
M	L	L	L	N	349
M	L	L	H	Y	13
M	L	L	H	N	64
M	L	LM	L	Y	9
M	L	LM	L	N	207
M	L	LM	H	Y	33
M	L	LM	H	N	72
M	L	UM	L	Y	12
M	L	UM	L	N	126
M	L	UM	H	Y	38
M	L	UM	H	N	54
M	L	H	L	Y	10
M	L	H	L	N	67
M	L	H	H	Y	49
M	L	H	H	N	43
M	LM	L	L	Y	2
M	LM	L	L	N	232
M	LM	L	H	Y	27
M	LM	L	H	N	84
M	LM	LM	L	Y	7
M	LM	LM	L	N	201
M	LM	LM	H	Y	64
M	LM	LM	H	N	95
M	LM	UM	L	Y	12
M	LM	UM	L	N	115
M	LM	UM	H	Y	93
M	LM	UM	H	N	92
M	LM	H	L	Y	17
M	LM	H	L	N	79
M	LM	H	H	Y	119
M	LM	H	H	N	59
M	UM	L	L	Y	8
M	UM	L	L	N	166
M	UM	L	H	Y	47
M	UM	L	H	N	91
M	UM	LM	L	Y	6
M	UM	LM	L	N	120
M	UM	LM	H	Y	74
M	UM	LM	H	N	110
M	UM	UM	L	Y	17
M	UM	UM	L	N	92
M	UM	UM	H	Y	148
M	UM	UM	H	N	100
M	UM	H	L	Y	6

Sex	SES	IQ	PE	CP	Frequency
M	UM	H	L	N	42
M	UM	H	H	Y	198
M	UM	H	H	N	73
M	H	L	L	Y	4
M	H	L	L	N	48
M	H	L	H	Y	39
M	H	L	H	N	57
M	H	LM	L	Y	5
M	H	LM	L	N	47
M	H	LM	H	Y	123
M	H	LM	H	N	90
M	H	UM	L	Y	9
M	H	UM	L	N	41
M	H	UM	H	Y	224
M	H	UM	H	N	65
M	H	H	L	Y	8
M	H	H	L	N	17
M	H	H	H	Y	414
M	H	H	H	N	54
F	L	L	L	Y	5
F	L	L	L	N	454
F	L	L	H	Y	9
F	L	L	H	N	44
F	L	LM	L	Y	5
F	L	LM	L	N	312
F	L	LM	H	Y	14
F	L	LM	H	N	47
F	L	UM	L	Y	8
F	L	UM	L	N	216
F	L	UM	H	Y	20
F	L	UM	H	N	35
F	L	H	L	Y	13
F	L	H	L	N	96
F	L	H	H	Y	28
F	L	H	H	N	24
F	LM	L	L	Y	11
F	LM	L	L	N	285
F	LM	L	H	Y	29
F	LM	L	H	N	61
F	LM	LM	L	Y	19
F	LM	LM	L	N	236
F	LM	LM	H	Y	47
F	LM	LM	H	N	88
F	LM	UM	L	Y	12
F	LM	UM	L	N	164
F	LM	UM	H	Y	62
F	LM	UM	H	N	85
F	LM	H	L	Y	15
F	LM	H	L	N	113
F	LM	H	H	Y	172
F	LM	H	H	N	50

Sex	SES	IQ	PE	CP	Frequency
F	UM	L	L	Y	7
F	UM	L	L	N	163
F	UM	L	H	Y	36
F	UM	L	H	N	72
F	UM	LM	L	Y	13
F	UM	LM	L	N	193
F	UM	LM	H	Y	75
F	UM	LM	H	N	90
F	UM	UM	L	Y	12
F	UM	UM	L	N	174
F	UM	UM	H	Y	91
F	UM	UM	H	N	100
F	UM	H	L	Y	20
F	UM	H	L	N	81
F	UM	H	H	Y	142
F	UM	H	H	N	77
F	H	L	L	Y	6
F	H	L	L	N	50
F	H	L	H	Y	36
F	H	L	H	N	58
F	H	LM	L	Y	5
F	H	LM	L	N	70
F	H	LM	H	Y	110
F	H	LM	H	N	76
F	H	UM	L	Y	12
F	H	UM	L	N	48
F	H	UM	H	Y	230
F	H	UM	H	N	81
F	H	H	L	Y	13
F	H	H	L	N	49
F	H	H	H	Y	360
F	H	H	H	N	98

Highlights of the data

1. The most dominant configuration is: (Gender, SES, IQ, PE, CP) = (Female, Low, Low, Low, No) with a frequency of 454.
2. The second most dominant configuration is: (Gender, SES, IQ, PE, CP) = (Male, High, High, High, Yes) with a frequency of 414.
3. The third most dominant configuration is: (Gender, SES, IQ, PE, CP) = (Female, High, High, High, Yes) with a frequency of 360.
4. The fourth most dominant configuration is: (Gender, SES, IQ, PE, CP) = (Male, Low, Low, Low, No) with a frequency of 349.

We want to find out the best Bayesian network model which summarizes the data. In this problem, we have five random variables and we can look at all possible network models. To illustrate the complexity involved, let us look at the case of three random variables X, Y, and Z and all possible network models.

Model 1

In this model, all random variables are independent.

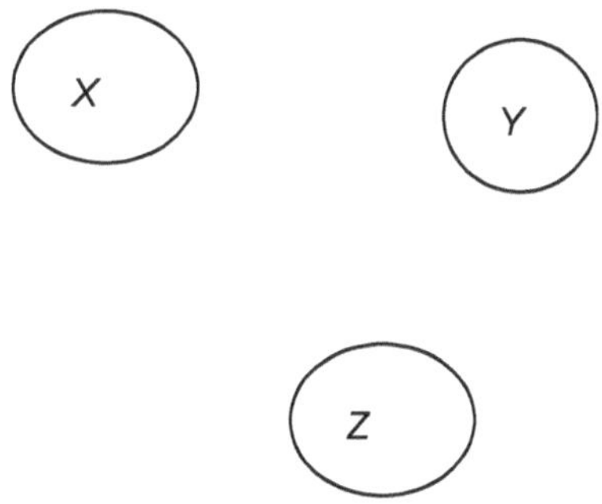

Model 2

In this model, we need the full information: the distribution of X, the conditional distribution of $Y|X$, and the conditional distribution of $Z|X, Y$.

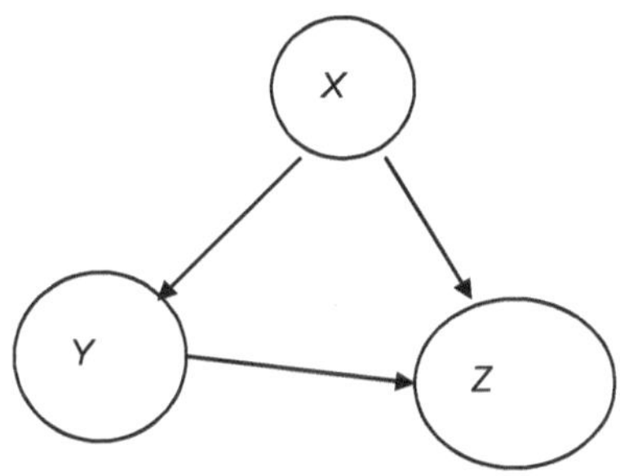

In Model 2, we can permute the positions of the random variables. We will have the same model. Model 1 and 2 are the extremes in the entire spectrum of network models. No. 1 is the simplest with least number of parameters.

Model 3

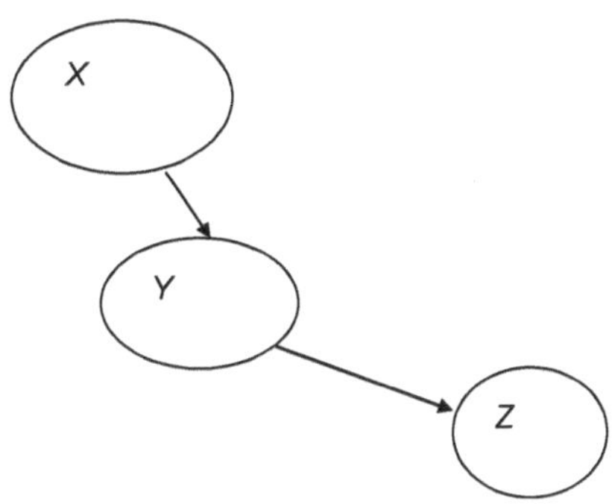

In this model, we need the distribution of X, the conditional distribution of $Y|X$, and the conditional distribution of $Z|Y$. In this model, Z and X are conditionally independent given Y. We can permute the positions of the random variables to get five other models.

Model 4

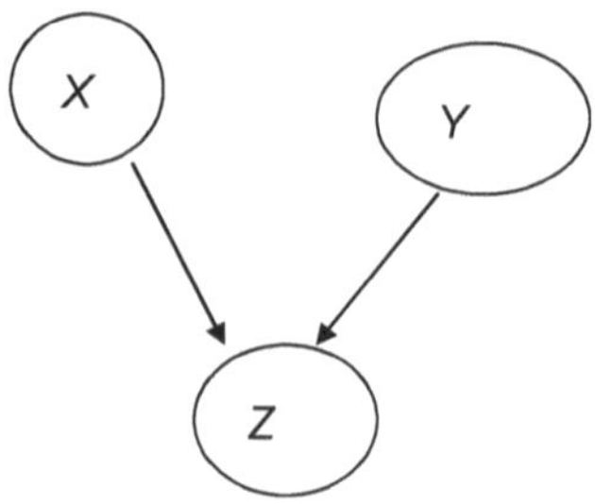

In this model, we need the distribution of X, the distribution of Y, and the conditional distribution of $Z|X, Y$. We can get two other models by permuting the positions of the random variables.

Model Selection

A Naïve Approach

Suppose we have data available on (X, Y, Z). We have a total of 11 models. Fit all these models to the data. Carry out a chi-squared goodness-of-fit test.

Let us start with Model 1. Using the data, estimate the distribution of each of the random variables X, Y, and Z. Using these marginal distributions, estimate all the joint probabilities and expected frequencies. For the sake of illustration, suppose each random variable is binary. Let

$$\hat{p}_1 = \hat{\Pr}(X=1) \text{ and } \hat{p}_0 = \hat{\Pr}(X=0);$$
$$\hat{q}_1 = \hat{\Pr}(Y=1) \text{ and } \hat{q}_0 = \hat{\Pr}(Y=0);$$
$$\hat{r}_1 = \hat{\Pr}(Z=1) \text{ and } \hat{r}_0 = \hat{\Pr}(Z=0).$$

The joint probabilities are estimated by (as per the model):

$$\widehat{\Pr}(X=1, Y=1, Z=1) = \hat{p}_1 \times \hat{q}_1 \times \hat{r}_1$$
$$\widehat{\Pr}(X=1, Y=1, Z=0) = \hat{p}_1 \times \hat{q}_1 \times \hat{r}_0$$
$$\widehat{\Pr}(X=1, Y=0, Z=1) = \hat{p}_1 \times \hat{q}_0 \times \hat{r}_1$$
$$\widehat{\Pr}(X=1, Y=0, Z=0) = \hat{p}_1 \times \hat{q}_0 \times \hat{r}_0$$
$$\widehat{\Pr}(X=0, Y=1, Z=1) = \hat{p}_0 \times \hat{q}_1 \times \hat{r}_1$$
$$\widehat{\Pr}(X=0, Y=1, Z=0) = \hat{p}_0 \times \hat{q}_1 \times \hat{r}_0$$
$$\widehat{\Pr}(X=0, Y=0, Z=1) = \hat{p}_0 \times \hat{q}_0 \times \hat{r}_1$$
$$\widehat{\Pr}(X=0, Y=0, Z=0) = \hat{p}_0 \times \hat{q}_0 \times \hat{r}_0$$

Multiply each estimated probability by the sample size N to get the expected frequencies under the model. Calculate the chi-squared statistic comparing the observed and expected frequencies.

The same tack is pursued with every model on the list. The chi-squared statistic value will be zero under Model 2.

Let us look at Model 4. Using the data, estimate the distribution of X, the distribution of Y, and the conditional distribution of $Z|X, Y$. Using these estimated distributions, estimate the joint distribution of X, Y, and Z. Calculate the expected frequencies as per the model. Do chi-squared statistic.

With the exception of the saturated Model 2, test the null hypothesis that the model is adequate.

Let us go back to our High School Data.

This is a multidimensional contingency table data. The variable College Plans (CP) is primary. Our favorite method is logistic regression model for an analysis of a binary response. The response variable is clearly identified. The model expresses the log odds of Pr(CP = Yes) as a linear combination of other factors. More precisely,

$$\ln\left(\frac{\Pr(\text{CP}=\text{Yes})}{\Pr(\text{CP}=\text{No})}\right) = \beta_0 + \beta_1 \times \text{Sex}_{\text{Male}} + \beta_2 \times \text{SES}_{\text{Low}} + \beta_3 \times \text{SES}_{\text{LowerMiddle}}$$
$$+ \beta_4 \times \text{SES}_{\text{UpperMiddle}} + \beta_5 \times \text{IQ}_{\text{Low}} + \beta_6 \times \text{IQ}_{\text{LowerMiddle}} + \beta_7 \times \text{IQ}_{\text{UpperMiddle}}$$
$$+ \beta_8 \times \text{PELow}$$

Since the possible levels of SES are four, one of the levels has to be the baseline. The level "High" is taken to be the baseline following the Graeco-numeric principle of choosing baselines in R. Likewise, the same level is taken to be the baseline followed for IQ.

What are the advantages of a logistic model?

1. We can quantify the impact of independent variables on the response variable CP.
2. Given information on the independent variables Sex, SES, IQ, and PE for a high school senior, we can predict the probability that he/she has plans for college.

R can be used to fit the logistic regression model. The data have to be prepared for R. The schematic arrangement of the data and commands is outlined below.

The data are in an EXCEL file. The data are copied into a clipboard. The data are downloaded into an R console.

```
> MB <- read.delim("clipboard")
# Dimension of the data
> dim(MB)
[1] 128 6
```

The top six rows of the data

```
> head(MB)
    Sex  SES  IQ  PE  CP  Frequency
1    M    L    L   L   Y          4
2    M    L    L   L   N        349
3    M    L    L   H   Y         13
4    M    L    L   H   N         64
5    M    L   LM   L   Y          9
6    M    L   LM   L   N        207
```

In order to fit a logistic regression model, we need to isolate the data pertaining to CP=Y and also to CP=N, to begin with.

```
> MB1 <- subset(MB, MB$CP == "Y")
> dim(MB1)
[1]  64   6
> MB2 <- subset(MB, MB$CP == "N")
```

The frequencies that show up in MB1 and MB2 are to be in the same data folder. We give a new identity to these columns of frequencies in the MB2 folder.

```
> MB2$CPNo <- MB2$Frequency
> MB2$CPYes <- MB1$Frequency
> head(MB2)
     Sex  SES  IQ  PE  CP  Frequency  CPNo  CPYes
2     M    L    L   L   N        349   349      4
4     M    L    L   H   N         64    64     13
6     M    L   LM   L   N        207   207      9
8     M    L   LM   H   N         72    72     33
10    M    L   UM   L   N        126   126     12
12    M    L   UM   H   N         54    54     38
```

Remove the unnecessary columns.

```
> MB3 <- MB2[ , -c(5, 6)]
> head(MB3)
     Sex  SES  IQ  PE  CPNo  CPYes
2     M    L    L   L   349      4
4     M    L    L   H    64     13
6     M    L   LM   L   207      9
8     M    L   LM   H    72     33
10    M    L   UM   L   126     12
12    M    L   UM   H    54     38
```

We fit the logistic regression model for the grouped data on hand.

```
> MB4 <- glm(cbind(CPYes, CPNo) ~ Sex + SES + IQ + PE,
+              family = binomial, data = MB3)
> summary(MB4)
```

```
Call:
glm(formula = cbind(CPYes, CPNo) ~ Sex + SES + IQ + PE,
    family = binomial, data = MB3)
```

Deviance Residuals:

```
    Min        1Q    Median       3Q      Max
-2.3647   -0.9461   -0.0309   0.7937   3.5510
```

Coefficients:

```
              Estimate   Std. Error   z Value      Pr(>|z|)
(Intercept)    1.56077      0.06683    23.354    <2e-16***
SexM           0.15943      0.05362     2.973    0.00295**
SESL          -1.41151      0.08942   -15.785    <2e-16***
SESLM         -0.87006      0.07117   -12.224    <2e-16***
SESUM         -0.71426      0.06840   -10.442    <2e-16***
IQL           -1.80854      0.08546   -21.163    <2e-16***
IQLM          -1.26446      0.07210   -17.539    <2e-16***
IQUM          -0.69080      0.06828   -10.117    <2e-16***
PEL           -2.47289      0.06827   -36.220    <2e-16***
—
Signif. codes: 0 '***' 0.001 '**' 0.01 '*' 0.05 '.' 0.1 ' ' 1
```

(Dispersion parameter for binomial family taken to be 1)

Null deviance: 4647.275 on 63 degrees of freedom
Residual deviance: 90.009 on 55 degrees of freedom
AIC: 401.35
Number of Fisher Scoring iterations: 4

Interpretation

1. Males and females do differ significantly with respect to "College Plans" (p-value = 0.003). A male is significantly more likely to entertain College Plans than a female (The regression coefficient is positive.) (Mind you, the data came from the 60 s.)
2. Under SES, the level "High" is significantly different from each of the other three levels. A student in the "High" status of SES is more likely to entertain College Plans than a student of any other status. (All regression coefficients are negative.)
3. Under IQ, the level "High" is significantly different from each of the other three levels. A student in the "High" status of IQ is more likely to entertain College Plans than a student of any other status. (All regression coefficients are negative.)
4. Parental Encouragement is a significant factor of College Plans. If the Parental Encouragement is Low, the student is less likely to entertain College Plans (p-value < 0.0001).

5. Even though the model fit is not good as exemplified by the "Residual Deviance," the "Residual Deviance" statistic is not indeed a good indicator of model fit. We conducted a proper chi-squared test, which gives a clean bill of health (adequate fit).
6. Even though the conclusions are common sense, the model provides numbers behind in the form of regression coefficients what we expect the results to be intuitively.

Conduct an "ANOVA" procedure to get a *p*-value for each predictor as a whole.

```
> anova(MB4, test = "Chisq")
```

Analysis of Deviance Table
Model: binomial, link: logit
Response: cbind(CPYes, CPNo)
Terms added sequentially (first to last)

```
        Df   Deviance   Resid. Df   Resid. Dev        Pr(>chi)
NULL                          63       4647.3
Sex      1      52.21         62       4595.1    4.985e-13***
SES      3    1667.77         59       2927.3     <2.2e-16***
IQ       3    1104.14         56       1823.1     <2.2e-16***
PE       1    1733.14         55         90.0     <2.2e-16***
—
Signif. codes: 0 ***"***" 0.001 "**" 0.01 "*" 0.05 "." 0.1 "·" 1
```

Comments: Every factor is significant. The most dominant factor is PE.
Odds Ratio Estimates

Effect	Point Estimate	95% Wald Confidence Limits	
Sex F vs. M	0.853	0.768	0.947
SES H vs. UM	2.043	1.786	2.336
SES L vs. UM	0.498	0.418	0.593
SES LM vs. UM	0.856	0.743	0.985
IQ H vs. UM	1.995	1.745	2.281
IQ L vs. UM	0.327	0.277	0.387
IQ LM vs. UM	0.563	0.489	0.649
PE H vs. L	11.857	10.372	13.554

Comments: The odds ratios are confirming the dominant effect of Parental Encouragement.

Let us see what a Bayesian network model does for us. With five categorical variables, one could write down hundreds of Bayesian network models. Many of them will not make sense. Let us assess the nature of each variable.

Sex: This variable cannot have any parents. No other variable affects or influences the gender of a high school student.

SES: This variable cannot have any parents. No other variable affects or influences the socioeconomic status of a high school student.

College Plans. This variable cannot have any children. It will not influence any of the other variables.

We are left with IQ and PE. We can see that SES directly affects IQ and PE. Sex affects PE. (We are in the 60 s.) We are not sure whether IQ has a direct effect on PE or PE has a direct effect on IQ. This leads to two plausible models.
Model 1

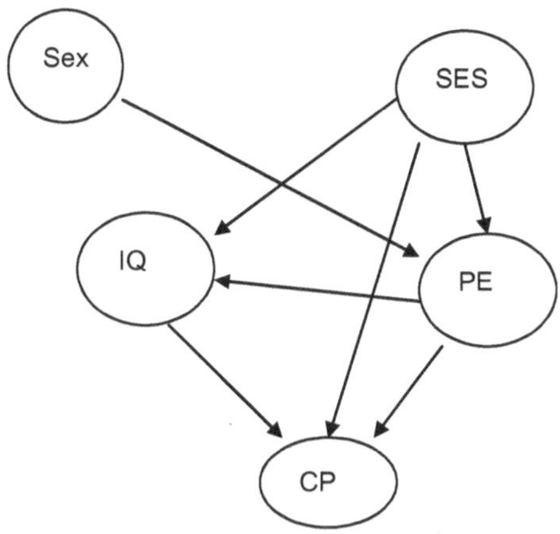

Model 2

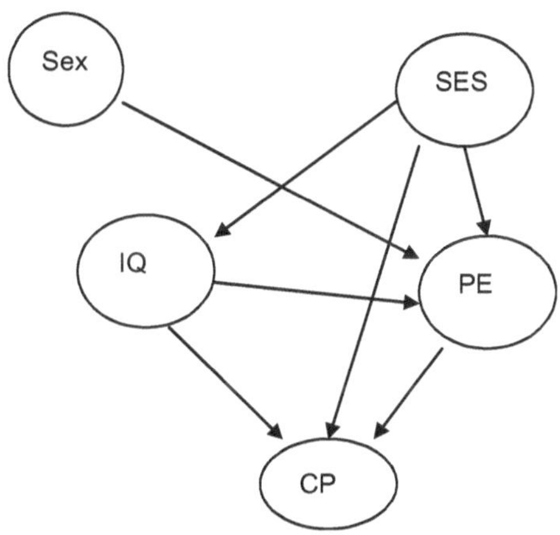

What is the difference between Models 1 and 2? In Model 1, there is a directed arc from PE to IQ, where as in Model 2, there is a directed arc from IQ to PE. All other directed arcs remain the same. The choice of these models is, basically, based on common sense.

Let us fit Models 1 and 2 to the data on hand. We will follow the frequentist approach as opposed to Bayesian approach.

For Model 1, we need to estimate

1. the distribution of Sex. (No. of probabilities to estimate = 1)
2. the distribution of SES. (No. of probabilities to estimate = 3)
3. the conditional distribution of IQ | SES, PE. (No. of probabilities to estimate = $3 \times 8 = 24$)

4. the conditional distribution of PE|SES, Sex. (No. of probabilities to estimate $= 1 \times 8 = 8$)
5. the conditional distribution of CP|IQ, PE, SES. (No. of probabilities to estimate $= 1 \times 32 = 32$)

Total no. of probabilities to estimate = 68

For Model 2, we need to estimate

1. the distribution of Sex. (No. of probabilities to estimate = 1)
2. the distribution of SES. (No. of probabilities to estimate = 3)
3. the conditional distribution of IQ|SES. (No. of probabilities to estimate $= 3 \times 4 = 12$)
4. the conditional distribution of PE|SES, Sex, IQ. (No. of probabilities to estimate $= 1 \times 32 = 32$)
5. the conditional distribution of CP|IQ, PE, SES. (No. of probabilities to estimate = 32)

Total no. of probabilities to estimate = 80

We fit these two models to our data in the next section.

6 NETWORK MODEL FITTING

The package "bnlearn" facilitates fitting of a network model. Download and activate the package. Our Wisconsin data on 10,418 high school students is grouped. We ungroup the data.

```
> MB5 <- MB[rep(1:nrow(MB), MB$Frequency), 1:5]
> dim(MB5)
[1] 10418 5
> head(MB5)
       Sex   SES   IQ   PE   CP
1        M     L    L    L    Y
1.1      M     L    L    L    Y
1.2      M     L    L    L    Y
1.3      M     L    L    L    Y
2        M     L    L    L    N
2.1      M     L    L    L    N
```

There were four students with Sex = M, SES = L, IQ = L, PE = L, and CP = Y. They were grouped in a single row of the original data folder "MB." In the new folder "MB5" they are separately listed. This is the so-called "ungrouping" operation.

We create an empty network with nodes, Sex, SES, IQ, PE, and CP with no arcs.

```
> MB6 <- empty.graph(nodes = c("Sex", "SES", "IQ", "PE", "CP"))
> MB6
  Random/Generated Bayesian network
```

```
  model:
    [Sex][SES][IQ][PE][CP]
   nodes:                          5
   arcs:                           0
     undirected arcs:              0
     directed arcs:                0
   average markov blanket size:    0.00
   average neighbourhood size:     0.00
   average branching factor:       0.00
   generation algorithm:           Empty
> plot(MB6, main = "Empty GraphWisconsin Study")
```

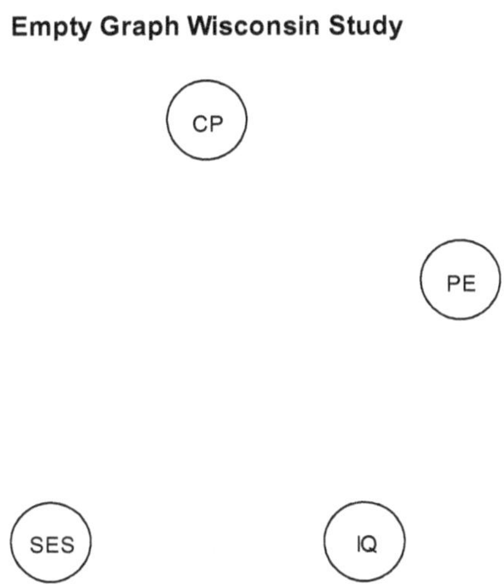

We create the Model 1 network.

```
> arcs(MB6) <- matrix(c("Sex", "PE", "SES", "IQ", "SES", "PE",
"SES", "CP", "PE", "IQ", "PE", "CP", "IQ", "CP"),
+       ncol = 2, byrow = T, dimnames = list(c(), c("from", "to")))
> MB6
  Random/Generated Bayesian network
  model:
    [Sex][SES][PE|Sex:SES][IQ|SES:PE][CP|SES:IQ:PE]
  nodes:                          5
  arcs:                           7
    undirected arcs:              0
    directed arcs:                7
  average markov blanket size:    3.20
  average neighbourhood size:     2.80
  average branching factor:       1.40
  generation algorithm:           Empty
> plot(MB6, main = "Model 1 Wisconsin Study")
```

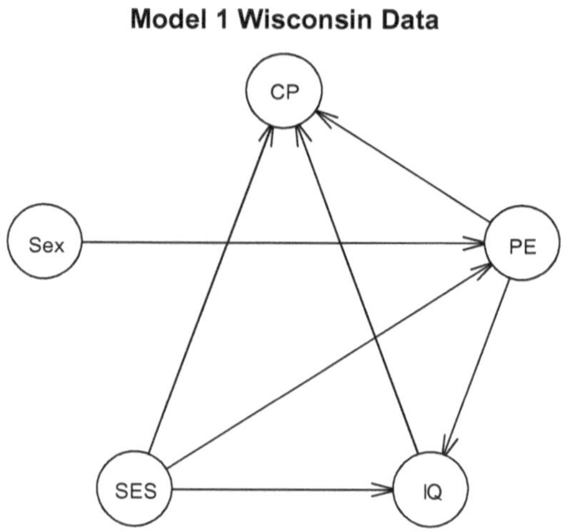


```
# Fit Model 1 to the data.
  > MB7Mle <- bn.fit(MB6, data = MB5, method = "mle")
  > MB7Mle
```

Bayesian network parameters
Marginal Distribution of Sex
Parameters of node Sex (multinomial distribution)

Conditional probability table:

```
       F          M
0.5209253  0.4790747
```

```
  # Marginal distribution of SES
```

Parameters of node SES (multinomial distribution)
Conditional probability table:

```
       H          L         LM         UM
0.2444807  0.2380495  0.2636782  0.2537915
```

```
  # Conditional distribution of IQ | PE
  Parameters of node IQ (multinomial distribution)
```

Conditional probability table:

```
  , , PE = H
    SES
  IQ          H           L          LM          UM
   H 0.43782506  0.24531516  0.32599837  0.32152231
   L 0.08983452  0.22146508  0.16381418  0.16141732
  LM 0.18865248  0.28279387  0.23960880  0.22900262
  UM 0.28368794  0.25042589  0.27057865  0.28805774
```

```
, , PE = L
     SES
IQ             H           L          LM          UM
 H  0.20138889  0.09825674  0.14736842  0.13303571
 L  0.25000000  0.42894876  0.34868421  0.30714286
LM  0.29398148  0.28156366  0.30460526  0.29642857
UM  0.25462963  0.19123085  0.19934211  0.26339286
```

Conditional distribution of PE | SES, Sex (8 conditional distributions)
Parameters of node PE (multinomial distribution)

Conditional probability table:

```
, , SES = H
  Sex
PE          F          M
 H  0.8056836  0.8562249
 L  0.1943164  0.1437751
, , SES = L
     Sex
PE          F          M
 H  0.1661654  0.3182609
 L  0.8338346  0.6817391
, , SES = LM
  Sex
PE          F          M
 H  0.4099379  0.4876733
 L  0.5900621  0.5123267
, , SES = UM
  Sex
PE          F          M
 H  0.5074294  0.6479199
 L  0.4925706  0.3520801
$ Conditional distribution of CP | IQ, PE, SES (32 distributions)
```

Parameters of node CP (multinomial distribution)

Conditional probability table:

```
, , IQ = H, PE = H
     SES
CP             H           L          LM          UM
 N  0.16414687  0.46527778  0.27250000  0.30612245
 Y  0.83585313  0.53472222  0.72750000  0.69387755
, , IQ = L, PE = H
     SES
CP             H           L          LM          UM
 N  0.60526316  0.83076923  0.72139303  0.66260163
 Y  0.39473684  0.16923077  0.27860697  0.33739837
, , IQ = LM, PE = H
     SES
```

```
CP           H            L           LM           UM
 N  0.41604010   0.71686747  0.62244898  0.57306590
 Y  0.58395990   0.28313253  0.37755102  0.42693410
, , IQ = UM, PE = H
   SES
CP           H            L           LM           UM
 N  0.24333333   0.60544218  0.53313253  0.45558087
 Y  0.75666667   0.39455782  0.46686747  0.54441913
, , IQ = H, PE = L
   SES
CP           H            L           LM           UM
 N  0.75862069   0.87634409  0.85714286  0.82550336
 Y  0.24137931   0.12365591  0.14285714  0.17449664
, , IQ = L, PE = L
   SES
CP           H            L           LM           UM
 N  0.90740741   0.98891626  0.97547170  0.95639535
 Y  0.09259259   0.01108374  0.02452830  0.04360465
, , IQ = LM, PE = L
   SES
CP           H            L           LM           UM
 N  0.92125984   0.97373358  0.94384449  0.94277108
 Y  0.07874016   0.02626642  0.05615551  0.05722892
, , IQ = UM, PE = L
   SES
CP           H            L           LM           UM
 N  0.80909091   0.94475138  0.92079208  0.90169492
 Y  0.19090909   0.05524862  0.07920792  0.09830508
```

The significance of each arc can be tested. The models with the arc present and the arc removed are compared with the resultant *p*-value reported under the column "strength." Every arc is significant.

```
> arc.strength(MB6, data = MB5, criterion = "x2")

   from  to      strength
1   Sex  PE  1.052422e-33
2   SES  IQ  5.881742e-48
3   SES  PE  0.000000e+00
4   SES  CP  9.608991e-58
5    PE  IQ 4.924133e-133
6    PE  CP  0.000000e+00
7    IQ  CP 5.359310e-120
```

The Bayesian Information criterion (BIC) is also computed for the model fitted.

```
> score(MB6, data = MB5, type = "bic")
[1] -46137.6
```

Model 2 is obtained by reversing the roles of PE and IQ.

```
> MB7 <- reverse.arc(MB6, "IQ", "PE")
```

```
> MB7
 Random/Generated Bayesian network
   model:
   [Sex][SES][IQ|SES][PE|Sex:SES:IQ][CP|SES:IQ:PE]
 nodes:                           5
 arcs:                            7
   undirected arcs:               0
   directed arcs:                 7
 average markov blanket size:     3.60
 average neighbourhood size:      2.80
 average branching factor:        1.40
 generation algorithm:            Empty
> plot(MB7, main = "Model 2 Wisconsin Study")
```

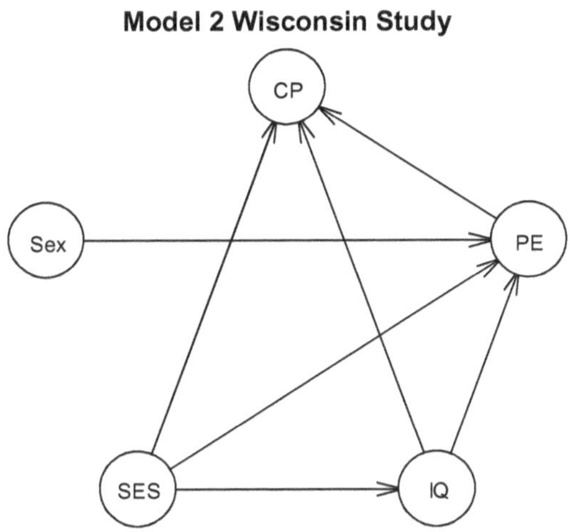

Model 2 is not fitted. However, the BIC score for Model 2 is computed.

```
> score(MB7, data = MB5, type = "bic")
[1] -46175.78
```

Models 1 and 2 are not nested. We cannot compare the models as to decide which one is a better fit.

7 LEARNING ALGORITHM

For a given set of random variables, the number of all possible network models rises rapidly with the set. Hill-climbing algorithm is one method akin to a model selection algorithm to identify the best model using the BIC score. The algorithm begins with an empty model and builds arcs one by one with significant improvement based on the BIC score. See Nagarajan et al. (2013). We have employed the R command “hc” on the Wisconsin expanded data.

```
> MB8 <- hc(MB5)
> MB8
 Bayesian network learned via Score-based methods
```

```
  model:
    [Sex][SES][PE|Sex:SES][CP|SES:PE][IQ|PE:CP]
  nodes:                                       5
  arcs: 6
    undirected arcs:                           0
    directed arcs:                             6
  average markov blanket size:                 2.80
  average neighbourhood size:                  2.40
  average branching factor:                    1.20
  learning algorithm:                          Hill-Climbing
score: BIC                                     (disc.)
 penalization coefficient:                     4.625645
 tests used in the learning procedure:         34
 optimized:                                    TRUE
```

The graph of this model is presented.

```
> MB9 <- empty.graph(nodes = c("Sex", "SES", "IQ", "PE", "CP"))
> arcs(MB9) <- matrix(c("Sex", "PE", "SES", "PE", "SES", "CP",
"PE", "CP", "PE", "IQ", "CP", "IQ"), ncol = 2, byrow = T,
+       dimnames = list(c(), c("from", "to")))
> MB9
 Random/Generated Bayesian network
 model:
   [Sex][SES][PE|Sex:SES][CP|SES:PE][IQ|PE:CP]
  nodes:                        5
  arcs:                         6
    undirected arcs:            0
    directed arcs:              6
  average markov blanket size:  2.80
  average neighbourhood size:   2.40
  average branching factor:     1.20
  generation algorithm:         Empty
> plot(MB4, main = "Best Model Hill-Climbing Algorithm BIC")
```

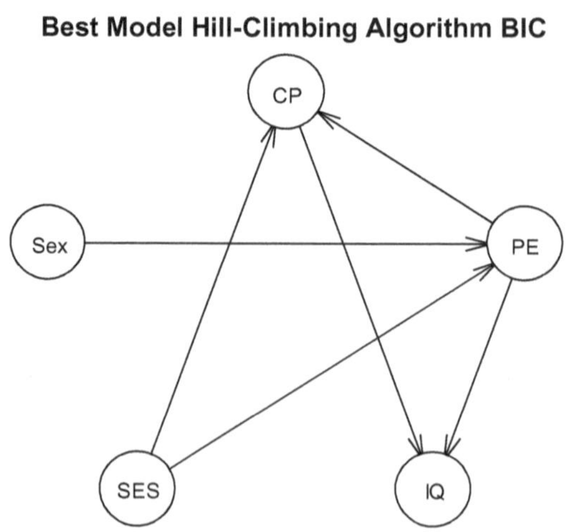

Comments: Models 1 and 2 that were postulated were different from the best model offered by the hill-climbing algorithm. The best model has six arcs compared the seven of Models 1 and 2. The BIC score is computed.

```
> score(MB4, data = MB2, type = "bic")
[1] -46056.02
```

Model	BIC Score	No. of Probabilities Involved in the Estimation
Best	-46056.02	32
1	-46137.6	68
2	-46175.78	80

REFERENCES

Almond, R., 1995. Graphical Belief Modeling. Chapman and Hall, New York.

Castillo, E., Guiterrez, J., Hadi, A., 1997. Expert Systems and Probabilistic Network Models. Springer, New York.

Cowell, R., Dawid, P., Lauritzen, S., 1999. Probabilistic Networks and Expert Systems. Springer, New York.

Edwards, D., 2000. Introduction to Graphical Modeling. Springer, New York.

Heckerman, D., 1995. A Tutorial on Learning with Bayesian Networks. Microsoft research, Technical Report: MSR-TR-95-06.

Hojsgaard, S., Edwards, D., Lauritzen, S., 2012. Graphical Models with R. Springer, New York.

Jensen, F., 2001. Bayesian Networks and Decision Graphs. Springer, New York.

Jordan, M. (Ed.), 1998. Learning in Graphical Models, NATO Science Series. In: Kluwer Academic Publishers, Dordecht, The Netherlands.

Lauritzen, S., 1996. Graphical Models. Oxford University Press, Oxford, UK.

Nagarajan, R., Scutari, M., Lebre, S., 2013. Bayesian Networks in R with Applications in Systems Biology. Springer, New York.

Neapolitan, R.E., 2004. Learning Bayesian Networks. Prentice Hall, Upper Saddle River, New Jersey.

Pearl, J., 1988. Probabilistic Reasoning in Intelligent Systems. Morgan Kaufmann, Burlington, Massachusetts.

Samatova, N., Hendrix, W., Jenkins, J., Padmanabhan, K., Chakraborty, A., 2013. Practical Graph Mining with R. Chapman and Hall, New York.

Sewell, W., Shah, V., 1968. Parent's education and children's educational aspirations and achievements. Am. Sociol. Rev. 33, 191–209.

Subject Index

Note: Page numbers followed by "*f*" indicate figures, and "*t*" indicate tables.

Kolibri 360 GmbH
Ferdinand-Jühlke-Straße 7
99095 Erfurt, Deutschland
produktsicherheit@kolibri360.de